Advanced SQL with SAS®

Christian FG Schendera

S.sas

sas.com/books

Contents

Preface[*]

SQL (Structured Query Language) is *the* programming language for the definition, query, and manipulation of very large data sets in relational databases. Worldwide. SQL is a quasi-industry standard. In June 2010, a web search for the term "SQL" with Google resulted in 135 million hits. In February 2021, the same Google search found 307 million hits. SQL has been a standard at ANSI (American National Standards Institute) since 1986 and at ISO since 1987. With PROC SQL, it is possible to query data from tables or views; create tables, views, new variables, and indexes; merge data from tables or views; create summary parameters or statistics; perform complex calculations (new variables); update values in SAS tables; or create and design reports with the Output Delivery System (ODS) (Schendera, 2011).

This volume shows what else PROC SQL can do. There are areas where PROC SQL goes far beyond the ANSI standard, for example:

- PROC SQL handles **missing values** differently than the ANSI standard. While entries represent the presence of information, missing values indicate the opposite, the absence of information. Chapter 2 presents the handling of missing values: defining (system- and user-defined), querying, as well as searching and replacing alphanumerical missing values.
- Automatic **data integrity checks** using integrity constraints and audit trails, as well as multiple value checks, outliers, filters, and unification of undesired strings (Chapter 3).
- PROC SQL also supports the use of the **SAS macro language**. This difference is of such importance that Chapter 4, the most extensive chapter, is dedicated to the interaction between SQL and the SAS Macro Facility. This chapter presents how to automate and accelerate processes using macro variables (automatic, user-defined) and macro programs including the listwise execution of commands and the use of loops.

There are some other "**SAS extras**" that go beyond the ANSI standard. However, no separate chapter covers them. These include countless SAS functions and function calls (see the overview in Subsection 10.2.), and other functionalities like CALCULATED, Boolean expressions, or remerging.

Chapters 6 through 9 extend PROC SQL focusing on **performance and efficiency**, which are covered from multiple angles: programming languages (hash objects, FedSQL, and DS2), programming environments (client/server, grid, or cloud), or special techniques (for example, in-memory or distributed processing). **SAS syntax** for PROC SQL, **SAS functions and function calls**, and **DBMS accesses** are covered in Chapter 10. PROC SQL supports many more evaluation functions than required by the ANSI standard. (See Chapter 10.)

In addition to functionality and programming, Base SAS users might need to get used to a slightly different **terminology.** (See Table 0.1.) In Base SAS, for example, a data file is called a SAS file, a data line is called an observation, and a data column is called a variable. In PROC SQL, however, a data file is called a table, a data row is called a row, and a data column is called a column. The names in RDBMS are again slightly different. Other models or modeling languages (ERM, UML) use different terminology.

Table 0.1 Terminology in PROC SQL, Base SAS, and RDBMS

Structure/Language	SQL	Base SAS	RDBMS
Data File	Table	SAS Data Set	Relation
Data Row	Row	Observation / Record	Tuple / Record
Data Column	Column	Variable	Attribute / Field
Data Cell	Value	Value	(Attribute) Value

Base SAS and PROC SQL elements are often used simultaneously in the same SAS program as this book will show in many places. Experience has shown that a strict terminological **differentiation** is therefore unnecessary. Although this book prefers SQL terminology, it will also use Base SAS terminology. This book therefore uses the terms "table", "SAS file", or "data set" interchangeably for a data set. In contrast to the RDBMS term, "data set" refers to the totality of all rows (even if a file should only contain one row).

PROC SQL creates empty tables, views, or "only" performs queries. What is the difference between tables, views, and queries?

- A **table** is nothing more than a SAS data set. SAS tables are often a consequence of queries. A SAS file (a) describes data, (b) contains data, and (c) requires storage space. The last two points are important; they serve to delimit views.
- A **view** (a) describes data, (b) contains no data itself, and therefore (c) does not require any storage space. SAS views are only a view on data or results, without having to physically contain the data itself. Views and tables have in common that they are each a result of a query or a sequence of queries. Tables and views could also be rewritten as "stored queries," for example, as the result of a SELECT statement or more complex queries.
- **Queries** are a collection of conditions (for example, "take all data from table A", "query only rows with a value in column X greater than 5 from view B", and so on). Queries in their basic form only query data and write their result immediately to the SAS output. Depending on the programming (including the addition of CREATE, AS, and QUIT), queries can output the result of their query in (a) tables, (b) views, (c) directly to the SAS output, or (d) to SAS macro variables. Queries can also contain one or more subqueries. A subquery is a query (in parentheses) that is embedded within another query. If there are several subqueries, the innermost one is executed first followed by all the others in the direction of the outermost query.

All concepts assume that a data row, observation, or tuple must be uniquely identifiable by at least one so-called **key** (also called a primary key or ID). A key is generally unique and thus uniquely identifies a row in a table, not its relative position in the table, which can always change by sorting. A key is generally used less for describing a data row than for managing it. A *primary* key is usually used to identify the rows in your own table, a *foreign* key is used to identify the rows in another table, usually its primary key. Keys are essential for joining two or more tables and therefore must not be deleted or changed.

These remarks so far about joins, scenarios, and keys were mainly concerned with data in the sense of valid or *existing* values. A whole chapter of this book is dedicated to the absence of information, the handling of *missing* values (also called NULL values, NULL, or missings). In contrast to the ANSI standard, the handling of **missing values** is based on a different definition. In the ANSI standard, expressions such as "5 > NULL", "0 > NULL" or "-5 > NULL" become *NULL*. In Boolean and comparison **operators**, however, these and other expressions become **true**, and this is also true when working with SAS.

About This Book

This book provides an in-depth look at working with SQL from SAS. SQL is, especially in the SAS variant, a very powerful programming language. This book presents the **topics** of missing values, data quality including integrity constraints and audit trails, macro variables and programs, the interplay of PROC SQL with geodata and even spherical distances, performance and efficiency, alternative programming languages, numerous other aids, tips and tricks, or the use of the numerous SAS functions and function calls. From time to time, users are explicitly warned of potentially **undesirable** results. Undesirable results when sorting, grouping, or joining tables with missing values often occur when one does not **know** how PROC SQL or SAS work in detail.

This book is intended as an in-depth consolidation and extension of ANSI SQL and is intended for advanced users. The chapters are organized so that they each illustrate different facets of programming and using PROC SQL. By the end of this book, you should be an advanced user of PROC SQL, capable of thinking conceptually beyond SQL topics and implementing data quality, performance, and automation including integrity constraints, performance (tuning), and programming using the SAS Macro Facility, hash objects, or FedSQL. You know what missing values are and can deal with missing values in queries, analyses, and joins. In addition to missing values, you know other criteria for data quality and SQL techniques to ensure it, including integrity constraints, audit trails, SAS functions, and queries. You will be able to accelerate and automate PROC SQL functionalities using macro variables and macro programs. You would also be no stranger to handling coordinate data using PROC SQL. If you need to handle very large amounts of data, you can also "translate" PROC SQL functionalities into alternative programming approaches using hash objects, FedSQL, or DS2 programs. Also, you know the respective advantages of these programming languages, not limited to the CAS environment. You can distinguish between performance and efficiency and know numerous options to speed up processing from programming languages to programming environments, as well as fundamental and SQL-specific processing optimization approaches. You will also know the benefits of SAS dictionaries and PROC SQL over the DATA step when updating tables with the UPDATE statement. Last but not least, the SAS functions and function calls will give you another impression of the power of SAS. Nevertheless, SAS can do so much more.

Many other exciting SQL topics had to be left out of the book due to lack of space and time, including the interaction of PROC SQL with SAS Studio, Enterprise Guide, or JMP, the strategic use of PROC SQL in Business Intelligence, the preparation of statistical data for complex statistical analyses using Enterprise Miner, or dynamic visualizations using JMP or other SAS applications, and so much more.

What Is SAS, for Me?

When I write about SAS, I like to start by saying that I have been working with SAS for more than 30 years. (See also Schendera, 2012/2011, 2004.) My focus is on Advanced Analytics and Business Intelligence with SAS. Many of the chapters and programs were written during long evenings, sometimes over weekends completely snowed in at a hotel, while preparing projects from all over Europe. Now, at the end of the last edits to this book, I see again and again and still with certain fascination, how many facets the use of PROC SQL from SAS can have. But **what is SAS** in the first place, personally speaking? SAS offers not just a wide range of products and services for almost every industry, role, or requirement. My personal answer to the question "What is SAS?" is fivefold. SAS is a product system, a company, a philosophy, a sophistication and an enthusiasm. Knowledge is power. Precise and timely knowledge is essential for information in performance and competition. SAS is *for* "The Power to Know." For the many solutions SAS offers, please refer to the SAS website. Readers can find information about topics, such as architecture, cloud computing, ETL, or data integration, on www.sas.com and through discussion groups (SAS-L, sasCommunity.org), a visit to one of the numerous SAS events (SAS Global Forum or local SAS Forums), or more specialized events (SAS@TDWI, SAS Government Executive Conference, or PhUSE).

*This is an updated version of the original 2011 preface.

Acknowledgments

Ten years ago, I published a **double volume** on SQL, a beginner's volume and an advanced volume. First of all, I would like to thank you for the overall positive reception and feedback on the German versions (Schendera, 2012/2011). Through writing these SAS books with heart and soul, I also made a great effort to make and keep them **relevant**. Consequently, this edition now contains a chapter introducing Cloud Analytic Services (CAS) and FedSQL (Chapter 7). Several chapters have been updated, including analysis of geodata (Chapter 5), programming with hash objects (Chapter 6, especially the section about fuzzy joins), and performance and efficiency (Chapter 8). I also updated the SAS functionalities, functions, and syntax to SAS 9.4. Some topics are fundamental as well as they are timeless. On the subject of missing values, for example, that miserable pitfall of programming and analysis, I managed to write 35 pages. I am fascinated again and again by the negative consequences empty data cells can have, and sometimes maybe even more about how clueless users sometimes are about this problem.

I have been working with SAS for over 30 years now (see also Schendera, 2004), but I still learn something new every day. If there's one thing I can say, it's that SAS is growing faster than you can ever learn it. When I was studying psychology at Heidelberg University, I quickly realized how important it is that you **understand exactly** where the data come from that you use for analysis and research. I had to know how SAS worked. I enrolled in SAS courses on main frames, still using punch cards. It was a plague! It took me two attempts to get my certificate. But then things took off. SAS got me hooked, and I developed an enthusiasm for SAS that sometimes kept me up all night. Although I wouldn't have called it work; it was, well, my passion. Which I would like to share with you and motivate you to get to know SAS—its power and versatility. Maybe to use it as a tool for good and make this world a better place.

I am grateful for advice, inspiration, or simply friendship to Prof.em. Gerd Antos (Martin-Luther-University Halle/S., Germany), Prof.em. Mark Galliker (University Bern, Switzerland), Prof. Kirk Lafler (San Diego State University), Roland Donalies (SAS Germany), Markus Grau (SAS Switzerland), Prof. Maria ReGester (Jacksonville, NC), Frank Ivis (CIHI, Canada), Suzanne Morgen and Siân Roberts (SAS Press), and Dr. Stefan Giesen (De Gruyter, Munich). I also want to thank the reviewers for their time and valuable comments: Lewis Church, Catherine Connolly, Vincent DelGobbo, Paul Grant, Mark Jordan, Suzanne Morgen, and Ross Richards. I want to especially highlight and appreciate Suzanne's efforts and contributions. Without you, this volume would not be on the level that it is now. I cannot thank you enough.

I translated and updated both volumes under rather challenging conditions. It is therefore a matter of the heart for me to dedicate this edition to **Captain Sir Tom Moore**. As a 99-year-old, he initially set out to raise £1,000 for NHS charities by walking 25m laps around his garden with the help of his walker. In the end, Capt. Sir Tom raised almost £33m ($45m) for NHS charities. His always positive role model and his "Tomorrow will be a good day" lifted people's spirits during the COVID-19 pandemic and beyond. My favorite quote of his is "Don't give in, just keep on going and things will certainly get better. That's the way to look at it." Capt. Sir Tom passed away in February 2021. The White House joined the United Kingdom and the world in honoring the memory of Captain Sir Tom Moore "who inspired millions through his life and his actions."

Again, I would like to thank Sigur Ros for their timeless artistic inspiration. I wish I could write as well as you can compose and create aural landscapes. Last, but not least, I thank my wife Xiao Yun Huang and my German and Chinese families and friends all over the world. If anything in this book should be still not clear or even incorrect, the responsibility lies solely with me.

Hergiswil NW, Switzerland
January 2022

Dr. CFG Schendera

Acknowledgments for the 2012 Edition

At this point I would like to take the opportunity to thank all those who have supported me in writing this book. First of all, I would like to thank you for the numerous feedback on "SQL with SAS: PROC SQL for Beginners" (Schendera, 2011), as well as "Data Management and Data Analysis with the SAS System" (Schendera, 2004). I was happy to implement the suggestion to include hash programming as an alternative to SQL.

I am grateful for i.a. technical advice and/or contribution in the form of syntax, data and/or documentation: Prof. Gerd Antos (Martin-Luther-University Halle/S., Germany), Prof. Wolfgang Auhagen (Martin-Luther-University Halle/S., Germany), Prof. Mark Galliker (University Bern, Switzerland), Prof. Jürgen Bock, Frank Ivis (Canadian Institute for Health Information, Toronto, Ontario, Canada), Kirk Lafler, Prof. Rainer Schlittgen, Mike Whitcher, as well as the exchange on SAS-L: e.g. Richard R. Allen, Michael Bramley, Dave Brewer, Nancy Brucken, David Carr, David Cassell, Laurel A. Copeland, Peter Crawford, Richard DeVenezia, Paul M. Dorfman, Harry "Bill" Droogendyk, Ron Fehd, Y. (Jim) Groeneveld, Gerhard Hellriegel, Sigurd Hermansen, Eric Hoogenboom, Ya Huang, Nick Longford, Gene Maguin, Dale McLerran, Lawrence H. Muhlbaier, Girish S. Patel, Prasad S. Ravi, Kattamuri Sarma, Howard Schreier, Karsten M. Self, Daniel Sharp, Erik Tilanus, Michel Vaillant, Kevin Viel, Ian Whitlock, John Whittington, Andre Wielki, Matthew M. Zack und last but not least the heroes of First Level Support at SAS in Heidelberg, Germany and Cary, NC.

My thanks go also to Wilhelm Petersmann and Markus Grau of SAS Switzerland (Wallisellen) for the generous provision of SAS software and technical documentation. A very special thanks goes to Roland Donalies (SAS Germany). I would also like to thank Mrs. Stephanie Walter, Mr. Thomas Ammon, and Mrs. Cornelia Horn of Oldenbourg Publishers (now: DeGruyter) for their confidence in also publishing Volume II and for their always generous support. My sincere thanks also go to Dr. Schechler and Kristin Beck. Volker Stehle (Mannheim) designed the artwork. I would like to thank Sigur Ros for their many years of artistic inspiration. If anything in this book should be unclear or incorrect, the responsibility lies solely with the author.

Hergiswil NW, Switzerland
August 2011

Dr. CFG Schendera

About This Book

What Does This Book Cover?

Structured Query Language (SQL) is the most widely used programming language and a quasi-industry standard. SQL is a standardized language that retrieves data from and updates data in tables and the views that are based on those tables. The SQL procedure implements SQL for SAS.

This book introduces the specifics of PROC SQL. Based on the premise that PROC SQL is developed for and executed in SAS, this book's mission is to present advanced SQL use. You will learn how to take advantage of the extra possibilities the power of SAS offers including:

- Essential topics like missing values and data quality with audit trails
- "Blind spots" like how missing values can affect even the simplest calculations and table joins
- SAS macro language and SAS macro programs
- SAS functions
- Integrity constraints
- SAS Dictionaries
- SAS Compute Server
- FedSQL on CAS

In addition to numerous tuning techniques, this book also touches on implicit and explicit pass-throughs, presents alternative SAS grid- and cloud-based processing environments, and compares SAS programming languages and approaches including FedSQL, CAS, DS2, and hash programming. Comparisons between SQL and FedSQL and overviews of SQL syntax and SAS functions help you see the possibilities at your disposal.

This book does not cover SQL basics like SQL logic, terminology, or how SQL works. It does not cover queries or joins, comparisons in programming and processing between Base SAS and PROC SQL, or calculating descriptive statistics, percentages, or working with weights. These fundamentals are specifically taught in a stand-alone volume that I wrote which is designed without SAS specifics and can be applied to systems other than SAS. Currently, the fundamental volume is only available in German.

This edition contains a chapter on FedSQL, which was not included in the "advanced" volume of the original German edition, the most comprehensive double-volume about PROC SQL worldwide at that time. The FedSQL part was written especially for this SAS Press edition. The chapter on geodata analysis has been updated and completely rewritten.

Is This Book for You?

This book is for SAS programmers, analysts, statisticians, and students who want to expand their SQL know-how. This book is also for users of other SQL variants who want to learn about the advantages of PROC SQL, the magic of macro programming, handle missing values, move from SQL to FedSQL, strive for more performance, or find alternatives to SQL and FedSQL.

Although the book addresses advanced topics, it is designed to progress from the simple and manageable to the complex and sophisticated. Some basic SQL and SAS programming skills might be helpful. You need to have access to licensed SAS software like SAS 9.4 or SAS Viya. You can also use SAS OnDemand for Academics or SAS Viya for Learners for some of the examples in this book.

What Should You Know about the Examples?

This book uses hundreds of SAS programs as examples, be it SQL, FedSQL, DS2, or others. The examples are chosen to cover hands-on needs like showing step-by-step how small adjustments to simple SQL programs turn them into powerfully accelerated SAS macro programs. All examples have been tested and are explained, sometimes code line by code line. Feedback from the SAS log or output is also explained, providing hundreds of explanations throughout the book. In more advanced programs, explanations detail even subtle differences in the structures of the generated tables. Some instances use "learning by mistake", in which mistakes are made on purpose to demonstrate which feedback a user might find in the SAS output and log, if any, and train the eye where to look at the result generated.

The chapters follow principles of instructional/text psychology. Typically, an Advance Organizer precedes the chapter, then the contents=examples are usually arranged from the fundamental and simple to the sophisticated and advanced topics. This enables readers=users to have an early understanding and make programming easy, wins=successes.

Software Used to Develop the Book's Content

Platforms: SAS 9.4; SAS Viya 3.5 plus Compute Server (SPRE), CAS.

Programming Environments: SAS Enterprise Guide 8.2, SAS 9.4 Editor Window; CAS WebApps: SAS Studio 5.2

Example Code and Data

With very few exceptions, this book uses SAS data sets from SASHELP to support self-study. A few exceptions are tiny data sets of a few rows especially tailored to explain specific SAS topics.

The list below compiles the most commonly used SAS data tables. Data is mainly from the SAS folder SASHELP and should be available in SAS when fully installed. The following short list does not contain dictionaries, views, and so on. The SAS tables are sorted alphabetically.

```
SASHELP.AIR
SASHELP.BUY
SASHELP.CLASS
SASHELP.ORSALES          FREQUENTFLYERS
SASHELP.PRDSALE
SASHELP.PRDSAL2
SASHELP.PRDSAL3
SASHELP.SHOES
SASHELP.SYR1001
SASHELP.ZIPCODE
```

SAS OnDemand for Academics

This book is compatible with SAS OnDemand for Academics. If you are using SAS OnDemand for Academics, then begin here: https://www.sas.com/en_us/software/on-demand-for-academics.html.

You can use the book's contents without major adjustments to the example code and data to run in SAS OnDemand for Academics. Further functionality depends on the scope of SAS ODA. At the moment, SAS ODA does not support CAS processing of PROC FEDSQL as described in Chapter 7.

We Want to Hear from You

SAS Press books are written *by* SAS Users *for* SAS Users. We welcome your participation in their development and your feedback on SAS Press books that you are using. Please visit sas.com/books to do the following:

- Sign up to review a book
- Recommend a topic
- Request information on how to become a SAS Press author
- Provide feedback on a book

Learn more about this author by visiting his author page at http://support.sas.com/schendera. There you can download free book excerpts, access example code and data, read the latest reviews, get updates, and more.

Author Feedback

Your feedback is encouraged, valued, and appreciated. If you have any suggestions for additions or improvements to this book, I would like to ask you to send your suggestions by email to the following address:

SAS_v2@method-consult.ch

Please use the keyword "Feedback SAS book" as the "Subject" and include at least the following information:

1. edition
2. page
3. keyword (for example, "typo")
4. description (for example, "for statistical analysis"). Please comment the program code.

About The Author

Dr. Christian FG Schendera is a Senior SAS Data Scientist and managing director at Method Consult in Switzerland. An avid SAS user for 30+ years, his experience ranges from scientific consulting, project management, and feature-engineering to statistical modeling using SAS. He studied at Heidelberg University and Martin Luther University Halle-Wittenberg. While still a student, he started consulting in statistics and research methods, lecturing on SAS, applied statistics, and the knowledge-shaping role of non-scientific methods. Christian views constructing knowledge as a two-way street: every step from data collection to applying methods will affect results. He has published several books about statistics, data quality, programming, and SAS. He is author of the most comprehensive double-volume about PROC SQL worldwide. Further information and downloads can be found at www.method-consult.ch.

Learn more about this author by visiting his author page at http://support.sas.com/schendera. There you can download free book excerpts, access example code and data, read the latest reviews, get updates, and more.

Chapter 1: Overview

This book is written for **advanced** users in PROC SQL. The handling of missing values (null values) in PROC SQL is an important topic since SAS SQL handles missing values differently from the ANSI standard. Failure to take these features into account can lead to potentially undesirable results when dealing with missing values. Other topics in this book include data quality, especially with integrity constraints, as well as special features in dealing with missing values and visual analysis of geodata and distances.

Three chapters in this book are particularly important as they help you to harness the power of SAS. The chapter on macro programming, for example, describes how the listwise execution of commands can speed up work with SAS many times over, both when programming and also executing programs. Two further chapters introduce two programming alternatives: programming with hash objects, as well as FedSQL and its possibilities to put it into practice using it in the procedures **FEDSQL**, CAS and DS2. The chapter on performance and efficiency compiles various possibilities of how to obtain even more performance, especially when handling large amounts of data. Further chapters and sections are reserved for overviews of SQL syntax, SAS functions and routines, as well as various special features of the SAS Pass-Through Facility for selected DBMS accesses. Because this book focuses on using many specific features in SAS, it is written primarily for SAS users, as well as anybody interested in gaining a deeper insight in the power of SAS.

This book also introduces a new processing platform, **Cloud Analytics Services** (CAS). Section 1.2 compares several programming languages and their advantages especially in the CAS environment. For details about when to use FedSQL and when to use SQL, please see Section 7.5.

1.1 Detailed Description of This Book

Table 1.1 Quick Finder

Chapter	Description
1	Overview of this book
2	Missing values: Definition, mode of operation, and conversion
3	Data quality: Integrity constraints (test rules) and audit trails, as well as finding duplicates and outliers.
4	Macro programming: From simple macro variables (SAS, SQL, user-defined), to helpful SAS macro programs including macros for list execution of commands or macros for retrieving system information
5	SQL for geodata, distances, and maps
6	Hashing as an alternative to SQL
7	FedSQL
8	Performance and efficiency: strategy and measures including narrowing, shortening, and compressing; sorting; and other tips and tricks including SAS dictionaries
9	Help, tips, and tricks
10	SAS syntax - PROC SQL, SAS functions, and SAS routines
11	References
12	Syntax index
13	Subject index

Chapter 1 is an *overview* and presents in a short summary the contents of this book.

Chapter 2 deals with the subject of **missing values**. While entries represent the presence of information, missing values indicate the opposite, the absence of information. In principle, missing values are unpleasant because they restrict the basis for deriving information. System-defined missing values are tricky because a user does not know why this data is missing. In the case of user-defined missing values, the user knows, but the data is still missing. What makes matters more difficult is that PROC SQL handles missing values differently than the ANSI standard. This chapter focuses on basic aspects of how PROC SQL handles missing data. These include: the *definition* of missing values, the *retrieval* of data from tables that contain missing values (Section 2.2), possibly undesirable effects of missing values on operations of *data analysis* and *data management* (Sections 2.3, 2.4, and 2.5), and fundamental *measures* for dealing with missing values (Section 2.6).

Chapter 3 introduces the topic of **data quality**. Data quality comes before analysis quality. (See Schendera, 2020/2007.) This chapter introduces SQL techniques for ensuring data quality in the problem areas of outliers, plausibility (Section 3.3), missing values (Section 3.1), uniformity (Section 3.4), and duplicates (Section 3.2). By integrating filters when accessing data via SQL, users can ensure that data is not entered incorrectly into the system or analysis in the first place. This chapter also introduces working with integrity constraints and audit trails (Section 3.1).

Chapter 4 introduces **macro variables** and **macro programs**. Using SAS macros, the scope of PROC SQL can be easily extended. The focus of this chapter is less on the introduction of programming SAS macros, but rather on their uncomplicated *application*. This chapter presents numerous examples of SAS macros in application-oriented sections, which extend the possibilities of working with PROC SQL, simplifying it through automation, and accelerating it. Because macros are based on SAS syntax (including PROC SQL), macros also embody all the advantages of syntax programming (see Schendera, 2005, 147ff), for example, validation, automation and reusability, speed, openness, clarity, and systematization. SAS syntax programming can be automated and used repeatedly with macros without the need to write or adapt new syntax commands in principle. This chapter is divided into four sections. Section 4.1 introduces *SAS macro variables*. Section 4.2 introduces programming *SAS macro programs*. Section 4.3 introduces the most important *elements of the SAS macro language*. Section 4.4 and all subsequent sections introduce interesting *SAS macro programs for special applications*.

Chapter 5 introduces the **analysis of geodata** with PROC SQL. The calculation of distances and related parameters such as time or costs is a common task. This section presents the calculation of **geographical distances** in two-dimensional and spherical space using different formats of coordinates (Sections 5.1 and 5.2). In addition, you will learn to calculate the shortest, fastest, or even cheapest distance. Further sections discuss projections, visualizations (Section 5.3), and so on.

Chapter 6 introduces the programming of PROC SQL or DATA step functionalities by means of **hash programming**. Two objects, the hash object and hash iterator, are introduced into the DATA step. These objects enable you to store, search, and query data from lookup tables. Since SAS 9.1 it has been possible to program hash objects specifically. Hash programs can be executed within a DATA step, the DS2 step (see also Section 7.3 and Table 1.2) and the FCMP procedure. One of the most important features of a hash object is that it resides completely in the physical memory of the DATA step. There are two reasons why a separate chapter is dedicated to hash programming: performance, especially with very large data volumes, and introducing terms and language elements for the DS2 section in the FedSQL chapter.

Chapter 7 introduces **FedSQL**. FedSQL is the next-level SQL. It is a vendor-neutral SQL dialect that offers a scalable, threaded, high-performance way to access, manage, analyze, and share nonrelational data in multiple data sources. FedSQL (Section 7.1) can perform federated queries (connect to multiple sources in one query); work with ANSI-1999 compliant data sources such as Google, Amazon, and Salesforce; use new ANSI data types; and work in a cloud environment like the SAS Cloud Analytics Services (CAS) framework. You can gradually expand your programming skills by moving from PROC SQL to PROC FEDSQL (Section 7.2), and from there to PROC CAS and also PROC DS2 (Section 7.3). Several overviews describe similar and different FedSQL functionalities compared to SQL and when executed on SAS versus CAS. Numerous examples illustrate where you can use PROC SQL and PROC FEDSQL and on which platforms (SAS 9.4, SAS Viya, and CAS).

Chapter 8 introduces the topics of performance and efficiency, especially in the context of programming with PROC SQL. **Performance** refers to the performance of users or systems. **Efficiency** describes the performance of users or systems taking into account investment, environmental factors, or sustainability. Therefore, *performant* is when the required amount of data is optimally processed in a minimal amount of time. On the other hand, *efficient* is when the required amount of data is optimally processed using a reasonable effort (time and money). Efficiency is therefore a measure of performance, taking into account the necessary costs. Not everything that is programmed quickly is also efficient in the sense that the system is able to process the program with high performance. Conversely, developing performant programs can be so costly that the cost of manpower is disproportionate to the performance gained on the system side. Next to techniques like reducing, shortening, or tuning (Sections 8.2–8.5), this chapter also touches on the specifics of the Implicit and Explicit Pass-Through (Section 8.6, and see also Section 10.3), and presents alternative grid- and cloud-based processing environments like SAS Grid Computing and SAS Cloud Analytic Services (Section 8.8).

Chapter 9 introduces a chapter full of further **help, tips,** and **hints** for various aspects of working with PROC SQL. Section 9.2, for example, introduces working with SAS dictionaries.

Chapter 10 primarily provides **SAS syntax** for PROC SQL, SAS functions and routines, and DBMS accesses. Section 10.3 features an in-depth discussion of the Implicit and Explicit Pass-Throughs as complementary approaches.

Chapters 11 to 13 contain **references** and the indexes for SAS **syntax** and **keywords**.

By the end of this book, you should know the advanced features and possibilities using PROC SQL, PROC FEDSQL, or FEDSQL in PROC CAS or DS2. You will know when to use which SAS language when it comes to macro or complex programming, multi-threading, or in-memory processing. From a more high-level point of view, you will have learned the basics of data quality and fundamentals in performance and efficiency, illustrated by SQL, but easily transferable to the other programming languages. By the end of this book, you should also understand that this is only an introduction. There is still so much more. Welcome to SAS.

1.2 Which SAS Language for the CAS Environment?

This book introduces several programming languages related to SQL and also a new processing platform, CAS. Table 1.2 aims to make clear from the outset which languages and language elements are suitable for CAS, where they can be found in this book, and what their strengths and weaknesses are.

Table 1.2 Programming Languages for CAS

CAS Support For ...	DATA Step	PROC SQL	FEDSQL (7.1-7.3)	DS2 (7.3)
Formats	●		●	●
In-Memory Processing (CAS)	●		●	●
Multi-Threading (SMP, MMP)	●		●	●
Precision (new integer types)			●	●
Pass-through (implicit, explicit)			●	●
Hash Objects (see Chapter 6)	●			
Unlimited use of SAS Macro Facility (see Chapter 4)	●			
Advanced package and methods development				●
Compatibility with special SAS hosting platforms				●
SAS Data Quality functions (see Chapter 10)	●			
BY-group processing using FIRST. and LAST.	●			●

Note: The **PROC SQL** column is **empty.** Table 1.2 highlights the fact that PROC SQL is not supported on CAS. On CAS, you can use SQL language elements, but not the procedure. Please see also Table 7.5-1 for details about when to use PROC FEDSQL and when to use PROC SQL.

PROC SQL programming expertise is a good starting point into possibilities of CAS by putting FedSQL to practice. For example, you can also use FedSQL statements in a DS2 program, which is another easy way to take advantage of the new integer data types.

I extended an overview originally provided by SAS Institute on how to choose the appropriate programming language for CAS by adding some more features and also a column for PROC SQL. Chapters 6 and 7 will discuss these programming languages (DS2, Hash Objects, FedSQL) from several angles.

Chapter 4 highlights some SAS enhancements of PROC SQL, which to date cannot be put into practice in the CAS environment.

Chapter 2: Missing Values

Data queries and analyses usually require the presence of entries such as numbers or strings in a table. Missing values indicate the absence of values. While entries represent the presence of information, missing values indicate absence of information. In data analysis, missing values are more of the rule than the exception. (See Schendera, 2020/2007, Chapter 6.)

A data set can be described as complete or incomplete by the extent to which the occurrence of missing values is observed. Complete data is observed when there is nothing missing within the data set, variable, or row. Incomplete data is observed when there are missing occurrences within the data set, variable, or row. If a data set, variable, or row contains missing data, it is described as empty. "Missing values," along with "completeness," "uniformity," and "duplicates" are among the basic criteria of data quality.

Data analysis also distinguishes between *system*-defined and *user*-defined missing values. The difference is this: with system-defined missing values, a cell is simply empty. It contains no data. With user-defined missing values, on the other hand, a cell contains a code for why no information is available, for example, "-1" for "Participant not found," "-2" for "Refused to answer," or even "-4" for "Question never asked." Such coding lists can easily reach a volume of several thousand codes. User-defined missing values allow statisticians to create information in the data record about why a value is missing. So, with user-defined missing values, statisticians know that an entry is missing and why a value is missing.

System-defined missing values do not initially contain any information except that values are missing. It is unknown why a value is missing. System-defined missing values can generally be converted to user-defined missing values. Since a table does not always consist exclusively of data, it should at least consist of data and user-defined missing values.

In principle, missing values are unpleasant because they restrict the basis for deriving information. System-defined missing values are tricky because a user does not know why this data is missing. What makes matters more difficult is that PROC SQL handles missing values slightly differently than the ANSI standard. This chapter focuses on basic aspects of how PROC SQL handles missing data. These include:

- the definition of missing values (that is, user-defined missing values)
- the retrieval of data from tables that contain missing values
- the effects of missing values on operations of data analysis such as aggregations and on data management (for example, joins)
- first fundamental measures for dealing with missing values, including deleting or searching and replacing missing values

The focus then shifts from user-defined missing values to system-defined missing values. However, *inferential statistical analyses* are not the subject of this chapter. Inferential statistical analyses of data containing missing values are extremely sensitive and require extra care.[1]

---

[1] For an initial introduction to aspects such as causes, patterns, consequences, and mechanisms of missing values, please refer to Schendera (2020/2007, Chapter 6), which presents further measures for dealing with missing values, including Hot Deck, reconstruction including imputation, or inclusion in an analysis via coding.

The handling of missing values in PROC SQL does not fully comply with the current ANSI standard. Failure to take these special features into account can lead to potentially undesirable results when dealing with missing values. (See, for example, the notes in Sections 2.4 and 2.5.)

According to the ANSI standard, the expressions "5 > NULL", "0 > NULL", or even "-5 > NULL" become *NULL*. In Boolean and comparison operators, however, this and other expressions become *true* in SAS, when working with SAS SQL. When sorting data, PROC SQL also places missing values before numeric or string data. Thus, if you specify an ascending sort order, missing values are placed at the top of all other nonmissing values. Sections 2.4 and 2.5 will present a selection of possible undesired results when handling missing values, including calculating, filtering, or joining.

The differences in the functionality of SAS and ANSI SQL are so central to working with missing data that all readers are strongly recommended to read this chapter *before* you deal with the other chapters. For example, it makes a difference whether you process missing values (NULL values) in SAS or in a non-SAS DBMS. For example, non-SAS DBMSes automatically remove missing values from a WHERE clause, whereas SAS does not. Users should be aware of these special features of SAS SQL to avoid possible programming errors or misinterpretations.

2.1 From the Start: Defining Missing Values

SAS defines missing values in different ways. Missing numerical values and missing strings are handled completely differently. A numeric variable can also have system- and user-defined missing values at the same time.

Please note that Program 2.1-1 creates a data set MISSINGS with blanks for STRING1 and STRINGX2 for ID 4 and Program 2.1-5 creates a data set MISSINGS_2 with a '_' for those same variables. The results and the MISSINGS table refer to the data set created by Program 2.1-1 below.

2.1.1 Missing Values in Numeric Variables

System-defined Numeric Missing Values

SAS defines a numerical missing as a point by default. Pay attention to the notes (◁).

Figure 2.1-1: System-defined Numeric Missing Values

VAR1	VAR2	
0	5	
-2	-6	
4	.	◁
4	7	
	B	
_		
.	◁ C	
A	8	

User-defined Numeric Missing Values

The user can also define numerical missing values after a point using an underscore or the letters from A to Z. Pay attention to the notes (◁).

Figure 2.1-2: User-defined Numeric Missing Values

VAR1		VAR2	
0		5	
-2		-6	
4		.	
4		7	
_	◁	B	◁
.		C	◁
A	◁	8	

PROC SQL treats all system-defined numerical missing values (".") the same. However, user-defined missing values are perceived and handled differently by PROC SQL; they have an internal order. All these numerical missing values are defined in the following order: " ._ " is interpreted as smaller as " . ", which again is interpreted smaller as

"A" (and so on: "A" < "B" < "C" < ... < "X" < "Y" < "Z"). If numerical missing values are compared with nonmissing numerical values, system- and user-defined missing values are interpreted as smaller than nonmissing values. Missing values that are inserted as time and date variables are perceived and handled like numerical missing values.

2.1.2 Missing Values in String Variables

SAS defines a missing letter or character string as "blank." A blank is ASCII character 20. So, if you compare blanks with each other, PROC SQL always perceives and handles them as the same. Pay attention to the notes (◁).

Figure 2.1-3: Missing Values in String Variables

```
STRING1     STRINGX2

ABC            12
DE             34
            ◁          ◁
FGH            56
            ◁   78
IJ                     ◁
KLE            90
```

Compare the sample data from MISSINGS with the results of your SAS program, especially if you use the PROC SQL version. If PROC SQL still outputs undesired results, it cannot be ruled out that your PROC SQL programming might still be handling missing values suboptimally. (See the notes in the next section.) If your PROC SQL produces the same results and you have used an older version of PROC SQL on your own data, it is possible that the new PROC SQL works *differently* and, for example, produces different results with the same data than the previous SQL version.

2.1.3 Example: Test Data for Handling Missing Values

The test data set MISSINGS contains an ID variable, four variables with missing values: two numeric variables VAR1 and VAR2 and two string variables, STRING1 and STRINGX2. Although STRINGX2 appears to be a numeric variable, it is defined as a string by the INPUT statement. Note the corresponding SAS output. The different missing values are explained under the test data; consequences are explained in the following sections.

The following examples apply to SAS 9.4 on Windows 10 Pro. On other platforms or with earlier PROC SQL versions, different results might occur if and because they do not fully comply with ANSI standards for SQL, including and especially when dealing with missing values.

Program 2.1-1: MISSINGS

```
data MISSINGS;
input VAR1 1-2 VAR2 4-5
STRING1 $7-9 STRINGX2 $11-12 ID ;
datalines;
0  5   ABC 12 1
-2 -6  DE 34 2
4          3
4  7   FGH 56 4
._  .B     78 5
.  .C  IJ    6
.A 8   KLE 90 7
;
run ;
proc print noobs ;
run ;
```

Figure 2.1-4: PROC PRINT Output

VAR1	VAR2	STRING1	STRINGX2	ID
0	5	ABC	12	1
-2	-6	DE	34	2
4	.			3
4	7	FGH	56	4
_	B		78	5
.	C	IJ		6
A	8	KLE	90	7

2.1.4 Defining Missing Entries when Creating Empty Tables

With CREATE TABLE (without AS), "empty" PROC SQL tables can be created. "Empty" means that the tables do not contain any data, but they can contain column attributes and check rules. One of the simplest check rules is the NOT NULL option. For more complex check rules, please refer to Chapter 3 on integrity constraints. NOT NULL applies equally to numeric and string variables.

2.1.5 SQL Approach: Creating an Empty Table Including Integrity Constraints for Missing Values

Using CREATE TABLE, the temporary table EMPTY is created with empty columns, named by the user as ID, STRING, and VALUE1. In the parenthesis expression, additional attributes are assigned to the columns, such as type, length, and label. NOT NULL is a check rule that prevents null values when updating, for example, when using INSERT VALUES.

Program 2.1-2: Create EMPTY table

```
proc sql ;
     create table EMPTY
         (ID num NOT NULL,
      STRING char (30) NOT NULL  label="String variable",
      VALUE1  num        NOT NULL  label="Numerical Variable") ;
quit ;
```

The specified NOT NULL conditions are logged in the specified order as integrity constraints _NM0001_, _NM0002_, etc.

If, for example, data rows without missing values in the monitored columns are added to the table EMPTY using INSERT INTO, the check rules allow the addition of the new data rows.

Program 2.1-3: Insert into EMPTY

```
proc sql ;
  insert into EMPTY
    values(1,'String', 123)
    values(2,'String2', 124) ;
quit ;
```

The SAS log for Program 2.1-3 provides the following feedback.

SAS Log 2.1-1: EMPTY (INSERT Successful)

```
NOTE: 2 rows were inserted into WORK.EMPTY.
```

If, on the other hand, an attempt is made to add data rows with missing values in the monitored columns to the table EMPTY, the check rules block the addition of the new data rows. The periods or blanks between the quotation marks represent missing values in numeric or string variables.

Program 2.1-4: Insert into EMPTY 2

```
proc sql ;
  insert into EMPTY
    values(1,'String', 123)
    values(1,' ', .)
    values(.,' ', .) ;
quit ;
```

The SAS Log for Program 2.1-4 provides the following feedback.

SAS Log 2.1-2: INSERT Failed
```
ERROR: Add/Update failed for data set WORK.EMPTY because data value(s) do not comply with
       integrity constraint _NM0002_.
NOTE: This insert failed while attempting to add data from VALUES clause 2 to the data set.
NOTE: Deleting the successful inserts before error noted above to restore table to a
consistent state.
NOTE: The SAS System stopped processing this step because of errors.
```

The note *Error:* means that the data to be inserted violates integrity constraint _NM0002_. The check rule _NM0002_ applies (due to its sequence) to the STRING column and requires that the data to be inserted there is nonzero, that is, no missing data. The first note indicates that the missing in STRING occurred in the second INSERT VALUES row. (See "VALUES clause 2".) This single missing prevented the insertion of all data rows. The *second* note indicates that the original state has been restored.

The next steps might be extremely iterative until all data is read in according to the rules. If, for example, only the missing in STRING is corrected, the missing in WERT1 (also in the second INSERT VALUES line) will trigger the check rule _NM0003_ in the next run. If this is corrected, the missing ID (in the third INSERT VALUES line) will trigger the checking rule _NM0001_ in the next run, and so on.

2.1.6 DATA Step Approach

By specifying MISSING_2, the DATA step enables you to assign different types of user-defined missing values, for example, A, B, C and "_", *even* before data is read into SAS. Due to the assigned underscore symbols in VAR1, STRING1, STRINGX2, the following example data set MISSING_2 differs from the example data set MISSINGS at the beginning of this chapter.

Program 2.1-5: Data MISSINGS_2	**Figure 2.1-5: PROC PRINT Output**

``` data MISSINGS_2 ; missing A B C _ ; input VAR1 1-2 VAR2 4-5 STRING1 $7-9 STRINGX2 $11-12 ID ; datalines; 0 5 ABC 12 1 -2 -6 DE 34 2 4 _ _ 3 4 7 FGH 56 4 _ B 78 5 . C IJ 6 A 8 KLE 90 7 ; run ; proc print noobs ; run ; ```	<table><tr><td>VAR1</td><td>VAR2</td><td>STRING1</td><td>STRINGX2</td><td>ID</td></tr><tr><td>0</td><td>5</td><td>ABC</td><td>12</td><td>1</td></tr><tr><td>-2</td><td>-6</td><td>DE</td><td>34</td><td>2</td></tr><tr><td>4</td><td>.</td><td>_</td><td>_</td><td>3</td></tr><tr><td>4</td><td>7</td><td>FGH</td><td>56</td><td>4</td></tr><tr><td>_</td><td>B</td><td></td><td>78</td><td>5</td></tr><tr><td>.</td><td>C</td><td>IJ</td><td></td><td>6</td></tr><tr><td>A</td><td>8</td><td>KLE</td><td>90</td><td>7</td></tr></table>

Note that the MISSING statement defines user-defined missing values only for numeric variables. It does not assign the underscore symbol ("_") to the missing string entries.

2.2 Queries for Missing Values

Before evaluating data or merging tables, it is recommended that you check for missing values in the tables. For example, a WHERE clause with IS NULL or IS MISSING queries all data that has a missing value in the corresponding column. Both IS conditions become true if a value in a column is missing and vice versa.

2.2.1 Query Missing Values in a Numeric Column

The following example queries all have missing values in column VAR1 of the MISSINGS table. Both IS conditions query system- and user-defined missing values at the same time.

Program 2.2-1: IS MISSING	Program 2.2-2: IS NULL
``proc sql ;
select ID, VAR1, VAR2
from MISSINGS
where VAR1 is missing ;
quit ;```	```proc sql ;
select ID, VAR1, VAR2
from MISSINGS
where VAR1 is null ;
quit ;```

SAS outputs all lines with missing values in column VAR1 of the MISSINGS table. You can tell from their formatting that these are system- and user-defined (with other characters) missing values.

Figure 2.2-1: Query Missing Values Output

```
     ID       VAR1       VAR2
  ----------------------------
      5         _          B
      6         .          C
      7         A          8
```

Note: Both IS conditions achieve the same result. SELECT changes the sequence of the columns. (See the position of ID.)

Note: A period represents system-defined missing values. Underscores or letters from A to Z represent user-defined missing values. The SAS documentation uses "special missing values" as standard term for user-defined missing values.

2.2.2 Exclusion of Missing Values from Two Numerical Columns

If the question allows it, users can take an alternative approach and filter out possible missing values from the data and query and include only the rows without missing values within the data set. The following example queries lines without missing data in columns VAR1 and VAR2 of the MISSINGS table. Both IS conditions are used simultaneously in the WHERE clause.

Program 2.2-3: Exclusion of Missing Values from Two Numerical Columns

```
proc sql ;
  select ID, VAR1, VAR2
    from MISSINGS
      where VAR1 is not missing and VAR2 is not null ;
quit ;
```

Figure 2.2-2: Output of Exclusion of Missing Values from Two Numerical Columns

```
     ID       VAR1       VAR2
  ----------------------------
      1         0          5
      2        -2         -6
      4         4          7
```

Note: Both IS conditions filter columns VAR1 and VAR2 so that so that only rows where both variables contain no missing values are selected.

Note: SELECT changes the sequence of the columns.

2.2.3 Query of User-defined Missing Values

The following example queries columns VAR1 and VAR2 of the MISSINGS table for user-defined missing values. An EQ and an IN-list are used as conditions in the WHERE clause. This approach assumes that it is known which codes are used to encode the user-defined missing values. Whether one knows which codes are used in which variable is secondary. The following example assumes that .A only appears in column VAR1 and .A, .B, and .C probably appear in column VAR2. The WHERE conditions only query for user-defined missing values.

Program 2.2-4: Query of User-defined Missing Values
```
proc sql ;
  select ID, VAR1, VAR2
    from MISSINGS
      where VAR1 eq .A or VAR2 in (.A, .B, .C) ;
quit ;
```

Figure 2.2-3: Output of Query of User-defined Missing Values
```
      ID      VAR1      VAR2
   --------------------------
       5        _         B
       6        .         C
       7        A         8
```

Note: The system missing value in VAR1 (ID 6) shows up because the condition for VAR2 is met.

2.2.4 Query of Missing Values in a String Column

If blanks are to be queried from a column of type string, the condition that is depicted subsequently is recommended. In the following example, all lines with missing values in the column STRING1 are queried.

Program 2.2-5: Query of Missing Values in a String Column
```
proc sql ;
  select ID, STRING1, VAR2
    from MISSINGS_2
      where STRING1 in ("    ");
quit ;
```

Note: Usually, only one (or no) blanks are used to query form missing character data.

Figure 2.2-4: Output of Query of Missing Values in a String Column
```
      ID   STRING1      VAR2
   --------------------------

       5                  B
```

SAS outputs all lines with missing values in column STRING1 of the MISSINGS_2 table. Missing values of type string, if they are not formatted, are generally recognized by the presence of a blank (ASCII 20), that is, the absence of any characters.

2.2.5 Variables, Accesses, and Possibly Undesired Results of Queries for Missing Values

PROC SQL interprets and handles system- or user-defined numerical missing values as well as string missing values differently. Both different data queries are required, as well as different results and outputs might follow. The following examples demonstrate how PROC SQL can produce different results depending on expression or missing values.

Logical Expression

Program 2.2-6: Logical Expression	**Figure 2.2-5: Output of Logical Expression**

```
proc sql;
create table LOGIC as
select not VAR1 as mis1_not,
       VAR1 or . as mis2_or,
       VAR1 and . as mis2_and
from MISSINGS;
quit;
```

mis1_not	mis2_or	mis2_and
1	0	0
0	1	0
0	1	0
0	1	0
1	0	0
1	0	0
1	0	0

In logical expressions (AND, OR, NOT), numerical variables and thus numerical missing values can take the values 1 (for "true") or 0 (for "false"). The created variables MIS1_NOT (when querying a variable) or MIS2_OR and MIS2_AND (each comparing two variables) demonstrate that PROC SQL perceives system- or user-defined numerical missing values equally. Missing values and the value 0 are interpreted as "false," nonmissing numerical values (positive or negative) as "true."

Arithmetic Expression

Program 2.2-7 Arithmetic Expression	**Figure 2.2-6: Output of Arithmetic Expression**

```
proc sql;
create table ARTHMTC as
select var1 + 1 as ADD_PLS1 from MISSINGS_2
where var1 < 0;
quit;
```

```
ADD_PLS1

   -1
    .
    .
    .
```

In arithmetic expressions (+, -, *, /), numerical missing values always lead to numerical missing values. If these numerical missing values are included in other expressions, the result will be another set of missing values.

String Expression

Program 2.2-8: STRING Expression	**Figure 2.2-7: Output of STRING Expression**

```
proc sql;
create table STRING as
select "PRFX_"||string1||stringx2
as STRING_1
from MISSINGS ;
quit ;
```

```
STRING_1
PRFX_ABC12
PRFX_DE 34
PRFX_
PRFX_FGH56
PRFX_   78
PRFX_IJ
PRFX_KLE90
```

Missing character variables consist of as many blanks (ASCII 20) as the number of bytes in the storage LENGTH of the variable. That's why many spaces appear in some of the values of STRING_1 (for example, the second, third, and fifth entry in STRING_1).

2.3 Missing Values in Aggregating Functions

Missing values are also a special topic in aggregating functions. Because many aggregation functions ignore missing values, it is quite possible that the results initially achieved do not correspond with what one would expect. For example, the AVG function, only includes the nonmissing values in the calculation of the mean. If, on the other hand, the divisor should also contain the number of missing values, a different approach must be chosen. Even small adjustments to aggregating functions can have a big impact. This can be illustrated when counting values or data rows using COUNT or calculating average values using AVG.

2.3.1 Counting Existing, Duplicate, and Missing Values with COUNT

Values of a column can be counted using the functions COUNT, NMISS, and MISS, among others. Clever combination with other keywords decides whether the desired query results are achieved.

Together with DISTINCT, the functions COUNT, NMISS, and MAX can be used in different ways to count the number of levels of the numeric column VAR1 in the MISSINGS table. Column VAR1 contains 7 rows, 3 of which are missing values, 4 values, and of the existing values, one value occurs twice, the 4. Many missing values can be handled as multiples and as unique, for example, counting every occurrence only once (see "count(distinct VAR1) + max(missing(VAR1)").

- "count(*)" returns the number of all rows of a data set including missing values and all other values (see "miss_in_else_in"): 3 + 4 = 7.
- "nmiss(VAR1)" returns the number of missing values in VAR1 without all other values (see "miss_in_else_out"): 3 + 0 = 3.
- "count(distinct VAR1) + max(missing(VAR1))" returns the number of all unique missing values in VAR1 plus the number of all available unique values. 1 + 3 = 4. So, DISTINCT reduces many missing values to 1 *unique* missing. In contrast, MISSING in combination the MAX function returns just 1. Although MISSING returns 1 three times (because a value is missing, otherwise 0), the MAX function cannot but return 1 as the highest value for a 1 for ID 5, a 1 for ID 6, and a 1 for ID 7.
- "count(VAR1)" counts only nonmissing values of VAR1, that is, excludes all missing values in VAR1, but returns the number of all existing values including double values: 0 + 4 = 4.
- "count(distinct VAR1)" counts unique, nonmissing values of VAR1, that is, it excludes all missing values and duplicates in VAR1, otherwise it returns the number of all unique existing values: 0 + 3 = 3.

These query variants can also be applied directly to string variables, for example, STRING1 or STRING2 from the MISSINGS table.

Program 2.3-1: Counting Valid and Missing Values
```
proc sql;
select
        count(*)                as miss_in_else_in,
        nmiss(VAR1)             as miss_in_else_out,
        count(distinct VAR1) +  max(missing(VAR1))
                                as miss_in_dups_out,
        count(VAR1)             as miss_out_else_in,
        count(distinct VAR1) as miss_out_dups_out
 from MISSINGS ;
quit ;
```

Figure 2.3-1: Counting Valid and Missing Values

miss_in_ dups_in	miss_in_ else_out	miss_in_ dups_out	miss_out_ else_in	miss_out_ dups_out
7	3	4	4	3

See also the concluding remarks on DISTINCT in Section 2.3.4.

2.3.2 Variations in Aggregating Functions

Under certain circumstances, missing values in aggregating functions can lead to conceptually different results. When specifying the divisor, you can choose to use a constant (for example, a theoretical maximum or target value) or the number of nonmissing values (Schendera, 2007, 129ff.). Confusing both approaches, the resulting error has two faces: 1. when divided by the number of theoretically possible values but *interpreted* as a mean based on nonmissing values, and 2. when divided by the number of nonmissing values but *interpreted* as a mean based on *all theoretically possible* values.

The challenge for the user is to decide which interpretation (division by number of nonmissing values or all theoretically possible values) is the most appropriate for his purposes. The following illustration shows the calculation of selected measures of location and dispersion.

Program 2.3-2: Missing Values in Aggregating Functions	Figure 2.3-2: Output of Missing Values in Aggregating Functions (Extract)
```proc sql ; select AVG(S0666) as MEAN_1 from SASHELP.SYR1001 ; select sum(S0666) as S0666_SUM, count(*) as S0666_N, calculated S0666_SUM/calculated S0666_N as MEAN_2 from SASHELP.SYR1001 ; select sum(S0666) as S0666_SUM, count(S0666) as S0666_N, calculated S0666_SUM/calculated S0666_N as MEAN_3 from SASHELP.SYR1001 ; select sum(S0666) as S0666_SUM, calculated S0666_SUM / 105 as MEAN_4 from SASHELP.SYR1001 ; select sum(S0666) as S0666_SUM, calculated S0666_SUM / 106 as MEAN_5 from SASHELP.SYR1001 ; quit ;```	``` MEAN_1 -------- 1.754444 MEAN_2 -------- 0.902285 MEAN_3 -------- 1.754444 MEAN_4 -------- 0.902285 MEAN_5 -------- 0.893773```

The variants MEAN_1 and MEAN_3 contain in the divisor all rows with values in S0666 (N=54) and achieve an average value of 1.754444. The variants MEAN_2 and MEAN_4 contain in the divisor all lines of the data set including the missing values in S0666 (N=105) and achieve an average value of 0.902285. The variant MEAN_5 contains the value 106 as divisor and achieves as mean value 0.893773. Mathematically, all three achieved mean values are correct; however, which divisor (54, 105, or 106) is correct (or also wrong) can only be judged with regard to which statement is to be made using the respective mean. Therefore, it must be checked whether it should be divided using a specific value (N=106), all rows (N=105) or all valid cases (N=54).

2.3.3 Adjusting an Aggregating Function Using CASE

As seen in the previous example, missing values in aggregating functions can lead to undesired results. For example, many aggregating functions ignore missing values.

In the following example, the average of all S0666 values is calculated again. For example, the first approach with the simple AVG function returns the arithmetic mean based on all nonmissing values only (N=54). The second approach, however, previously assigns the value 0 to each missing value in S0666 via CASE and stores it in the new variable S0666_0. The arithmetic mean is now calculated on the basis of *all* data rows (N=105).

Program 2.3-3: Syntax with Simple AVG

```
proc sql ;
      create table MYDATA as select
      T, S0666, avg(S0666) as S0666_MEAN
      from SASHELP.SYR1001 ;
quit ;
```

SAS Log 2.3-1: Syntax with Simple AVG

```
NOTE: Table WORK.MYDATA created, with 105 rows and 3 columns.
```

Figure 2.3-3: Output for Simple AVG (Extract)

```
                       S0666_
   T     S0666         MEAN
   1     1.80000       1.75444
   2     1.52000       1.75444
   3     1.51000       1.75444
   4     1.80000       1.75444
   5     1.85000       1.75444 ...
```

The arithmetic mean (1.75444) is calculated on the basis of the sum of all nonmissing (valid) values (94.740) divided by the number of all nonmissing (valid) values (N=54). If, however, you want to divide by the number of all data rows (N=105) (or, from another perspective, interpret a missing value as the value 0), you can choose the above-mentioned approach using CASE.

Program 2.3-4: Syntax with Extended AVG

```
proc sql;
     create table MYDATA
        as select T, S0666, case
            when S0666 is missing then 0
                    else S0666
                    end as S0666_0 ,
              avg(calculated S0666_0) as S0666_MEAN
        from SASHELP.SYR1001 ;
quit;
```

SAS Log 2.3-2: Syntax with Extended AVG

```
NOTE: Table WORK.MYDATA created, with 105 rows and 4 columns.
```

Figure 2.3-4: Output for Extended AVG (Extract)

```
                                S0666_
    T     S0666      S0666_0     MEAN
   50     1.61000    1.61000    0.90229
   51     1.74000    1.74000    0.90229
   52     1.52000    1.52000    0.90229
   53     1.84000    1.84000    0.90229
   54     1.96000    1.96000    0.90229
   55     .          0.00000    0.90229
   56     .          0.00000    0.90229
   57     .          0.00000    0.90229
   58     .          0.00000    0.90229 ...
   59     ...
```

PROC SQL uses CASE to assign the value 0 to each missing in S0666 and stores it in the new variable S0666_0. The arithmetic mean is now calculated on the basis of the sum of all nonmissing (valid) values (94,740), but also the number of all data rows (N=105). The arithmetic mean is now 0.90 compared to 1.75 in the first variant. See also the notes on grouping for missing values in the next chapter.

Numeric and alphanumeric missing values can also be converted into each other using the INPUT or PUT functions. (See Schendera, 2004.) The next example shows how PROC SQL converts numerals stored in a character data type variable into a number in a numeric type variable before aggregation. At the same time, a blank in the original variable STRINGX2 from the MISSINGS table is converted to the value 0 of the newly created variable MIN_ZEROS. (See output MYDATA1, ID 3.)

Program 2.3-5: Syntax for Missing Values Using INPUT or PUT

```
proc sql;
 create table MYDATA1 as
  select ID, STRINGX2,
    min(input(STRINGX2, 2.)) as MIN_VALIDS
label="Minimum nonmissing values",
    min(sum(0, input(STRINGX2, 2.))) as MIN_ZEROS
label="Minimum with 0 instead of missing"
      from MISSINGS
group by ID;
 create table MYDATA2 as
  select
      mean(MIN_VALIDS) as MEAN_VALIDS
label="Average of nonmissing values",
      mean(MIN_ZEROS) as MEAN_INCZERO
label="Average of nonmissing values including 0",
      sum(MIN_VALIDS) as SUM_VALIDS
label="Sum of nonmissing values",
      sum(MIN_ZEROS) as SUM_INCZERO
label="Sum of nonmissing values including 0"
      from MYDATA1 ;
quit ;
proc print data=MYDATA1 noobs;
run;
proc print data=MYDATA2 noobs;
run;
```

Figure 2.3-5 Output MYDATA1 (Extract)				Figure 2.3-6 Output MYDATA2 (Extract)			
ID	STRINGX2	MIN_VALIDS	MIN_ZEROS	MEAN_VALIDS	MEAN_INCZERO	SUM_VALIDS	SUM_INCZERO
1	12	12	12	54	38.5714	270	270
2	34	34	34				
3		.	0				
4	56	56	56				
5	78	78	78				
6		.	0				
7	90	90	90				

2.3.4 Final Remarks on DISTINCT

DISTINCT causes only unique values of the SQL expression to be included in the calculation (in contrast to ALL). What is often overlooked is that DISTINCT also works for missing values. For example, DISTINCT reduces many system-defined missing values to 1 system-defined missing value, which can have fatal consequences when filtering or joining tables. (See, for example, Program 2.3-1.) This effect also applies to user-defined missing values.

DISTINCT also differentiates between types of missing values. For example, if data contains system- and user-defined missing values, aggregations or joins are performed including *each type of missing value*.

DISTINCT also determines the feedback of results: if DISTINCT is applied together with one or more aggregation functions to data that itself contains only missing values, only two different results can be returned. COUNT(*), COUNT(DISTINCT), and NMISS(DISTINCT) return a 0; all other aggregation functions return a missing value.

Only an appropriate use of DISTINCT will achieve the desired results in dealing with missing values, with different types of missing values, and in the form of feedback of results.

2.4 Possibly Undesirable Results with WHERE, GROUP, and ORDER

Missing values can lead to undesired results while working with only one table, for example, when sorting data (ORDER BY), in WHERE conditions, and when grouping data (GROUP BY). (See the following examples.) When grouping data, other, possibly undesired effects can occur. For example, cells with missing values might be treated as if they belong *together* or calculated values might be returned even for rows that did not provide any data for the calculation. Subsequently, Section 2.5 will present possibly undesired results when working with two tables.

2.4.1 Unwanted Results with WHERE

If a column in a WHERE condition contains missing values, a simple WHERE condition can lead to undesired results under certain circumstances. In the following example, the following query should output all S0666 values that are smaller than 1.5. The intuitively obvious SQL programming, simply specifying a WHERE with a < character, can lead to undesired results in case of missing values in the data set.

Program 2.4-1: Syntax with Simple WHERE

```
proc sql ;
      create table MYDATA as select *
      from SASHELP.SYR1001
      where S0666 < 1.5 ;
quit ;
```

SAS Log 2.4-1: Syntax with Simple WHERE

```
NOTE: Table WORK.MYDATA created, with 54 rows and 3 columns.
```

Figure 2.4-1: Output for Simple WHERE

T	S0502	S0666
16	411	1.19000
26	344	1.30000
27	294	1.48000
55	342	.
56	405	.
57	389	.
58	364	.
59	361	

Because SAS defines missing values as smaller than any nonmissing value, the simple WHERE condition causes the created table to contain not only the S0666 values smaller than 1.5, but also all missing values in S0666.

If you want to keep only nonmissing S0666 values (only values less than 1.5, but without missing values), you have to extend the WHERE condition by adding IS NOT MISSING.

Program 2.4-2: Syntax with Extended WHERE

```
proc sql ;
      create table MYDATA as select *
      from SASHELP.SYR1001
      where S0666 < 1.5
            and S0666 is not missing ;
quit ;
```

SAS Log 2.4-2: Syntax with Extended WHERE

```
NOTE: Table WORK.MYDATA created, with 3 rows and 3 columns.
```

Figure 2.4-2: Output for Extended WHERE

```
  T    S0502    S0666
 16      411  1.19000
 26      344  1.30000
 27      294  1.48000
```

If, however, missing values in S0666 values are not desired (if only all values less than 1.5 are to be stored but without missing values), the WHERE condition must be extended by the additional condition IS NOT MISSING.

> **Important note:** When you program WHERE clauses, you must take into account the system and the way in which any missing values are processed. The important point is: Processing data stored in a SAS data set recognizes non-NULLs and SAS special missing values as valid; in contrast, processing data stored in a relational database system like Oracle recognizes only NULLs as missing. If the data in SAS is processed in a non-SAS DBMS, these missing values are often removed from a WHERE clause by default. However, if the data is processed in SAS, SAS keeps any missing values in the WHERE clause by default. Undesired different results can be the consequence indeed. (See the following examples.)

2.4.2 Undesired Results when Sorting Data

Missing values can possibly also lead to undesired results when sorting one or more columns. As mentioned in the introduction, SAS SQL handles missing values *differently* than the ANSI standard for SQL. PROC SQL follows the SAS convention for handling missing values: If numeric missing values are compared with nonmissing (valid) values, the numeric missing values are always interpreted as *less or smaller* than all nonmissing (valid) values. However, if missing values of type character are compared with nonmissing characters, the missing values of type character are interpreted as a string of blanks. When sorting data, PROC SQL arranges missing values before data of type numeric or string. If you specify an ascending sort sequence (default, ASC, ascending), missing values are placed at the top of all other nonmissing values and can therefore appear in the rows of a query or table. In this respect, SAS SQL can also differ from other SQL variants.

Program 2.4-3: Syntax with Simple ORDER BY

```
proc sql ;
      create table MYDATA as select *
      from SASHELP.SYR1001
      order by S0666 ;
quit ;
```

SAS Log 2.4-3: Syntax with Simple ORDER BY

```
NOTE: Table WORK.MYDATA created, with 105 rows and 3 columns.
```

Figure 2.4-3: Output for Simple ORDER BY

```
   T    S0502   S0666
 100      839       .
  80      846       .
  79      755       .
  56      405       .
  72      764       .
```

Because S0666 contains missing values, the ORDER BY places them before all other values. Because ORDER BY does not contain any other sorting statement, for example, by other variables, the values in columns T and S0502 are not sorted. In T, for example, the value 56 is between 79 and 72. If the values are to be explicitly ordered, especially in the presence of missing values, the ORDER BY must be supplemented by further sorting variables, for example, T and S0502.

Program 2.4-4: Syntax with Augmented ORDER BY

```
proc sql ;
      create table MYDATA as select *
      from SASHELP.SYR1001
      order by S0666, T, S0502 ;
quit ;
```

SAS Log 2.4-4: Syntax with Augmented ORDER BY

```
NOTE: Table WORK.MYDATA created, with 105 rows and 3 columns.
```

Figure 2.4-4: Output for Simple ORDER BY

```
    T      S0502     S0666

   55       342        .
   56       405        .
   57       389        .
   58       364        .
   59       361        .
   60       396        .
   61       478        .
   62       348        .
```

If the ORDER BY was augmented by T and S0502 as sorting variables, the values are ordered in ascending order despite the missing values in S0666. First the rows were sorted by the values of S0666, then by T, and finally by S0502. Because the sorting of S0666 and T has priority, column S0502 only appears to be unsorted. For example, if S0666 and T each contained two identical (valid, missing) values, then a smaller value would be sorted before a larger value in S0502.

2.4.3 Undesirable Results when Grouping Data

When grouping data, two variants of possibly undesired results can occur. The first variant is of a more mathematical nature and is primarily observed with aggregating functions. The second is more content-related and can occur primarily with categorical data.

For the mathematical variant, an SQL example already presented in another context is used for demonstration purposes. The arithmetic means of S0666 and S0666_0 are calculated based on the sum of all available values divided by the number of nonmissing values data rows.

Program 2.4-5: Grouping when Calculating with Missing Values I

```
proc sql;
      create table MYDATA as select
      T, S0666,
          avg(S0666) as S0666_MEAN,
             case
               when S0666 is missing then 0
                     else S0666
             end as S0666_0,
        avg(calculated S0666_0) as S0666_0_MEAN
      from SASHELP.SYR1001 ;
quit;
```

Figure 2.4-5: Output for Grouping with Missing Values in Aggregating Functions (Extract)

T	S0666	S0666_ mean	S0666_0	S0666_0_ mean
52	1.52000	1.75444	1.52000	0.90229
53	1.84000	1.75444	1.84000	0.90229
54	1.96000	1.75444	1.96000	0.90229
55	.	1.75444	0.00000	0.90229
56	.	1.75444	0.00000	0.90229
57	.	1.75444	0.00000	0.90229
58	.	. . .		

It might be undesirable for users that PROC SQL returns means for S0666 and the calculated S0666_0 even for rows that did not provide any values for its calculation, for example, the rows of T-values 55, 56, and so on. Here, "grouping" does not mean using GROUP (for example,, "group by T"; observe the dramatically different results though); it means a group (subset) of values changes from having missing values to no missing values.

To better illustrate what this can mean from a content-related point of view, an example with string data from the SAS documentation is used for demonstration purposes. (See Figure 2.4-6.) The SAS table COUNTRIES contains countries (NAME), their area in square miles (AREA), and the continent to which they are assigned. The data set contains three countries that are not assigned to a continent, that is, entries that have missing values in CONTINENT. These are Bermuda, Iceland, and Kalaallit Nunaat (not shown in the example output). The aim of this analysis is to determine the total area of all countries per continent (C_AREA) and to return this value for each country.

Figure 2.4-6: Content of COUNTRIES (Extract)

NAME	AREA	CONTINENT
Afghanistan	251825	Asia
Albania	11100	Europe
Algeria	919595	Africa
Andorra	200	Europe
Angola	481300	Africa
Antigua and Barbuda	171	Central America
Argentina	1073518	South America
Armenia	11500	Asia
Australia	2966200	Australia ...

Program 2.4-6: Grouping when Calculating with Missing Values II

```
proc sql ;
     create table MYDATA as
    select NAME format=$25., CONTINENT,
          sum(AREA) format=comma12. as C_AREA
    from COUNTRIES
    group by CONTINENT
    order by CONTINENT, NAME ;
quit ;
```

Figure 2.4-7: Output for Grouping with Missing Values in Category Data (Extract)

NAME	CONTINENT	C_AREA
Bermuda		876,800
Iceland		876,800
Kalaallit Nunaat		876,800
Algeria	Africa	11,299,595
Angola	Africa	11,299,595
Benin	Africa	11,299,595
Botswana	Africa	11,299,595 ...

In the output, the user can see from the table COUNTRIES that in the countries Bermuda, Iceland, and Kalaallit Nunaat, there are missing values in the column CONTINENT. Similar to how PROC SQL calculates the total area for the continent Africa and interprets all CONTINENT entries in "Africa" as belonging to one group, SQL interprets all entries with missing values in CONTINENT as belonging to the continent "Missings."

Arithmetically speaking, it might be undesirable for the user that PROC SQL returns the calculated value even in rows that did not provide the variable CONTINENT with any values for an explicit grouping. On the other hand, Bermuda, Iceland, and Kalaallit Nunaat do *not* belong to the same continent *geographically* speaking. The fact that PROC SQL treats these countries in such a way is exclusively due to the missing values in CONTINENT. This is an obvious mistake and can only be understood if this incorrect grouping was also recognized as incorrect in terms of content, which can be considerably more complicated, especially in the case of more complex groupings.

2.4.4 Undesired Results when Joining Tables

> **Important note:** SAS SQL treats missing values differently than the ANSI standard for SQL. By default, the set operators automatically exclude multiple rows from their result tables. You can use ALL to keep multiple rows. (See also Section 2.5.) DISTINCT has the effect that only the unique values of the relevant columns are kept in the join. However, since DISTINCT also differentiates between types of missing values, it is advisable to check whether different, especially *user-defined*, missing values occur in the data and to check the *consistency* of their respective definitions in several tables.

2.5 Possibly Undesirable Results with Joins

When tables are joined, undesired results can occur if the ID variables (keys) of the tables to be joined not only have missing values, but also if the missing values are different (system- versus user-defined). Section 2.5.1 presents several examples of the effect of missing values for self-joins. Section 2.5.2 presents examples of the effect of missing values when joining two tables.

2.5.1 Examples I: Self-Joins on a Single Table

Example 1: Inner Join

A self-join is performed on the table MISSINGS using the inner join method. Using ON and table aliases, the system selects all the rows whose values in VAR1 and VAR2 match. Here, the variables VAR1 and VAR2 are the key variables (not the variable ID!). The missing values in VAR1 and VAR2 are both system- *and* user-defined missing values.

In the inner join example, the filter ON A.VAR1=B.VAR2 is used to select only the VAR1-VAR2 value pairs that match. The created table JOIN_INN contains only rows with the key variables matching (not necessarily) nonmissing values in both data supplying tables. Since only the system-defined missing values of rows 3 and 6 in columns VAR1 and VAR2 of table MISSINGS match, that is, the value of VAR1 in row 6 ('.') matches the value of VAR2 in row 3 ('.') (but not the two user-defined missing values in row 5, or the other system- and user-defined pairings!), only these are stored in table JOIN_INN.

All other lines are excluded from JOIN_INN. The data is not explicitly sorted. If user-defined missing values were to match *exactly*, they would also be output in PROC SQL (not necessarily in other SQL versions).

Program 2.5-1: Syntax for Inner Join	**Figure 2.5-1: Output: Syntax for Inner Join**
``proc sql ;
create table JOIN_INN as
select a.ID,
b.ID as ID_B,
a.var1, b.var2
from MISSINGS as a inner join
MISSINGS as b
on a.var1=b.var2;
quit;```	ID ID_B VAR1 VAR2 6 3 . .

The COALESCE function together with ID leads to what appears to be the same result, but in a completely different way. COALESCE returns the first nonmissing value from a list of numeric arguments, whereas COALESCEC returns it from a list of character arguments. Previously filtered with ON, the COALESCE function replaces the '._' with '.B' in row 5, '.' with '.C' in row 6, and '.A' with '8' in row 7. Still the only match is the '.' in rows 6 (VAR1) and 3 (VAR2).

Program 2.5-2: Syntax for Inner Join (COALESCE)	**Figure 2.5-2 Output: Syntax for Inner Join (COALESCE)**
``proc sql;
create table JOIN_INN
as select a.ID, b.ID as ID_B,
coalesce(a.VAR1, b.VAR2)
as VAR1, b.VAR2
from MISSINGS as a inner join
MISSINGS as b
on a.var1=b.var2
order by a.ID ;
quit;```	ID ID_B VAR1 VAR2 6 3 . .

Example 2: Outer Join (Left)

A self-join is performed on the table MISSINGS using the outer join method "left." Outer joins are basically inner joins that have been extended by rows that do not match *any* row of the other table. A left outer join (LEFT JOIN with ON) keeps all rows of the Cartesian product of the two tables for which the SQL expression is true, and also the rows from the left table that do not match any row in the second table. Using ON and table aliases, all rows whose VAR1 and VAR2 values match each other exactly are selected. Here, the variables VAR1 and VAR2 are the key variables (not the variable ID!).

Program 2.5-3: Syntax for Outer Join (Left)	**Figure 2.5-3: Output: Outer Join (Left)**
``proc sql ;
create table JOIN_LEFT as
select a.ID,
b.ID as ID_B,
a.var1, b.var2
from MISSINGS as a left join
MISSINGS as b
on a.var1=b.var2
order by a.ID ;
quit ;```	ID ID_B VAR1 VAR2 1 . 0 . 2 . -2 . 3 . 4 . 4 . 4 . 5 . _ . 6 3 . . 7 . A .

Since no b.VAR2 value in the MISSINGS table corresponds to the values of VAR1, with the exception of the system-defined missing (see IDs 3 and 6), a complete a.VAR1 column and an empty VAR2 column are output.

The Outer Join Method Left keeps the system-defined missing values (ID 6) in a.VAR1 and (ID 3) in b.VAR2 because they correspond to each other. In addition, the left outer join keeps all rows of the "left" table (that is, a.VAR1) that do not match any row of the second table (b.VAR2). All other b.VAR2 values (except for the value in row ID3) are system-defined missing values ('.') because that is the value that matches the '.' of a.VAR1 in ID 6. That value is then paired with all the non-matching values of a.VAR1. In other words, the output is the Cartesian product of the matching rows and the non-matching rows of a.MISSINGS. In the outer join example, the missing values in column VAR2 in the output table JOIN_LEFT are therefore created in two ways: The value in row 6 is created because it matches the value in VAR1; all other VAR2 values because they do *not* match the values in VAR1. In addition, this result also shows that SAS SQL treats *differently user-defined* missing values as non-matching (see ID 5), whereas *user- and system-defined* missing values are treated as matching (see ID 6).

If the COALESCE function is used, the result is *not* the same.

Program 2.5-4 Syntax with COALESCE function	Figure 2.5-4: Output: COALESCE

```
proc sql;
create table JOIN_LEFT
   as select a.ID,
      b.ID as B_ID,
      a.VAR1 as VAR1_A,
      a.VAR2 as VAR2_A,
   coalesce(a.VAR1, b.VAR2) as VAR1_C,
      b.VAR2
from MISSINGS as a left join
   MISSINGS as b
on a.var1=b.var2
order by a.ID ;
quit;
```

ID	ID_B	VAR1_A	VAR2_A	VAR1_C	VAR2
1	.	0	5	0	.
2	.	-2	-6	-2	.
3	.	4	.	4	.
4	.	4	7	4	.
5	.	_	B	.	.
6	3	.	C	.	.
7	.	A	8	.	.

The data filtered by ON contain nonmissing values and missing values of different types. COALESCE checks the specified variables VAR1 and VAR2 (in "different" tables) and returns the first nonmissing value found. COALESCE returns a missing value when user-defined missing values do not match system-defined missing values or when they were found before the nonmissing values (see 'A': if 8 were in VAR1 and '.A' in VAR2, the result would be 8). Because there are no matching user-defined missing values, the JOIN_LEFT table does not contain any user-defined missing values (see VAR1_C in ID 5 to 7; cf. 'A': if '.A' were in VAR1 and VAR2, VAR1_C would be '.A').

2.5.2 Examples II: Missing Values in Multiple Tables

The two tables ONE and THREE are combined using different join methods. The column ID in both tables is the key variable (with the same name). Each of the missing IDs are system-defined missing values.

To avoid possible misunderstandings, the two key variables should not have the same name. Most of the programs presented here also work if the key variables have different names.

The Demo Tables

The data set ONE contains the IDs 1 to 5 and the variables A, B, and C. Data set THREE contains the variables E, F, G, and also the variable ID. However, the two tables differ in their values in ID. In ONE, the value 6 is missing (the value 4 occurs instead). In THREE, the value 4 is missing (the value 6 occurs instead). JOIN operators are used to combine these tables. The following JOIN examples highlight the effect of missing values in two key variables (here: "ID").

Program 2.5-5: Syntax ONE

```
data ONE ;
   input ID A B C ;
datalines;
01 1   2  3
02 1   2  3
03 99 99 99
04 1   2  3
05 1   2  3
;
```

Program 2.5-6: Syntax THREE

```
data THREE ;
   input ID E F G ;
datalines;
01 4   4  4
02 4   4  4
03 99 99 99
05 4   4  4
06 4   4  4
;
run ;
```

Program 2.5-7 SQL Syntax INNJOIN: FROM with AS

```
proc sql ;
create table INNJOIN as
select a.ID, A, B, E, F
from ONE as a inner join THREE as b
on a.ID=b.ID ;
quit ;
```

Figure 2.5-5: Output: INNJOIN (see ID 4 and 6)

ID	A	B	E	F
1	1	2	4	4
2	1	2	4	4
3	99	99	99	99
5	1	2	4	4

Program 2.5-8: SQL Syntax: Short Form (without AS)

```
proc sql ;
create table INNJOIN as
select a.ID, A, B, E, F
from ONE a inner join THREE b
on a.ID=b.ID ;
quit ;
```

The table INNJOIN created using the inner join method only contains rows with IDs with nonmissing (valid) values in both data supplying tables. If an ID value only occurs in one table, this row is excluded from INNJOIN. For example, the rows with IDs 4 and 6 were excluded. The data is not explicitly sorted.

Program 2.5-9: LEFTJOIN

```
proc sql ;
create table LEFTJOIN as
select a.ID, A, B, E, F
from ONE a left join THREE b
on a.ID=b.ID
order by ID ;
quit ;
```

Figure 2.5-6: Output: LEFTJOIN (see ID 6)

ID	A	B	E	F
1	1	2	4	4
2	1	2	4	4
3	99	99	99	99
4	1	2	.	.
5	1	2	4	4

The table LEFTJOIN created using the left join method only contains rows for the IDs for the structure-providing table ONE. If an ID value only occurs in table THREE, this row is excluded from LEFTJOIN. For example, the row with the ID 6 was excluded. The data is explicitly sorted. Since table THREE does not contain ID 4, the values in E and F are filled with missing values for this row.

Program 2.5-10: FULLJOIN

```
proc sql ;
create table FULLJOIN as
select a.ID, A, B, E, F
from ONE a full join THREE b
on a.ID=b.ID ;
quit ;
```

Figure 2.5-7: Output: FULLJOIN (See Data Row for ID 6)

ID	A	B	E	F
1	1	2	4	4
2	1	2	4	4
3	99	99	99	99
4	1	2	.	.
5	1	2	4	4
.	.	.	4	4

The table FULLJOIN created using the full join method contains all rows for the structure-providing table ONE. However, the ID 6 does not occur in table ONE; therefore, ID of FULLJOIN contains a missing entry at this point. The corresponding data row is specified according to its order in table THREE. Since table ONE does not contain ID 6, the values in A and B are filled with missing values for the row with the missing in ID. Since table THREE does not contain ID 4, the values in E and F are filled with missing values for this line. The data is not explicitly sorted.

Program 2.5-11: FULLJOIN with ORDER		Figure 2.5-8: Output: FULLJOIN (See Data Row for ID 6)			

```
proc sql ;
create table FULLJOIN as
select a.ID, A, B, E, F
from ONE a full join THREE b
on a.ID=b.ID
order by ID ;
quit ;
```

ID	A	B	E	F
.	.	.	4	4
1	1	2	4	4
2	1	2	4	4
3	99	99	99	99
4	1	2	.	.
5	1	2	4	4

If the data is explicitly sorted using ORDER BY, the original physical position (in the data set) is changed into a sequence according to the value of ID. Since ID 6 does not occur in table ONE, ID of FULLJOIN still contains a missing value at this newly acquired: "first" position. If there are several missing values in an ID variable, sorting by ID can be counterproductive under certain circumstances, since the rows can be distinguished from each other *before* sorting, at least by their position in the table. Sorting would possibly make the lines with missing values in the ID or sorting variable indistinguishable.

Program 2.5-12: FULLJOIN (by ID)		Figure 2.5-9: Output: FULLJOIN (See Data Row for ID 4)			

```
proc sql ;
create table FULLJOIN as
select b.ID, A, B, E, F
from ONE a full join THREE b
on a.ID=b.ID ;
quit ;
```

ID	A	B	E	F
1	1	2	4	4
2	1	2	4	4
3	99	99	99	99
.	1	2	.	.
5	1	2	4	4
6	.	.	4	4

If the data is sorted by the ID from THREE, ID of FULLJOIN (at the position of the value 4) contains a missing value because the value 4 does not occur in THREE but only in ONE. The data is not explicitly sorted.

Program 2.5-13: FULLJOIN (by ID, ORDER)		Figure 2.5-10: Output: FULLJOIN (See Data Row for ID 4)			

```
proc sql ;
create table FULLJOIN as
select b.ID, A, B, E, F
from ONE a full join THREE b
on a.ID=b.ID
order by ID ;
quit ;
```

ID	A	B	E	F
.	1	2	.	.
1	1	2	4	4
2	1	2	4	4
3	99	99	99	99
5	1	2	4	4
6	.	.	4	4

If the data is explicitly sorted using ORDER BY, the original physical position (in the data set) is changed into a sequence according to the value of ID. Since ID 4 does not occur in table THREE, ID of FULLJOIN still contains a missing value at this (now: first) position.

Program 2.5-14: FULLJOIN (COALESCE)		Figure 2.5-11: Output (See Data Row for ID 4)			

```
proc sql ;
create table FULLJOIN as
select coalesce (a.ID, b.ID)
    as ID, A, B, E, F
from ONE a full join THREE b
on a.ID=b.ID ;
quit ;
```

ID	A	B	E	F
1	1	2	4	4
2	1	2	4	4
3	99	99	99	99
4	1	2	.	.
5	1	2	4	4
6	.	.	4	4

If the COALESCE function is used, the FULLJOIN table contains all ID values, even if an ID value only occurs in one table. Since table ONE does not contain ID 6, the values in A and B are filled with missing values within the row. Since table THREE does not contain ID 4, the values in E and F are filled with missing values for this row. The data is not explicitly sorted.

Program 2.5-15: FULLJOIN (a.ID as ID_ONE)

```
proc sql ;
create table FULLJOIN as
select a.ID as ID_ONE,
       b.ID as ID_THREE, A, B, E, F
from ONE a full join THREE b
on a.ID=b.ID ;
quit ;
```

Figure 2.5-12: Output

ID_ONE	ID_THREE	A	B	E	F
1	1	1	2	4	4
2	2	1	2	4	4
3	3	99	99	99	99
4	.	1	2	.	.
5	5	1	2	4	4
.	6	.	.	4	4

Program 2.5-16: FULLJOIN (COALESCE variant)

```
proc sql ;
create table FULLJOIN as
select coalesce(a.ID) as ID_ONE,
       coalesce(b.ID) as ID_THREE,
                     A, B, E, F
from ONE a full join THREE b
on a.ID=b.ID ;
quit ;
```

The COALESCE variant achieves the same result.

The two preceding approaches differ from all the approaches presented thus far in that *two* ID variables are created, which is appropriate in applications where security has priority over efficiency. The FULLJOIN table does not contain an ID variable that contains all the values (from 1 to 6 in the example) of all data-providing tables. In a further step, a (composite) index variable could be created from two (or more) ID variables.

Program 2.5-17: INNJOIN (COALESCE variant)

```
proc sql ;
create table INNJOIN as
select a.ID,
   coalesce(a.A, b.F) as NEW_AF,
   coalesce(a.B, b.G) as NEW_BG
from ONE as a inner join THREE as b
on a.ID=b.ID ;
quit ;
```

Figure 2.5-13: Output

ID	NEW_AF	NEW_BG
1	1	2
2	1	2
3	99	99
5	1	2

In this approach, the table INNJOIN created using the inner join method only contains rows with IDs with nonmissing values in both data-providing tables. If an ID value only existed in one table, this row was excluded from INNJOIN. The two COALESCE functions select the first nonmissing value from the two variable pairs A and F or B or G and store it in the newly created variables NEW_AF or NEW_BG.

Program 2.5-18: FULLJOIN (MISSING, ID Variant I)

```
proc sql ;
create table FULLJOIN as
select coalesce (a.ID, b.ID)
       as ID, A, B, E, F
from ONE a full join THREE b
on a.ID=b.ID
where a.ID is missing or b.ID is missing ;
quit ;
```

Figure 2.5-14: Output: ID Variant I

ID	A	B	E	F
4	1	2	.	.
6	.	.	4	4

This approach retains the "negative intersection," that is, data rows with IDs that only occur in one, but not both data-supplying tables. The original variable ID is used for identification.

Program 2.5-19: FULLJOIN (MISSING, ID Variant II)	**Figure 2.5-15: Output: ID variant II**

```
proc sql ;
create table FULLJOIN as
select a.ID as ID_ONE,
       b.ID as ID_THREE, A, B, E, F
from ONE a full join THREE b
on a.ID=b.ID
where a.ID is missing or b.ID is missing ;
quit ;
```

ID_ONE	ID_THREE	A	B	E	F
4	.	1	2	.	.
.	6	.	.	4	4

This concluding approach retains the "negative intersection," that is,, data rows with IDs that only occur in one, but not both data-supplying tables. For identification purposes, two ID variables are created, whose names identify the source of the delivered data (ID_ONE, ID_THREE).

2.6 Searching and Replacing Missing Values

The replacement of missing values has several advantages:

- Information is not discarded.
- Further analysis is based on a complete data set.
- Missing values in one variable can possibly be explained by other variables in the SAS table, which are probably the cause of the missing values.

Provided that the reconstruction itself is plausible, replacing missing values is always better than deleting rows that contain missing values.

Section 2.6.1 introduces two approaches to finding missing values using PROC SQL. Section 2.6.2 compiles different approaches to search for and replace numeric and alphanumeric missing values. These approaches can be summarized under the generic term "single imputation" (SI). With SI, one value replaces one missing value. Multivariate estimation procedures such as multiple imputation are not presented because they are somewhat distant from the core topic of this book, SQL with SAS. Furthermore, the complexity of the reconstruction of missing values should not be underestimated as it is also influenced by the cause, patterns, and statistical concepts of the applied reconstruction methods. (See for example, Schendera, 2007, 119-162.)

2.6.1 Searching for Missing Values (Screening)

Searching for missing values in SAS tables in various ways is simple when using PROC SQL , to include:

- Approach 1: This SQL approach based on a WHERE clause in which the IS MISSING condition outputs the cases that have user- *or* system-defined missing values.
- Approach 2: This approach is based on a macro and analyzes a single variable for different types of missing values. In contrast to approach 1, no cases are output; instead, the respective frequencies of occurrence of the missing values found in the examined variable are output.

Approach 1: Searching for Missing Values (Univariate or Multivariate)

The following two approaches enable you to select rows that contain user- or system-defined missing values in one or more variables. Approach 1 can be helpful limiting the query to a few rows or columns.

Missing Values in One Variable (Univariate)

This SQL approach outputs all rows where VAR1 contains system- or user-defined missing values.

Program 2.6-1: Missing Values (Univariate)	Figure 2.6-1: Output: Missing Values (Univariate)

```
proc sql;
   create table MISSDATA as
   select VAR1, VAR2
   from MISSINGS
   where VAR1 is missing ;
quit;
```

VAR1	VAR2
	B
_	C
.	
A	8

Missing Values with Several Variables (Multivariate)

This SQL approach outputs all rows where VAR1 contains system- or user-defined missing values.

Program 2.6-2: Missing Values (Multivariate)	Figure 2.6-2: Output: Missing Values (Multivariate)

```
proc sql;
   create table MISSDATA
   as select ID, VAR1, VAR2, STRING1,
STRINGX2
   from MISSINGS
   where VAR1 is missing or
         VAR2 is missing or
         STRING1 is missing or
         STRINGX2 is missing ;
   quit;
```

ID	VAR1	VAR2	STRING1	STRINGX2
3	4	.		
5	_	B		78
6	.	C	IJ	
7	A	8	KLE	90

This SQL approach returns cases that have user- *or* system-defined missing values in either VAR1, VAR2, STRING1, or STRINGX2. If the cases are to have missing values in *all* specified variables, the AND operator must be used.

Approach 2: Screen a Variable for Types of Missing Values (Macro)

The CHECK macro analyzes a single variable for different types of missing values simultaneously. The result is the frequency of occurrence of the missing values found in the examined variable. Approach 2 can be helpful with a few rows but very large amounts of data.

Program 2.6-3: Macro %CHECK

```
%macro CHECK (data_in=, mis_var=, out_data=);
proc sql;
title "Screening of &mis_var.
      in SAS Dataset &data_in. for Missings" ;
  create table &OUT_DATA. as select
    count(*) as ROW_DATA
      label= "Total number of rows in dataset &data_in.",
    count(&mis_var.) as COUT&mis_var.
      label="Number of nonmissing values in &mis_var.",
    nmiss(&mis_var.) as NMISS&mis_var.
      label="Missing or invalid values in &mis_var.",
    count(case when &mis_var.=.B then "count me" end)
                                  as CD_CHCK1
      label="Entry coded as 'Response denied' in &mis_var.",
    count(case when &mis_var.=.C then "count me" end)
                                  as CD_CHCK2
      label="Entry coded as 'invalid' in &mis_var.",
    count(case when &mis_var.=. then "count me" end) as CD_CHCK3
      label="Number of missings in &mis_var."
  from &DATA_IN. ;
     quit;
```

```
proc print label;
   run ;
%mend CHECK ;

%CHECK (data_in=MISSINGS, mis_var=var2, out_data=MYDATA);
```

%CHECK is the actual call of the macro. After DATA_IN=, MIS_VAR= and OUT_DATA=, only the input data set (MISSINGS), the variable to be checked (VAR2), and the data set to be created (MYDATA) need to be specified. After initializing the macro, SAS outputs the following result.

Figure 2.6-3: Output: Macro %Check

```
Screening of var2 in SAS Dataset MISSINGS for Missings
```

Obs	Total number of rows in dataset MISSINGS	Number of non-miss ing values in var2	Missing or invalid values in var2	Entry coded as 'Response denied' in var2	Entry coded as 'invalid' in var2	Number of missings in var2
1	7	4	3	1	1	1

With minor extensions such as the PARMBUFF option, this macro can also be enhanced to analyze lists of variables. (See Chapter 4.)

2.6.2 Searching and Replacing of Missing Values (Conversion)

There are numerous approaches for replacing missing values (Schendera, 2020/2007; Little and Rubin, 2002; Wothke, 1998). Missing values can be replaced by logically correcting data and by reliable estimates. The approaches presented above can be summarized as "single imputation" (SI) in which a value replaces a missing value. Multivariate estimation approaches such as multiple imputation, cluster analysis, or hot deck are not presented because they are thematically rather distant from the core topic of this book, SQL. However, the interested SAS user is referred to the SAS procedures MI and MIANALYZE. PROC MI performs a so-called multiple imputation (hence the acronym MI) of missing data. With multiple imputation, a missing value that is not replaced by one but by one of a number of possible values until a certain optimum is reached. Conversely, PROC MIANALYZE summarizes results of the analyses of the imputations and derives reliable statistical conclusions.

This section summarizes different approaches to search for and also replace or convert numeric and character missing values. These include:

- Approach 1: Converting valid values into missing values (CALL MISSING)
- Approach 2: Writing missing values into macro variables
- Approach 3: Finding and replacing missing values with constants (COALESCE)
- Approach 4: Replacement of missing values by calculated values (for example, mean substitution)
- Approach 5: Replace with valid values from another table (UPDATE)
- Approach 6: Replacement of missing values by a join (COALESCE)
- Approach 7: Filtering out lines with missing values (COALESCE)

All approaches assume the plausibility of the other entries in the SAS table. Thus, if a value in a source variable is incorrect for determining the missing value, the substitute value determined is also incorrect. In practice, several approaches are often combined to replace missing values with reliable values. (See Little and Rubin, 2002.)

Approach 1: Converting Valid Values into Missing Values (CALL MISSING)

The application of the routine CALL MISSING in a DATA step operates in the opposite direction compared to the approaches presented in the following: not missing values are converted into values, but valid values are converted into missing values. CALL MISSING is extremely convenient and converts the values of all variables in parentheses into missing values. String, numeric, date, and time variables can be specified simultaneously in the parentheses.

Program 2.6-4: Syntax Approach 1	Figure 2.6-4: Output: Syntax Approach 1

```
data MYDATA ;
set SASHELP.CLASS ;
call missing
     (SEX, HEIGHT, WEIGHT) ;
run ;
proc print noobs ;
run ;
```

Name	Sex	Age	Height	Weight
Alfred		14	.	.
Alice		13	.	.
Barbara		13	.	.
Carol		14	.	.
Henry		14	.	.
James		12	.	...

CALL MISSING converts string entries into a blank and numeric values into system-defined missing values. In the example, the length of the created missing values corresponds to the original lengths of the variables to be converted. The SAS documentation is not clear whether this is always the case, and it is recommended that you check the length of the missing values created if they are relevant for further processing.

Approach 2: Writing Missing Values into Macro Variables

Approach 2 uses the principle of SAS macros that entries in macro variables are always text values. (See Section 4.1.) Even values that appear to be numbers are text values strictly speaking. Numeric missing values can be easily written as alphanumeric characters using the INTO clause and macro variables. System- or user-defined numeric missing values could be converted to strings in the form of a period ("."), underscore ("_"), or the respective codes for the user-defined numeric missing values (from "A" to "Z"). (See the first three SELECT statements in Program 2.6-5.) Alphanumeric missing values (blanks) are converted to a period (".") in this approach. (See the last SELECT statement.)

Program 2.6-5: Syntax Approach 2

```
proc sql;
select var1 into :val_var1 from MISSINGS
     where var1 = .;
select var1 into :val_var2 from MISSINGS
     where var1 = ._;
select var1 into :val_var3 from MISSINGS
     where var1 = .A;
select string1 into :valstring1 from MISSINGS
     where string1=' ';
quit;

%put val_var1=||&val_var1|| is of length %length(&val_var1);
%put val_var2=||&val_var2|| is of length %length(&val_var2);
%put val_var3=||&val_var3|| is of length %length(&val_var3);
%put valstring1=||&valstring1|| is of length %length(&valstring1);
```

Figure 2.6-5: Output in SAS Log: Syntax Approach 2

```
val_var1=||      _|| is of length 1
val_var2=||      .|| is of length 1
val_var3=||      A|| is of length 1
valstring1=||    || is of length 0
```

The log displays the result of that approach. The names assigned identify the macro variables in which the values were stored. Note that different types of missing values within one variable are stored in *different* macro variables. The numeric missing values of variable VAR1 were stored in the macro variables VAL_VAR1 to VAL_VAR3. Blanks in STRING1 were stored in the macro variable VALSTRING1. The chapter on macro programs contains further approaches to handling missing values.

Approach 3: Finding and Replacing Missing Values with Constants (COALESCE)

The COALESCE function checks the value of one or more columns and returns the first nonmissing value. If the values in all columns are missing, that is, if there are only missing values in a row across several columns, the COALESCE function returns one missing value. In some SQL DBMSes, the COALESCE function is also known as an IFNULL function. For example, the COALESCE function can be used to search for missing values in a variable and replace them with a user-defined value in the same variable. The CASE expression can be used, for example, to search for missing values in a first variable and replace them with a user-defined value in a second variable.

In the following example, the COALESCE function can be used to easily convert missing values of type numeric or string into user-defined values or codes for missing string values, respectively. The missing values in the two numeric variables VAR1 and VAR2 are replaced by the value 0; the missing values in the character variable STRING1 are converted into the string "Missing!".

Program 2.6-6: Syntax Approach 3 (COALESCE)
```
proc sql;
create table REPLACED1 as
   select ID, coalesce(VAR1,0) as VAR1,
             coalesce(VAR2,0) as VAR2,
             coalesce(STRING1, 'Missing!') as STRING1
      from MISSINGS ;
quit;
proc print data= REPLACED1 noobs;
run ;
```

Figure 2.6-6: Output: Syntax Approach 3 (COALESCE)

ID	VAR1	VAR2	STRING1
1	0	5	ABC
2	-2	-6	DE
3	4	0	Missing!
4	4	7	FGH
5	0	0	Missing!
6	0	0	IJ
7	0	8	KLE

Note: The COALESCE function is used to search for missing values and replace them with a user-defined *value in the same variable*. Please note the slightly different programming of the COALESCE function when replacing numeric versus string missing values (that is, the quotation marks around the code for the missing string value).

The following example for converting string type missing values can easily be adapted into an example for numerical missing values with user-defined missing codes.

Program 2.6-7: Syntax Approach 3 (CASE/WHEN)
```
proc sql;
create table REPLACED2 as
   select ID, STRING1, case
                 when STRING1  is missing then 'Missing!'
                 else STRING1
                    end as STRING3
      from MISSINGS ;
quit;
proc print data= REPLACED2 ;
run ;
```

Figure 2.6-7: Output: Syntax Approach 3 (CASE/WHEN)

```
ID     STRING1     STRING3

 1      ABC         ABC
 2      DE          DE
 3                  Missing!
 4      FGH         FGH
 5                  Missing!
 6      IJ          IJ
 7      KLE         KLE
```

Note: The COALESCE function searches for missing values in the variable STRING1, replaces them with a user-defined value and stores them modified in this way as the new variable STRING3.

Approach 4: Replacement of Missing Values by Calculated Values

A common approach is the "mean substitution." The desired parameter (for example, mean value, if metric data; mode, if categorical) is calculated for the available values and used instead of the missing values.

Scenario: In a data set, approximately 1% of the data is missing in the variable "age"; other variables are not used for the estimate. Strictly speaking, the proportion of missing data is already too large for this example.

Options for location measures include MEAN, MEDIAN, SUM, EUCLEN (Euclidean length), USTD (standard deviation about the origin), STD, RANGE, MIDRANGE (Range/2), MAXABS (maximum absolute value), IQR, or the MAD (median absolute deviation from the median).

As alternative solutions, the procedures STDIZE (together with the REPONLY option) or STANDARD can be chosen. In PROC STDIZE, the missing values in a distribution are replaced by the selected location measure (to be specified after METHOD=). In PROC STANDARD, missing values can only be replaced by a specified mean value.

Program 2.6-8: Syntax Approach 4: Data

```
data MEANSUB ;
input ID AGE ;
datalines ;
1 05
2 10
3 .
4 60
5 75
 ;
run ;
```

Program 2.6-9: Syntax Approach 4: Query

```
proc sql ;
  select *,
    case
      when AGE = .
         then mean(AGE)
      else AGE
    end as AGE_new
from MEANSUB ;
quit;
```

Figure 2.6-8: Output: Syntax Approach 4 (Query)

```
ID        AGE     AGE_new
---------------------------

 1          5          5
 2         10         10
 3          .       37.5
 4         60         60
 5         75         75
```

This example uses PROC SQL to calculate an average for the AGE values (based on nonmissing AGE values), and uses this value to replace any missing values when writing the new column AGE_new using a CASE-WHEN-THEN approach.

The speculative nature of this procedure manifests itself in the inhomogeneity of the initial data; whether the average value used is the correct one appears somewhat doubtful without including the information from other variables. Using a mean based on grouped data might be more appropriate, but this would require a second variable, the grouping variable.

The univariate calculation and imputation of a parameter is not recommended for several reasons. (See Schendera, 2020/2007, Chapter 6.) The main reason is that the univariate approach unrealistically excludes other variable interactions. (See example.) The variance in the replaced variable and the covariance with other variables are artificially reduced by this process. Furthermore, an estimated value is equated with an observed value (uncertainty). Especially for metrically scaled values with a large range, this approach must be considered speculative and artificial because no checks were made for interactions with other variables. Since speculatively inserted values can already turn out to be counterproductive in bivariate analyses, this approach should only be used if no other method is available or the proportion of missing values is really very small (< 5%).

Approach 5: Replace with Valid Values from Another Table (UPDATE)

UPDATE and CASE enable you to replace missing values in a first table with the *correct* values from a second table. Replacing missing values with correct values is best demonstrated using tables with identical structures.

Example: Table RECH_ALT contains incomplete invoice data in the string variable ART_BEZ and in the numeric variable PREIS, while table BILL_NEW contains the correct values. The task now is to replace the missing values in RECH_ALT with the valid values from BILL_NEW. The two tables RECH_ALT and BILL_NEW have the same number of rows and columns, but also have three differences: (a) the column names in the first table are in English and those in the second table are in German, (b) the two columns ART_BEZ and PREIS in RECH_ALT are incomplete, whereas in BILL_NEW they are complete, and (c) the row with RECH_NR 0104 only appears in table RECH_ALT, but not in BILL_NEW. The task now is to replace the missing values in RECH_ALT with the valid values in BILL_NEW and only if the same invoice numbers actually exist.

Program 2.6-10: Syntax Approach 5: Data

```
data RECH_ALT ;
    input RECH_NR KUNDE $ 7-20 KUNDEN_NR
          MA_NR ART_BEZ $ 34-44 MENGE PREIS ;
    format PREIS euro. ;
    datalines ;
0101  SportHalle          16  101                  20      .
0102  SportHalle          16  102                  20      65
0103  SportHalle          16  103   Skating        20      .
0104  SportHalle          15  102                  15      54
0105  SportHalle           5  103   Skating        15      .
;
run ;

data BILL_NEW ;
    input BILL_NO CLIENT $ 7-20 CLIENT_NR
          MA_NR ART_BEZ $ 34-44 MENGE PRICE ;
    format PRICE dollar. ;
    datalines ;
0101  SportHalle          16  101   Kindersport 20      7
0102  SportHalle          16  102   Fitness      20     33
0103  SportHalle          16  103   Skating      20     27
9999  SportHall           67  103   Skating      15     67
0105  SportHalle           5  103   Skating      15     33
;
run ;
```

The following SQL program passes through two phases. In the first UPDATE statement, the missing values in the string column ART_BEZ are replaced by valid strings from BILL_NEW. In the second UPDATE statement, the missing values in the numerical column PRICE are replaced by valid numerical values from BILL_NEW only if the same invoice numbers (RECH_NR = BILL_NO) actually exist. The SQL programming is the same for replacing numeric values and strings.

Program 2.6-11: Syntax Approach 5: UPDATE (Subquery)

```
proc sql;
update RECH_ALT as old
   set ART_BEZ=(select ART_BEZ from BILL_NEW as new
                 where old.RECH_NR=new.BILL_NO)
                   where old.RECH_NR in (select BILL_NO from BILL_NEW);
select   RECH_NR, KUNDE, KUNDEN_NR, MA_NR,
  ART_BEZ, MENGE, PREIS format=euro. from RECH_ALT ;
update RECH_ALT as old
   set PREIS=(select PRICE from BILL_NEW as new
                 where old.RECH_NR=new.BILL_NO)
                   where old.RECH_NR in (select BILL_NO from BILL_NEW);
select   RECH_NR, KUNDE, KUNDEN_NR, MA_NR,
  ART_BEZ, MENGE, PREIS format=euro. from RECH_ALT ;
quit;

proc print data=RECH_ALT noobs ;
run;
```

Since the SQL programming for replacing numeric values and strings is the same, the approach will only be explained for replacing string missing values. For later comparison with the table containing the current values, the table to be updated (RECH_ALT) is first given the alias OLD. After SET, you specify that the variable to be updated is the column ART_BEZ, which is to be replaced by the identically named ART_BEZ values from BILL_NEW (with the alias NEW), where invoice numbers in both tables are to match (RECH_NR = BILL_NO). In addition, the other variables RECHN_NR, KUNDE, and so on are to be transferred to the updated table. The Euro format is then assigned to the transferred values. PROC PRINT is used to check the success of the update.

Figure 2.6-9: Output: Syntax Approach 5: UPDATE (Subquery)

RECH_NR	KUNDE	KUNDEN_ NR	MA_NR	ART_BEZ	MENGE	PREIS
101	SportHalle	16	101	Kindersport	20	€7
102	SportHalle	16	102	Fitness	20	€33
103	SportHalle	16	103	Skating	20	€27
104	SportHalle	15	102		15	€54
105	SportHalle	5	103	Skating	15	€33

If this check output is compared with the contents of RECH_ALT, it can be seen that the ART_BEZ values were successfully replaced by the ART_BEZ values from BILL_NEW, and always when the same invoice numbers are present. Therefore, row 0104 was *not* updated because this invoice number did not appear in the table that was to provide the valid values.

Approach 6: Replacement of Missing Values by a Join (COALESCE)

If there are a large number of columns with missing values that need to be replaced with values, the missing values within them can also be replaced by a full join with the table with the *correct* values. The strategy in the subsequent example is to use the COALESCE function. The COALESCE function returns the first nonmissing value from a list of columns. If two or more columns are joined with valid but also missing values, the COALESCE function enables you to replace the missing values with nonmissing (valid) values in the created table (provided that they exist).

The subsequent example not only demonstrates the effect of COALESCE on three columns to be joined, but also includes a small extra: when tables are to be joined, in one table often the information is distributed over one column and in a second table, over two columns. In the example, the missing values in *one* column KUNDE ("client") from the table RECH_KND are to be replaced by the information from the *two* columns CLIENT and SUBSIDIARY from the table BILL_SUB. In other words, before replacing the strings in KUNDE in RECH_KND, the two substrings CLIENT and SUBSIDIARY must be joined together.

Example: Table RECH_ALT contains incomplete invoice data in the string variable ART_BEZ and in the numeric variable PREIS, while table BILL_NEW contains the correct values. The task now is to replace the missing values in RECH_ALT with the valid values from BILL_NEW. The two tables RECH_ALT and BILL_NEW have the same number of rows and columns, but also have three differences: (a) the column titles in the first table are in English and those in the second table are in German, (b) the two columns ART_BEZ and PREIS in RECH_ALT are incomplete, whereas in BILL_NEW they are complete, and (c) the row with RECH_NR 0104 only appears in table RECH_ALT, but not in BILL_NEW. The task now is to replace the missing values in RECH_ALT with the valid values in BILL_NEW and only if the same invoice numbers actually exist.

Program 2.6-12: Syntax Approach 6: Data

```
data RECH_KND ;
   input RECH_NR KUNDE $ 7-22 KUNDEN_NR
         MA_NR ART_BEZ $ 36-44 MENGE PREIS ;
   datalines ;
0101  SportHalle, Köln   16  101              20    .
0102                      16  102              20    65
0103  SportHalle, Köln   16  103  Skating     20    .
0104                      15  102              15    54
0105  SportHalle, Köln    5  103  Skating     15    .
;
run ;
proc print data=RECH_KND ;
run ;

data BILL_SUB ;
   input BILL_NO CLIENT $ 7-17 SUBSIDIARY $ 18-21 CLIENT_NR
         EMP_NR ARTICLE $ 35-45 AMOUNT PRICE ;
   datalines ;
0101  SportHalle Köln   16  101  Kindersport 20    14
0102  SportHalle Köln   16  102  Fitness     20    65
0103  SportHalle Köln   16  103  Skating     20    54
0104  SportHalle Köln   15  102  Fussball    15    54
0105  SportHalle Köln    5  103  Skating     15    65
;
run ;
proc print data=BILL_SUB ;
run ;
```

Program 2.6-13: Syntax Approach 6: UPDATE (COALESCE)

```
proc sql;
   create table MYDATA as
   select RECH_NR,
coalesce(empty.KUNDE,trim(filled.CLIENT)
            ||' '||filled.SUBSIDIARY)as KUNDE format=$18.,
coalesce(empty.ART_BEZ,filled.ARTICLE)as ART_BEZ format=$14.,
coalesce(empty.PREIS,filled.PRICE)as PREIS, MA_NR, MENGE
      from RECH_KND as empty full join BILL_SUB as filled
          on empty.RECH_NR=filled.BILL_NO
      order by RECH_NR ;
quit;

proc print data=MYDATA noobs ;
run ;
```

Figure 2.6-10: Output: Syntax Approach 5: UPDATE (COALESCE)

RECH_NR	KUNDE	ART_BEZ	PREIS	MA_NR	MENGE
101	SportHalle, Köln	Kindersport	14	101	20
102	SportHalle Köln Fitness	65 102	20		
103	SportHalle, Köln	Skating	54	103	20
104	SportHalle Köln	Fussball	54	102	15
105	SportHalle, Köln	Skating	65	103	15

> **Important note:** Using FORMAT= you may have to create enough space for the created strings. Often you should also pay attention to correct punctuation. In the example, no comma was intentionally placed between the two joined substrings. In the output, for example, this small difference makes it easier to see which entries already existed (with comma) and which were created during the join (without comma). Depending on how the tables are joined, a sorting order in variables is not reliable if they are specified as value-contributing variables in the COALESCE function. It is therefore recommended to specify, for example, an ID variable or another variable that does not have duplicate values under ORDER BY.

Approach 7: Filtering out Rows with Missing Values (COALESCE)

The COALESCE function is known to be useful for replacing missing values with a code 0. The following SQL code replaces user- and system-defined missing values equally with the code 0.

Program 2.6-14: Syntax Approach 7 I	**Figure 2.6-11: Output: Syntax Approach 7 I**
```proc sql ; select coalesce(VAR1,0) as VAR1_0, coalesce(VAR2,0) as VAR2_0 from MISSINGS ; quit ;```	``` VAR1_0 VAR2_0 ---------------- 0 5 -2 -6 4 0 4 7 (... abbreviated)```

With COALESCE, the missing values in several numerical columns (VAR1, VAR2) can be converted into zeros before a query. The trick is to multiply the resulting numeric columns with each other and use the finally emerging product ZEROS_NE_MISS for filtering.

Program 2.6-15: Syntax Approach 7 II a	**Figure 2.6-12: Output: Syntax Approach 7 II a**
```proc sql ; select VAR1, VAR2, (coalesce(VAR1,0)*coalesce(VAR2,0)) as ZEROS_NE_MISS from MISSINGS ; quit;```	``` ZEROS_NE_ VAR1 VAR2 MISS --------------------------------- 0 5 0 -2 -6 12 4 . 0 4 7 28 _ B 0 . C 0 A 8 0```

The trick is based on the assumption that the product of several columns achieves the filter value 0 if at least one of the columns included in the multiplication contains a missing value.

Program 2.6-16: Syntax Approach 7 II b	**Figure 2.6-13: Output: Syntax Approach 7 II b**
```proc sql ; select VAR1, VAR2, (coalesce(VAR1,0)*coalesce(VAR2,0)) as ZEROS_NE_MISS from MISSINGS where calculated ZEROS_NE_MISS ne 0 ; quit;```	``` ZEROS_NE_ VAR1 VAR2 MISS --------------------------------- -2 -6 12 4 7 28```

This approach treats rows with zeros in the same way as rows with missing values. For example, the first line with 0 and 5 should not have been filtered out. To prevent unwanted results from occurring, you must make sure that the rows do *not* contain zeros.

2.7 Predicates

PROC SQL allows checking of logical conditions by means of so-called predicates. The check of logical conditions can only take on the events "true" (output: 1) or "false" (output: 0). PROC SQL provides general predicates including ALL/ANY/SOME, EXISTS, IN, IS, and LIKE and special predicates for missing values including IS NULL and IS MISSING. This section illustrates how predicates work when handling with valid values and user- and system-defined missing values. The handling of missing values in string format is demonstrated using LIKE. The predicates are presented in alphabetical order. This section builds on a NESUG paper by Danbo Yi and Lei Zhang. Yi and Zhang (1998) worked out the different effects of predicates on the work with system- and user-defined missing values. This and other sections of this book were inspired by this. A sequence phenomenon of the [NOT] LIKE predicate is interesting.

2.7.1 ALL, ANY, and SOME

ALL, ANY, and SOME are used in subqueries in connection with comparisons (for example, <, =, >). All three predicates adopt the values that are output by the subquery. ANY and SOME behave the same in PROC SQL. In earlier PROC SQL versions, ALL could produce quite different results from ANY and SOME.

Program 2.7-1: Syntax: Three Subqueries with ALL   **Figure 2.7-1: Output: Three ALL Outputs**

```
proc sql;
select var2 from MISSINGS
     where var2 > ALL
     (select var1 from MISSINGS
     where var1 > 2);
select var2 from MISSINGS
     where var2 < ALL
     (select var1 from MISSINGS
     where var1 > 2);
select var2 from MISSINGS
     where var2 NE ALL
     (select var1 from MISSINGS
     where var1 > 2);
quit;
```

VAR2	VAR2	VAR2
----	----	----
5	-6	5
7	.	-6
8	B	.
	C	7
		B
		C
		8

Program 2.7-2: Syntax: Three Subqueries with ANY   **Figure 2.7-2: Output: Three ANY Outputs**

```
proc sql;
select var2 from MISSINGS
     where var2 > ANY
     (select var1 from MISSINGS
     where var1 > 2);
select var2 from MISSINGS
     where var2 < ANY
     (select var1 from MISSINGS
     where var1 > 2);
select var2 from MISSINGS
     where var2 NE ANY
     (select var1 from MISSINGS
     where var1 > 2);
quit;
```

VAR2	VAR2	VAR2
----	----	----
5	-6	5
7	.	-6
8	B	.
	C	7
		B
		C
		8

In previous versions, ALL behaved differently from ANY and SOME if the subquery did not return values for a comparison with ANY or SOME and ALL. In this case, the comparison with ALL automatically returned 1 ("true"), but the comparison with ANY or SOME returned 0 ("false").

2.7.2 EXISTS or NOT EXISTS

Missing values are normal comparison values in PROC SQL, which can be included in the evaluation of a subquery using EXISTS or NOT EXISTS. With EXISTS, only the output of nonmissing values would be expected, with NOT EXISTS only the output of missing values.

Program 2.7-3: Syntax: Predicate EXISTS	**Figure 2.7-3: Output: Predicate EXISTS**

<pre>proc sql;
create table PRED_EXISTS
    as select var1 from MISSINGS as a
where exists
   (select var1 from MISSINGS as b
where a.var1=b.var1);
quit;
proc print noobs;
run;</pre>
	<pre>VAR1
	
	 0
	-2
	 4
	 4
	
	 _
	 .
	 A</pre>

In fact, a query with PROC SQL and EXISTS produces a counterintuitive result in case of missing values. Actually, with the EXISTS predicate, only all nonmissing values would be expected. However, the counterintuitive result is that the output for VAR1 also contains all missing variants (user-defined, system-defined), and NOT EXISTS (not displayed) does not output any values at all.

2.7.3 IN and NOT IN

IN or NOT IN are predicates for the evaluation of values or missing values in lists. The use of IN and NOT IN for alphanumeric missing values treats missing strings independently of their length, as if they had the same length. For example, regardless of the number of blanks in the "in (" ")" expression, IN filters out all missing strings regardless of their length, including missing strings whose number of characters exceeds the number of blanks in the IN expression. However, the LIKE predicate (see below) can be used to filter out missing strings that take into account the maximum character length using the "(" ")" expression. IN is a more general predicate than LIKE. LIKE, on the other hand, is a more precise predicate because it also takes the maximum character length into account.

Program 2.7-4: Syntax: Predicate IN	**Figure 2.7-4: Output: Predicate IN**

<pre>proc sql;
create table PRED_IN as
select var1, string1 from MISSINGS
where string1 in ("   ");
quit;
proc print noobs;
run;</pre>
	<pre>VAR1 STRING1
	
	 4
	
	 _</pre>

Program 2.7-5: Syntax: Predicate NOT IN	**Figure 2.7-5: Output: Predicate NOT IN**

<pre>proc sql;
create table PRED_IN as
select var1, string1 from MISSINGS
where string1 not in ("   ");
quit;
proc print noobs;
run;</pre>
	<pre>VAR1 STRING1
	
	 0 ABC
	-2 DE
	 4 FGH
	 . IJ
	 A KLE</pre>

2.7.4 IS NULL, IS NOT NULL, IS MISSING, and IS NOT MISSING

IS NULL and IS NOT NULL or IS MISSING and IS NOT MISSING are special predicates for dealing with missing values. IS NULL and IS NOT NULL are SQL standard while or IS MISSING and IS NOT MISSING are SAS standard. They can be used for numerical and alphanumerical missing values.

Program 2.7-6: Syntax: Predicates IS [NOT] NULL / MISSING	Figure 2.7-6: Output: Predicates IS [NOT] NULL / MISSING

```
proc sql;
create table PRED_ISNOT as
select var1, string1 from MISSINGS
where var1 is
            not null
and string1 is
            not missing;
quit;
```

```
VAR1     STRING1

  0        ABC
 -2        DE
  4        FGH
```

For example, the IS NOT example queries all values that have *no* missing values in VAR1 (numeric) and STRING1 (string) at the same time. Exchanging NULL and MISSING leads to the same result.

2.7.5 LIKE and NOT LIKE

LIKE or NOT LIKE, like IN and NOT IN, are also predicates for the listwise evaluation of values or missing values. However, LIKE and NOT LIKE differ in one central point. While IN and NOT IN filter out all alphanumeric missing values, LIKE and NOT LIKE only filter out those whose character length does not exceed the number of blanks in the "like (" ")" expression. Thus, if there are three blanks in the "(" ")" expression, "in (" ")" would output missing values with a length of more or less than three characters, but "like (" ")" would only output missing values up to a maximum of three characters. LIKE expressions connected in series have then exactly the same effect as IN expressions when the length of the longest missing string does not exceed the length of the blanks in the "like (" ")" expression.

Program 2.7-7: Syntax: Predicate [NOT] LIKE	Figure 2.7-7: Output: Predicate [NOT] LIKE

```
proc sql;
create table PRED_LIKE as
select var1, string1 from MISSINGS
where string1 like "   ";
quit;
```

```
VAR1     STRING1

  4

  —
```

With the LIKE and NOT LIKE predicates, there is an interesting sequence phenomenon in comparisons (for example, STRING1=STRINGX2). If both variables have missing values, missing values always lead to 1 ("true"). However, the direction of the comparison is also decisive. The order STRING1=STRINGX2 produces a result, whereas the reverse order STRINGX2=STRING1 produces none.

Program 2.7-8: Syntax: Sequence Phenomenon (See WHERE)	Figure 2.7-8: Output: Sequence Phenomenon

```
proc sql;
create table PRED_WHERE1 as
select var1, string1, stringx2
from MISSINGS
   where string1 like stringx2 ;
quit ;

proc sql;
create table PRED_WHERE2 as
select var1, string1, stringx2
from MISSINGS
   where stringx2 like string1 ;
quit ;
```

```
VAR1      STRING1     STRINGX2
  4




NOTE:
No observations in data set WORK.PRED_WHERE2.
```

Chapter 3: Data Quality with PROC SQL

Data quality comes before the quality of the analysis. This section introduces SQL techniques for ensuring data quality in three problem areas: outliers, uniformity, and duplicates. The previous chapter, Chapter 2, was dedicated to a fourth problem area: missing values. By integrating filters when accessing data via SQL, users can ensure that data is not entered incorrectly into the system or analysis in the first place.

Section 3.1 introduces working with integrity constraints and audit trails. In simple terms, integrity constraints are check rules for ensuring the data quality of a SAS table.

Section 3.2 deals with the handling, finding, and filtering of multiple values (duplicates). Multiple occurrences of data generally lead to various problems. First, it should be clarified whether the SAS table should contain only one case per row or whether the same rows are allowed to occur more than once. This section presents a variety of approaches, including how to display or filter duplicate IDs or individual values, how to create lists for duplicate rows, or how to identify duplicates in multiple tables.

Section 3.3 presents different approaches to identifying quantitative outliers. Outliers can massively undermine the robustness of statistical methods. The results of data analyses can be completely distorted by only a few outliers. Several approaches for dealing with outliers are presented, including checking for outliers by means of descriptive statistics, by means of statistical tests, and by filtering outliers through conditions such as limits or intervals. Not every value that is suspicious from a formal point of view is automatically false, though. An outlier is always relative to expectations. Expectations vary according to time and place, and might not necessarily coincide with empirical variability. At the end of the section, there are some hints for the further handling of outliers.

Section 3.4 deals with the identification, filtering, standardizing, and replacement of character values or strings. The uniformity of numeric and alphanumeric values is one of the basic criteria of data quality. The consequences of inconsistent strings can be serious. Several possibilities for checking and filtering are presented, including checking for longer character strings, checking for several strings and substrings, and checking for single characters. Finally, different possibilities for searching and replacing strings, substrings, or characters are presented.

The implied concept of the DQ Pyramid is based on Schendera (2020/2007); there is also an unpublished SAS version. The following chapter presents only a fraction of what could be said about data quality with SQL.

3.1 Integrity Constraints and Audit Trails

Integrity constraints are, in simple terms, check rules to ensure the data quality of a SAS table. Integrity constraints serve first and foremost to *identify* possible errors, but not to interpret or correct them. (See Schendera, 2020/2007.) For an iterative check of larger data volumes, especially when the integrity rules are complex and the data is continuously deployed, the use of integrity constraints is highly recommended.

Integrity constraints can be created with PROC SQL, PROC DATASETS, and in SAS Component Language (SCL). Check rules cannot be defined for views. The following section introduces you to working with integrity constraints with SAS SQL and points out the advantages of other procedures if necessary.

Section 3.1.1 introduces working with integrity constraints. Section 3.1.2 introduces working with audit trails. Audit trails do not just log changes to a table and store them in a separate file. It is important to know that an audit trail is also currently the only technical possibility in SAS to save cases from failed APPEND operations or cases rejected by integrity constraints.

3.1.1 Integrity Constraints (Check Rules)

Integrity constraints provide feedback if certain criteria are met. If a rule is violated, it is generally interpreted as an error. However, rules can be formulated negatively, but also positively, for example, with regard to reference values such as 95, 96, 97, and so on. In the case of a negative rule, any other value (for example, 91, 92, and 93) will trigger an error message. If a *positive* rule *meets* the reference values (for example, 96 and 97) an error message is triggered as well. Violating a negative rule or meeting a positive rule equally results in the return code "1", and depending on the other settings, an additional "0" and a missing value. When defining rules, programming them, and interpreting the result, it is essential to consider the difference between positive and negative rules.

Test rules are used in a wide variety of contexts such as for data access, for checking or replacing data, or for determining derived measures, values, and calculations. The reliability of check rules is therefore of fundamental importance, since in each case, further values, analyses, or results depend on them. An incorrect check rule leads to incorrect values, analyses, or results. The problem with dysfunctional check rules and accesses is that the data itself might be fine, but the data quality problems are caused by the incorrect rules or accesses. Check rules and queries should therefore be checked for several aspects, ideally before programming them in SAS syntax. These aspects include semantics, logic, and computer science.

- Is the check rule positive or negative?
- For complex check rules, are there logically realizable "true" and "false" events or only a justified selection of "relevant" events?
- For complex check rules, review the logically correct check rules against the complexity and dynamics of empirical reality. Is everything that is logically possible also empirically useful or probable?
- Which SAS syntax or procedure allows the programming of the rules or conditions? How do several syntax possibilities differ in detail? Pay attention to the actuality of syntax, the associated range of functions and changes, as well as the smooth interaction of the data with software and hardware from different manufacturers, systems, and versions in general.
- Last but not least: Evaluate the correctness of the programming. Is the check rule or condition formulated negatively as planned? Does it take into account the relevant, logically realizable "true" and "false" events including a check for a possible programming error?

Check rules or conditions should always be checked first using manageable test data before integration into a workflow. Incorrect results will otherwise not be detected at all or might be discovered by chance because they are only implausible in a certain context. An often overlooked source of errors is simple programming errors such as the incorrect handling of missing values. (See Chapter 2.)

Due to possible false positives, errors should be interpreted with caution at first. A high proportion of missing values does not necessarily have to be an error. In the case of online surveys, for example, questions that are not applicable might have been skipped. Even apparently absent error messages should be interpreted rather cautiously, as there is the possibility of false negatives. Such a result does not necessarily mean that the tested variable does not contain any further errors in the form of differently formulated test rules, settings (limit values), or criteria (duplicates).

On closer inspection, integrity constraints themselves require compliance with various other criteria for data quality such as completeness. (See Schendera, 2020/2007, Chapter 9.) The results of integrity constraints must be seen in the context of the variables examined, the checks made, and the criteria specified. This central aspect will be explicitly referred to repeatedly in the description of the exemplary checks carried out. The interpretation of the results should consider the dual problem of false positives and false negatives. False positives are error messages that, on closer examination, turn out to be false alarms. False negatives are apparently absent error messages. A very simple cause for false positives or false negatives can be the wrong programming of check rules. This can happen if check rules are supposed to indicate a deviation from criteria as an error, but inadvertently display the adherence to certain criteria due to an incorrectly implemented check logic.

In the rest of this section, I will present the following three examples of working with integrity constraints:

- The first example shows how integrity constraints can be created for a still empty table and displayed in the log.
- The second example shows how users can define their own feedback for the SAS log and interpret that feedback when errors might be present in the data.
- The third example shows what happens when integrity constraints are applied retroactively to an already filled table that already contains errors.

Example 1: Creating and Displaying Integrity Constraints (Empty Table)

This example demonstrates how integrity constraints can be created for a table that is still empty and displayed in the SAS log.

Program 3.1-1: Example 1: Syntax for the Definition of the Table and the Check Rules

```
proc sql ;
   create table EMPTY
     (ID num,
      STRING char (30) label="String variable" ,
      VALUE1  num label="Numerical variable" ,
      DATE_VAR date format=ddmmyy8. label="Date variable" ,
        constraint DATE_VAR_c   check(DATE_VAR >= "02JAN70"D) ,
        constraint ID_0      not null(ID) ,
        constraint ID_pkey   primary key(ID) ,
        constraint STRING_0  not null(STRING) ,
        constraint STRING_c  check(STRING in ('Banking',
                                              'Finance')) ,
        constraint VALUE1_0   not null(VALUE1) ,
        constraint VALUE1_c   check(VALUE1 >= 0.1) ,
        constraint VALUE1_unq unique(VALUE1) ) ;
        describe table constraints EMPTY ;
        quit ;
```

Note: Using CREATE TABLE, the temporary file EMPTY is created with empty columns with the names ID, STRING, VALUE1, and DATE_VAR. The next step is to conceptually define the criteria to be checked. In the example, the following should be checked or guaranteed (sorted alphabetically):

- **DATE_VAR**
  - The values in the DATE_VAR column should be greater (more current) or equal to the reference date 2nd of January, 1970 ("02Jan1970" in date 9 format).

- **ID**
  - ID should contain only valid values or no missing values.
  - ID is to be the primary key when connecting to other tables.

- **STRING**
  - STRING should contain only valid values or no missing values.
  - STRING should contain the entries "Banking" or "Finance".

- **VALUE1**
  - VALUE1 should contain only valid values or no missing values.
  - The values in the VALUE1 column should be greater or equal than 0.1.
  - The values in the VALUE1 column should be unique.

After CONSTRAINT, the check rules are then created for the respective variables. A separate CONSTRAINT statement is created for each check rule. Each check rule must be assigned a different name. Identical names for integrity constraints trigger an error message and SAS aborts further processing. One of the following check modes can be specified after each CONSTRAINT (arranged alphabetically):

- **CHECK:** User-defined checking rules are applied to the checked column(s). Users can filter the values of a column with user-defined WHERE conditions, which can contain one or more variables.
- **DISTINCT:** The checked column contains only unique values and must not contain duplicates (multiple values). Missing values are allowed but might also occur only once. DISTINCT corresponds to the UNIQUE check mode and will not be introduced further.
- **NOT NULL:** The checked column contains only valid values and no missing values.
- **PRIMARY KEY:** The checked column is defined as the primary key when connecting to other tables. PRIMARY KEY also automatically checks for unique values. If check rules are applied to primary keys and unique values for one and the same variable, the primary key check rule replaces the check rule for unique values.
- **UNIQUE:** The checked column contains only unique values and must not contain duplicates (multiple values). Missing values are permitted but might also occur only once. UNIQUE corresponds to the DISTINCT check mode (not further introduced).

Once check rules have been created, they can be output to the SAS log with the command DESCRIBE TABLE CONSTRAINTS. The check rules are displayed even if the table itself is still empty.

Figure 3.1-1: Example 1: Output of Integrity Constraints (DESCRIBE TABLE CONSTRAINTS)

```
-----Alphabetic List of Integrity Constraints-----

     Integrity                                  Where
  #  Constraints    Type          Variables     Clause
  --------------------------------------------------------------
  1  DATE_VAR_c     Check                        DATE_VAR>='02JAN1970'D
  2  ID_0           Not Null      ID
  3  ID_pkey        Primary Key   ID
  4  STRING_0       Not Null      STRING
  5  STRING_c       Check                        STRING in ('Banking', 'Finance')
  6  VALUE1_0       Not Null      VALUE1
  7  VALUE1_c       Check                        VALUE1>=0.1
  8  VALUE1_unq     Unique        VALUE1
```

The check rules (see "Integrity Constraints" column) are output alphabetically. The extenders "_c", "_pkey", "_0" and "_unq" assigned in the example indicate the rule type ("Check", "Primary Key", "Not Null", "Unique") that is applied to the variable in question. Users are free to choose their own names for test rules.

For example, the check rule "DATE_VAR_c" is of type "Check" and contains the WHERE condition "DATE_VAR>='02JAN1970'D". The values in the DATE_VAR column should be greater (more current) or equal than the reference date 2nd of January, 1970. For example, the check rule "ID_0" is of type "Not Null" and refers to the variable ID. ID might contain only valid values or no missing values. For example, the check rule "ID_pkey" is of the type "Primary Key" and refers to the variable ID. ID thus becomes the primary key when connecting to other tables and so on. The separate check rules "DATE_VAR_c", "STRING_c" and "VALUE1_c" could be combined into a single check rule called MULTIRULE, for example:

Program 3.1-2: Syntax Example for a Multi-rule Constraint

```
constraint MULTIRULE check(DATE_VAR >= "02JAN70"D &
                      STRING in ('Banking', 'Finance') & VALUE1 >= 0.1).
```

Alternatively, integrity constraints can be viewed using the CONTENTS and DATASETS procedures (the output is not presented). These procedures also enable you to store information about integrity constraints in a file using the OUT2 option.

Program 3.1-3: Two Approaches to Retrieve Audit Information

```
proc datasets library=work ;              proc contents data=EMPTY ;
   contents data=EMPTY ;                     run ;
  run ;
quit ;
```

SCL provides the ICTYPE, ICVALUE, and ICDESCRIBE functions for querying information about integrity constraints (not further introduced).

Example 2: Integrity Constraints for Errors in Data of a Row- or Tablewise Update

The following example shows how users can define their own feedback for the SAS log. Then a table is updated, which contains errors. Example 3 demonstrates what happens when extended integrity constraints are applied to a table that already contains errors.

Program 3.1-4: Example 2: Syntax for Defining Table, Check Rule, and User Messages

```
proc sql ;
  create table EMPTY
   (ID num,
    STRING char (30) label="String variable" ,
    VALUE1  num label="Numerical variable" ,
    DATE_VAR date format=ddmmyy8. label="Date variable" ,
     constraint DATE_VAR_c   check(DATE_VAR >= "02JAN70"D)
      message="DATE_VAR contains values older than the date."
                                       msgtype=newline ,
     constraint ID_0      not null(ID)
      message="ID contains missings."      msgtype=newline ,
     constraint ID_pkey   primary key(ID)
      message="ID is not the primary key."
                                       msgtype=newline ,
     constraint STRING_0  not null(STRING)
      message="STRING contains missings." msgtype=newline ,
     constraint STRING_c  check(STRING in ('Banking',
                                      'Finance'))
      message="STRING does not contain 'Banking' or 'Finance'."
                                       msgtype=newline ,
     constraint VALUE1_0   not null(VALUE1)
      message="VALUE1 contains missings."  msgtype=newline ,
     constraint VALUE1_c   check(VALUE1 >= 0.1)
      message="VALUE1 contains values not greater than or equal to 0.1."
                                       msgtype=newline ,
     constraint VALUE1_unq unique(VALUE1)
      message="VALUE1 contains duplicates."  msgtype=newline ) ;
describe table constraints EMPTY ;
      quit ;
```

The syntax for Example 2 differs from Example 1 only in that each CONSTRAINT statement also contains a text for a user-defined error message (MESSAGE= "User Message"), as well as the type of its response in the SAS log (MSGTYPE=).

An error message is then displayed in the log as soon as the delivered data violates one of the check rules. In the event of data errors, SAS not only aborts the inserting process immediately, but if data was successfully inserted before the displayed error occurred, SAS also deletes the data to restore the state before the update. The text for the response must not exceed 250 characters. The text for the feedback should be as short and concise as possible.

For the type of user-defined feedback, users can choose between USER and NEWLINE. With USER, if a check rule is violated, only the error message defined by the user is written to the SAS log.

SAS Log 3.1-1: Failed Insert (Example 2)

```
ERROR: STRING does not contain 'Banking' or 'Finance'.
```

With NEWLINE, if a check rule is violated, the automatic response from SAS is also output, which in this case outputs the name of the check rule in the log.

```
ERROR: STRING does not contain 'Banking' or 'Finance'. Add/Update failed for data set
WORK.EMPTY because data value(s) do not comply with integrity constraint STRING_c.
```

Figure 3.1-2: Example 2: Output of Integrity Constraints (DESCRIBE TABLE CONSTRAINTS)

```
     -----Alphabetic List of Integrity Constraints-----

   Integrity                        Where      User
#  Constraints  Type     Variables  Clause     Message
-----------------------------------------------------------------------
1 DATE_VAR_c  Check                 DATE_VAR>=' DATE_VAR contains values older
                                    02JAN1970'D          than the date.
2 ID_0        Not Null   ID                    ID contains missings.
3 ID_pkey     Primary Key ID                   ID is not the primary key.
4 STRING_0    Not Null   STRING                STRING contains missings.
5 STRING_c    Check                 STRING in  STRING does not contain 'Bank-
                                    ('Banking',          ing' or 'Finance'.
                                    'Finance')
6 VALUE1_0    Not Null   VALUE1                VALUE1 contains missings.
7 VALUE1_c    Check                 VALUE1>=0.1 VALUE1 contains values not greater
                                                         than or equal to 0.1.
8 VALUE1_unq  Unique     VALUE1                VALUE1 contains duplicates.
```

Compared to Example 1, the table "Alphabetic List of Integrity Constraints" has been extended by the column "User Message". This column shows the respective text of the user-defined feedback messages.

Syntax for the Rowwise Update (VALUES)

Using INSERT, four cases are now added to the empty table, EMPTY. The values of each case are specified in a separate line after VALUES. The order of the entries in the parentheses after VALUES must correspond to the order in the target table (in this case, EMPTY). Because four cases are to be inserted after EMPTY, INSERT contains four VALUES lines.

Program 3.1-5: Syntax for the Rowwise Update (VALUES)

```
proc sql ;
insert into EMPTY
       values (1,"Banking",123,"03JAN70"d)
       values (3,"Finance",321,"29JAN08"d)
       values (3,"Telco",321,"15OCT94"d)
       values (4,"Finance",321,"23OCT04"d);
quit ;
```

To make the operation of the SQL check rules a bit more transparent, errors were intentionally added to the lines. In line 3, for example, the entry "Telco" violates the check rule "STRING_c", and the date entry violates the check rule "DATE_c". The multiple values "321" in column VALUE1 violate check rule "VALUE1_unq". It is now interesting that SAS initially reports only *one* error and then aborts. However, this feedback does *not* mean that this is the *only* error.

SAS Log 3.1-2: Failed Insert 2

```
ERROR: STRING does not contain 'Banking' or 'Finance'. Add/Update failed for data set
WORK.EMPTY because data value(s) do not comply with integrity constraint STRING_c.
NOTE: This insert failed while attempting to add data from VALUES clause 3 to the data set.
NOTE: Deleting the successful inserts before error noted above to restore table to a
consistent state.
NOTE: The SAS System stopped processing this step because of errors.
```

The feedback in the SAS log is largely self-explanatory. The VALUES condition means that an error was identified in the third VALUES line. The first error is that in the third line neither the entries "Banking" nor "Finance" appear in the column STRING, but the entry "Telco".

If this error is corrected in the third VALUES line by replacing "Telco" with "Finance", and the syntax for the line-by-line update is submitted again, SAS returns the following feedback.

SAS Log 3.1-3: Failed Insert 3

```
ERROR: VALUE1 contains duplicates. Add/Update failed for data set WORK.EMPTY because
data value(s) do not comply with integrity constraint VALUE1_unq.
NOTE: This insert failed while attempting to add data from VALUES clause 3 to the data set.
NOTE: Deleting the successful inserts before error noted above to restore table to a
consistent state.
NOTE: The SAS System stopped processing this step because of errors.
```

The feedback in the SAS log is self-explanatory. The error is the first duplicate of a value that already occurred in the third line in the column VALUE1. (See 321 in line 2.) If the value is corrected (and also the value that still occurs in row 3), the INSERT VALUES syntax that was submitted again now returns the following feedback:

SAS Log 3.1-4: Failed Insert 4

```
ERROR: ID is not the primary key. Add/Update failed for data set WORK.EMPTY because
data value(s) do not comply with integrity constraint ID_pkey.
NOTE: This insert failed while attempting to add data from VALUES clause 3 to the
data set.
```

This feedback means that duplicate values occur in the primary key ID. If this error is also corrected, the INSERT VALUES syntax outputs the following message:

SAS Log 3.1-5: Successful Insert

```
NOTE: 4 rows were inserted into WORK.EMPTY.
```

This feedback means three things:

- Four cases have been inserted into the EMPTY table.
- The *formulated* check rules did not identify any other errors. The data quality meets the specified criteria.
- The result does *not* mean that the data in EMPTY contains no other errors. If relevant rules are missing, or if the rules are formulated too tolerantly or even programmed incorrectly, the quality of the data in EMPTY might still be doubtful.

Syntax for a Tablewise Update (SELECT, FROM)

Despite its advantages, the repeated checking of data updated line by line can be quite tedious for larger amounts of data. A more efficient approach might be to update the data table by table. Instead of four separate VALUES lines, the same incorrect data is created as file NEW_DATA.

Program 3.1-6: Test Data NEW_DATA for Tablewise Update

```
data NEW_DATA ;
  input ID STRING $2-9 VALUE1 DATE_VAR date7. ;
datalines ;
1 Banking 123 03JAN70
3 Finance 321 29JAN08
3 Telco   321 15OCT94
4 Finance 321 23OCT04
  ;
run ;
```

The new data in NEW_DATA is transferred to the EMPTY file using an INSERT INTO.

Program 3.1-7: Syntax for the Tablewise Update (SELECT, FROM)

```
proc sql;
      insert into EMPTY (ID, STRING, VALUE1, DATE_VAR)
      select ID, STRING, VALUE1, DATE_VAR
from NEW_DATA  ;
quit ;
```

Despite a more efficient transfer in the form of one update file, even with this approach SAS initially reports only *one* error and then aborts.

SAS Log 3.1-6: Failed Insert (Tablewise)

```
ERROR: STRING does not contain 'Banking' or 'Finance'. Add/Update failed for data set
WORK.EMPTY because data value(s) do not comply with integrity constraint STRING_c.

NOTE: Deleting the successful inserts before error noted above to restore table to a
consistent state.

NOTE: The SAS System stopped processing this step because of errors.
```

What was said for the rowwise approach also applies to the tablewise approach. SAS aborts after the first error found in NEW_DATA. If this error is corrected in NEW_DATA and this file is submitted again, SAS aborts after the next error.

To prevent SAS from aborting after each incorrect row passed, each new data row must be transferred to SAS in a separate INSERT statement. This approach is presented in Section 3.2 (second example) in conjunction with an audit trail.

Example 3: Applying Integrity Constraints Subsequently to an Already Filled Table

Check rules can be added to an empty table or to a table that already contains values. If the check rules are added to an empty table, the rules are activated only when values are inserted. (See Example 1.) If the columns concerned have such check rules, they are activated each time values are inserted (INSERT), deleted (DELETE), or changed (ALTER), and they perform their criteria-based checks accordingly. If the check rules are added to a filled table, the first check is whether the contents of the table match the assigned integrity constraints.

This example demonstrates what happens when integrity constraints are applied subsequently to a table that has already been filled and which itself already contains errors. The table with errors is represented by table DEMO. Its contents correspond to the VALUES lines from Example 2 with all their errors. The integrity constraints including their user-defined feedback are also taken from Example 2. Finally, it is shown how check rules can be deactivated (deleted) and reactivated.

Program 3.1-8: Example 3: Test Data DEMO

```
data DEMO ;
   input ID STRING $2-9 VALUE1 DATE_VAR date7. ;
   format DATE_VAR date7. ;                          proc print ;
   datalines ;                                       run ;
1 Banking 123 03JAN70
3 Finance 321 29JAN08
3 Telco   321 15OCT94
4 Finance 321 23OCT04
   ;
run ;
```

If an already filled table DEMO is to be checked subsequently using integrity constraints, you only need to specify the required integrity constraints after an ADD in the ALTER TABLE statement in an SQL step. When SQL is executed, these integrity constraints are applied to the table in question, that is, its incorrect contents are checked to see whether they comply with the rules of the integrity constraints passed. The result after DESCRIBE TABLE CONSTRAINTS is correspondingly informative. (See below.)

Program 3.1-9: Example 3: Syntax for Defining Table, Check Rule, and User Messages

```
proc sql ;
  alter table DEMO
    add constraint DATE_VAR_c   check(DATE_VAR >= "02JAN70"D)
        message="DATE_VAR contains values older than the date."
                                    msgtype=newline ,
      constraint ID_0       not null(ID)
       message="ID contains missings." msgtype=newline ,
      constraint ID_pkey    primary key(ID)
       message="ID is not the primary key."
                                    msgtype=newline ,
     constraint STRING_0  not null(STRING)
      message="STRING contains missings." msgtype=newline ,
     constraint STRING_c   check(STRING in ('Banking',
                                   'Finance'))
       message="STRING does not contain 'Banking' or 'Finance'."
                                    msgtype=newline ,
     constraint VALUE1_0   not null(VALUE1)
      message="VALUE1 contains missings."  msgtype=newline ,
     constraint VALUE1_c    check(VALUE1 >= 0.1)
       message="VALUE1 contains values not greater than or equal to 0.1."
                                    msgtype=newline ,
     constraint VALUE1_unq unique(VALUE1)
      message="VALUE1 contains duplicates." msgtype=newline ;
describe table constraints DEMO ;
      quit ;
run;

proc contents data=DEMO ;
run ;

proc print data=DEMO noobs ;
run ;
```

Figure 3.1-3: Example 3: Output of Integrity Constraints (DESCRIBE TABLE CONSTRAINTS)

```
       -----Alphabetic List of Integrity Constraints-----

  Integrity                      Where              User
# Constraints  Type   Variables  Clause             Message
-------------------------------------------------------------------------
1 DATE_VAR_c  Check              DATE_VAR>='02JAN1970'D  DATE_VAR contains values
                                                         older than the date.

2 ID_0        Not Null  ID                          ID contains missings.
```

Only two of the eight check rules passed to SAS are output in the list. This is because the data in the table only complies with these two integrity constraints. If integrity constraints are subsequently applied to a filled table, only those constraints are successfully passed to the table in question and output in this list that are met by the data in the table. If the data in the table meets all integrity constraints, all check rules are listed. PROC SQL aborts when it hits the first constraint that is violated by the data. (See constraint VALUE1_unq.) If a check rule is *not* listed, it might be either because the constraint itself was violated or the procedure aborted because of an earlier constraint. We learn more about this when we take a closer look at the data. The information about integrity constraints displayed by PROC CONTENTS is the same except for minor differences in the formatting of the output and will be not repeated.

The following PROC PRINT output describes the contents of DEMO.

Figure 3.1-4: Example 3: PROC PRINT Output

```
ID    STRING    VALUE1    DATE_VAR
 1    Banking    123      03JAN70
 3    Finance    321      29JAN08
 3    Telco      321      15OCT94
 4    Finance    321      23OCT04
```

Table DEMO still contains the errors that were added intentionally. In ID 3, for example, the entry "Telco" still violates the unlisted check rule "STRING_c" and the date entry still violates the unlisted check rule DATE_VAR_c. The multiple values "321" in the column VALUE1 still violate the check rule VALUE1_unq, and so on.

However, this result also means that the subsequent application of check rules only serves to *identify* possible errors, but not to interpret or correct them. The table still contains even more incorrect data. How do we know that? We know that *all* integrity constraints are listed only when *all* data meet *all* their rules. As some rules are *not* listed, we know that these check rules have identified further errors. This result also does not rule out the possibility that the data contains other errors.

Removing Integrity Constraints

If integrity constraints are to be deleted again, they need to be specified after DROP in an ALTER TABLE statement. In the following example, all validation rules are deleted except the three check rules VALUE1_0, VALUE1_c, and VALUE1_unq.

Program 3.1-10: Removing Integrity Constraints

```
proc sql ;
  alter table DEMO
   drop constraint DATE_VAR_c
   drop constraint ID_0
   drop constraint ID_pkey
   drop constraint STRING_0
   drop constraint STRING_c ;
     quit ;
```

Reactivating Integrity Constraints of the Type "Foreign Key"

Under certain circumstances, integrity constraints of the type "foreign key" must be reactivated again. For example, if tables containing check rules are processed further using a COPY, CPORT, CIMPORT, UPLOAD, or DOWNLOAD procedure, the default setting is to suppress the copying of integrity constraints. (See also PROC APPEND and PROC SORT.) After copying using PROC COPY, any previously existing check rules of the type "foreign key" are inactive. To actively copy the foreign keys, you can either use a CONSTRAINT=YES (or a similar option) in these procedures or a subsequent PROC DATASETS.

Program 3.1-11: Reactivating Integrity Constraints

```
proc datasets library=WORK ;
   modify DEMO ;
      ic reactivate FRN_KEY references FPATH ;
   run ;
quit ;
```

After LIBRARY=, the current location of the file is specified. After REFERENCES (default), the previous location from where a file was copied is specified (for example, using PROC COPY). FRN_KEY represents a fictitious foreign key in this case.

Compared to PROC SQL, PROC DATASETS has a similar range of functions for creating, changing, or deleting integrity constraints. Users are advised to consider the potential gain of higher performance against the effort of the programming required.

3.1.2 Audit Trails

This section introduces working with audit trails. Audit trails log changes to a table and store them in a separate file. It is important to note that an audit trail is also currently the only technical option in SAS to save cases from failed APPEND operations or cases rejected by integrity constraints.

A Conceptual Perspective

An audit trail is a SAS file that can be created to log changes to a SAS table and store them in a separate file. Each time a case is added, deleted, or updated, the log shows who made the change, what change was made, and when the change was made. An audit trail therefore logs the "history" of data and tables, ideally from the time they enter a system until the time they leave it.

Audit trails are therefore relevant in business intelligence because they support the following four areas:

- Transparency of data flows within a company
- Security of data and companies (documentation, control)
- Efficiency of data flows and companies (performance, Green IT)
- Quality of data and data operations within a company (completeness, "backup" function)

A Technical Perspective

An audit trail is created by the default Base SAS engine and has the same path (libref) and name as the data table, but is of the file type AUDIT. An audit trail works similarly in local and remote environments, possibly with the exception of the time stamp. The logged time when writing in the remote SAS session might differ from the user's SAS session. This file replicates the variables from the data table and automatically creates so-called _AT*_ variables in it. _AT*_ variables automatically store information about changes to data as shown in Table 3.1-1.

Table 3.1-1: _AT*_ Variables

_AT*_ Variable	Description
ATDATETIME	Saves the date and time of a change (except in remote environments, if applicable).
ATUSERID	Saves the login user ID in connection with a change.
ATOBSNO	Saves the case number that is affected by the change (except when REUSE= YES).
ATRETURNCODE	Saves the so-called Event Return Code.
ATMESSAGE	Saves the SAS log at the time of the change (except in remote environments).
ATOPCODE	Saves a code that describes the type of change: • AL: Auditing continues. • AS: Auditing is suspended. • DA: Image of the added data row. • DD: Image of the deleted data row. • DR: Image of the data row before the update. • DW: Image of the data row after the update. • EA: Adding observation failed. • ED: Deleting observation failed. • EU: Updating observation failed.

Besides the _AT*_ variables, user-defined variables can also be created. User-defined variables are variables for which the information about the changes to the data can be transferred to the audit file by the user. A user-defined variable is associated with the values in the data table and is updated by changes to these values in the data table. For example, data entry staff can specify a reason or other special circumstances for the respective update in the form of free text.

An audit trail logs successful updates as well as failed updates. The logs are in the form of an audit file.

Note: An audit trail is currently the only technical possibility in SAS to store cases from failed APPEND operations or rejected by integrity constraints.

By means of a DATA step, the rejected cases can be extracted from the audit file and the information from _AT*_ and user-defined variables can be used for root cause analysis and error correction to successfully pass the originally rejected cases to the data table.

Due to the large number of row-level processes and the fact that each update of the data table is also passed to the audit file, an audit trail can make heavy use of the performance of a system. For large, regularly running batch updates, it can therefore make sense to suspend an audit trail. An audit trail is not recommended for tables that undergo one of the following operations: copy, move, replace, or transfer to another operating environment. These operations do not preserve the audit trail. If the audit file is damaged, the data table can only be processed further when the audit trail is terminated. The data table can then be processed without an audit trail. If necessary, a new audit trail can be initiated for the data table.

Audit trails cannot be initialized for generation data sets.

Two examples of working with audit trails are presented below.

Example 1: Initiating and Interpreting an Audit Trail (Example without Integrity Constraints)

Audit trails log changes to a table and store them in a separate file called an audit file. This example shows how an audit file is initialized and how user-defined variables can be created for this purpose. After four cases have been processed in the audit trail, the audit file is viewed using the CONTENTS and SQL procedures. Differences to SAS data tables are also pointed out, including the user-defined variables and the so-called _AT*_ variables.

Program 3.1-12: Example 1: Demo Data

```
data DEMO ;                                   proc print ;
  input ID STRING $2-9 VALUE1 DATE_VAR date7. ;   format DATE_VAR date7. ;
  datalines ;                                 run ;
1 Banking 001 03JAN70
2 Finance 002 29JAN08
3 Banking 003 15OCT94
  ;
run ;
```

An audit trail is initialized with PROC DATASETS. The audit file is automatically stored under the same directory and name as the table to be monitored. In this case, the audit file is created in the directory WORK and the name DEMO. In addition, two user-defined variables named REASON_1 and REASON_2 are created after USER_VAR. The user-defined variables are created only in the audit file, not in the data table.

Program 3.1-13: Initiating an Audit Trail

```
proc datasets lib=WORK ;
  audit DEMO ;
     initiate ;
     user_var REASON_1 $ 30 REASON_2 $ 30 ;
run ;
quit ;
```

SAS Log 3.1-7: Feedback in the SAS Log

```
WARNING: The audited data file WORK.DEMO.DATA is not password protected.
         Apply an ALTER password to prevent accidental deletion or replacement of it
         and any associated audit files.
NOTE: The data set WORK.DEMO.AUDIT has 0 observations and 12 variables.
```

Four cases are now added to table DEMO using INSERT. The values of each case are specified in a separate line after VALUES. The order of the entries in the parentheses after VALUES must correspond to the order in the target table (DEMO). Since four cases are to be inserted in DEMO, INSERT contains four VALUES lines.

Program 3.1-14: Syntax for the Rowwise Update (VALUES)

```
proc sql ;
insert into DEMO
      values (5,"Banking",123,"03JAN70"d, "Supplement", "---")
      values (6,"Finance",321,"29JAN08"d, "Supplement", "---")
      values (6,"Telco",321,"15OCT94"d,   "Supplement", "---")
      values (8,"Finance",321,"23OCT04"d, "Supplement", "---") ;
quit ;
```

SAS Log 3.1-8: Feedback in the SAS Log

```
NOTE: 4 rows were inserted into WORK.DEMO.
```

PROC CONTENTS allows an insight into the content of the audit file. Alternatively, an audit file can be viewed with any other SAS procedure for reading files, for example, PROC PRINT, PROC SQL, and so on. Here too, the file type must be explicitly requested with TYPE=AUDIT. The audit file is read-only, meaning that its entries cannot be changed or overwritten.

Program 3.1-15: Syntax for Looking into an Audit File

```
proc contents data=DEMO (type=audit) ;
run ;
```

In the following, the special features of an audit file are in bold in the PROC CONTENTS output.

Figure 3.1-5: Contents of the Audit File in the SAS Output

```
                        The CONTENTS Procedure
```

Data Set Name	WORK.DEMO.**AUDIT**	Observations	4
Member Type	**AUDIT**	Variables	12
Engine	V9	Indexes	0
Created	12/14/2020 17:07:27	Observation Length	158
Last Modified	12/14/2020 17:09:22	Deleted Observations	0
Protection		Compressed	NO
Data Set Type	**AUDIT**	Sorted	NO
Label			
Data Representation	WINDOWS_64		
Encoding	wlatin1 Western (Windows)		

```
                   Engine/Host Dependent Information
```

Data Set Page Size	65536
Number of Data Set Pages	1
First Data Page	1
Max Obs per Page	385
Obs in First Data Page	4
Number of Data Set Repairs	0
ExtendObsCounter	YES
Filename	C:\...\demo.sas7baud
Release Created	9.0401M7
Host Created	X64_10PRO
Owner Name	MC-T410-1\cfgsc
File Size	65KB
File Size (bytes)	66560

```
                Alphabetic List of Variables and Attributes
```

#	Variable	Type	Len	Format
4	DATE_VAR	Num	8	
1	ID	Num	8	
5	**REASON_1**	**Char**	**30**	
6	**REASON_2**	**Char**	**30**	
2	STRING	Char	8	
3	VALUE1	Num	8	
7	**_ATDATETIME_**	**Num**	**8**	**DATETIME19.**
12	**_ATMESSAGE_**	**Char**	**8**	
8	**_ATOBSNO_**	**Num**	**8**	
11	**_ATOPCODE_**	**Char**	**2**	
9	**_ATRETURNCODE_**	**Num**	**8**	
10	**_ATUSERID_**	**Char**	**32**	

The output looks the same as a SAS data file, but in the bold lines, SAS refers several times to the fact that it is the content of an *audit* file. However, this does not rule out confusion with the contents of a SAS table with certainty. The bold elements of the list of variables and attributes are characteristic of audit files, including _AT*_ variables as well as user-defined variables (REASON_1, REASON_2). The information after "Data Set Name" in the first line of the output indicates that the audit file is automatically created by SAS in the same path and with the same name as the data table.

Using PROC SQL, the intermediate status of the audit trail is now queried by viewing the created audit file. The following are requested: the four data columns from DEMO (ID, STRING, VALUE1, DATE_VAR), two user-defined variables (REASON_1, REASON_2), and three _AT*_ variables (_atopcode_, _atuserid_, _atdatetime_). With TYPE=AUDIT, the file type is explicitly requested.

Program 3.1-16: Retrieving the Intermediate Status of an Audit Trail (SQL)

```
proc sql;
  select ID, STRING, VALUE1, DATE_VAR format=date7., REASON_1, REASON_2,
         _atopcode_,
         _atuserid_ format=$6.,
         _atdatetime_
         from DEMO(type=audit) ;
quit ;
```

Figure 3.1-6: Feedback in SAS Output

```
ID STRING  VALUE1 DATE_VAR   REASON_1    REASON_2 _ATOPCODE_  _ATUSERID_   _ATDATETIME_
-------------------------------------------------------------------------------------
5  Banking   123   03JAN70   Supplement    ---       DA         cfgsc     14DEC2020:17:09:22
6  Finance   321   29JAN08   Supplement    ---       DA         cfgsc     14DEC2020:17:09:22
6  Telco     321   15OCT94   Supplement    ---       DA         cfgsc     14DEC2020:17:09:22
8  Finance   321   23OCT04   Supplement    ---       DA         cfgsc     14DEC2020:17:09:22
```

Rowwise reading: The SAS output shows the number of lines *processed*. The first four lines are not displayed because they already existed, and no processing was necessary.

Columnwise reading: The values passed to ID, STRING, VALUE1, and DATE_VAR are displayed in the processed lines under the four data columns (for DATE_VAR after a corresponding formatting). Under the two user-defined variables REASON_1 and REASON_2, the information specified by the user in free text can be viewed in the processed rows. The user-defined variables are not stored in DEMO. The three _AT*_ variables show the code of the change ("DA", image of the added data row) (_atopcode_), the login user ID (_atuserid_, this was a single user each, "cfgsc"), and the date and time of the change (_atdatetime_). Also, the _AT*_ variables are not stored in DEMO.

With PROC PRINT, the intermediate status of the created SAS data file is now queried. This query could also have been made with PROC SQL.

Program 3.1-17: Retrieving the Intermediate Status of the Created SAS Data Set

```
proc print data=DEMO noobs ;
format DATE_VAR date7. ;
run ;
```

Figure 3.1-7 Feedback in SAS Output

```
ID     STRING     VALUE1       DATE_VAR

 1     Banking       1         03JAN70
 2     Finance       2         29JAN08
 3     Banking       3         15OCT94
 4     Finance      321        23OCT04
 5     Banking      123        03JAN70
 6     Finance      321        29JAN08
 6     Telco        321        15OCT94
 8     Finance      321        23OCT04
```

Columnwise reading: PROC PRINT shows all columns from the SAS file DEMO. The two user-defined variables REASON_1 and REASON_2 are not stored in DEMO. No _AT*_ variable is stored in DEMO either.

Rowwise reading: The SAS output shows the number of all available rows. In this form, no distinction is made between existing rows and subsequently processed rows.

Example 2: Audit Trails in Conjunction with Integrity Constraints (Display of Rejected Cases)

Audit trails log changes to a table and store them in a so-called audit file. An audit file therefore logs not only the accepted data that the SAS table could also display, but also all rejected data. The following example demonstrates how audit trails work in conjunction with integrity constraints. For simplicity, this approach uses elements already known from the previous sections. To prevent SAS from aborting after each incorrect line passed, two ways for passing data to SAS are presented. In contrast to all previous examples, an update is *not* aborted if data errors occur. Successfully inserted data is kept; data with errors is stored in a separate file.

Table DEMO2 is created in the first step. The contents of DEMO2 fulfill the criteria of the integrity constraints added in the subsequent step.

Program 3.1-18: Example 2: Demo Data

```
data DEMO2 ;
  format DATE_VAR date9.;
  input ID STRING $2-9 VALUE1 DATE_VAR date7.;
  datalines ;
1 Banking 123 03JAN70
2 Finance 321 29JAN08
3 Banking 322 15OCT94
4 Finance 323 23OCT04
  ;
run ;
```

In the second step, the integrity constraints already known are applied to table DEMO2. Since the contents of the table fulfill the criteria of the integrity constraints, all the check rules are transferred. No further result after DESCRIBE TABLE CONSTRAINTS is displayed.

Program 3.1-19: Applying Integrity Constraints to Demo Data

```
proc sql ;
  alter table DEMO2
    add constraint DATE_VAR_c    check(DATE_VAR >= "02JAN70"D)
        message="DATE_VAR contains values older than the date."
                                      msgtype=newline ,
      constraint ID_0      not null(ID)
      message="ID contains missings." msgtype=newline ,
      constraint ID_pkey    primary key(ID)
       message="ID is not the primary key."
                                    msgtype=newline ,
      constraint STRING_0   not null(STRING)
       message="STRING contains missings." msgtype=newline ,
      constraint STRING_c   check(STRING in ('Banking',
                                        'Finance'))
       message="STRING does not contain 'Banking' or 'Finance'."
                                    msgtype=newline ,
      constraint VALUE1_0    not null(VALUE1)
       message="VALUE1 contains missings."  msgtype=newline ,
      constraint VALUE1_c    check(VALUE1 >= 0.1)
       message="VALUE1 contains values not greater than or equal to 0.1."
                                      msgtype=newline ,
      constraint VALUE1_unq unique(VALUE1)
       message="VALUE1 contains duplicates."  msgtype=newline ;
  describe table constraints DEMO2 ;
      quit ;
```

The audit trail is initialized by means of PROC DATASETS. The audit file is automatically stored under the same directory and name as the table to be monitored. In this case, the audit file is created in the WORK directory and named DEMO2. In addition, two user-defined variables named REASON_1 und REASON_2 are created after USER_VAR.

Program 3.1-20: Initializing Audit Trail

```
proc datasets lib=WORK ;
  audit DEMO2 ;
     initiate ;
     user_var REASON_1 $ 30 REASON_2 $ 30 ;
run ;
quit ;
```

To prevent SAS from aborting after each incorrect row passed, each new data row is passed to SAS in a *separate* INSERT statement. This approach is shown in two variants. The first approach passes each new row of data to SAS in a separate INSERT statement. The second approach does exactly the same, but in the form of a more efficient macro. Both approaches lead to absolutely identical results.

It is important to know that all new data rows contain at least one error.

Program 3.1-21: Approach I: Multiple INSERT Statements

```
proc sql;
    insert into DEMO2 (ID, STRING, VALUE1, DATE_VAR, REASON_1, REASON_2 )
    set ID=5, STRING="Banking", VALUE1=123, DATE_VAR="01JAN70"d, REASON_1="Supplement",
REASON_2="---";
    insert into DEMO2 (ID, STRING, VALUE1, DATE_VAR, REASON_1, REASON_2 )
    set ID=6, STRING="Finance", VALUE1=321, DATE_VAR="29JAN08"d, REASON_1="Supplement",
REASON_2="---";
    insert into DEMO2 (ID, STRING, VALUE1, DATE_VAR, REASON_1, REASON_2 )
    set ID=6, STRING="Telco", VALUE1=321, DATE_VAR="15OCT94"d, REASON_1="Supplement",
REASON_2="---";
    insert into DEMO2 (ID, STRING, VALUE1, DATE_VAR, REASON_1, REASON_2 )
    set ID=8, STRING="Finance", VALUE1=321, DATE_VAR="23OCT04"d, REASON_1="Supplement",
REASON_2="---";
 quit;
```

Program 3.1-22: Approach II: Macro AUDIDAT

```
%macro AUDIDAT(ID, STRING, VALUE1, DATE_VAR, REASON1, REASON2 );
proc sql;
    insert into DEMO2 (ID, STRING, VALUE1, DATE_VAR, REASON_1, REASON_2)
    set ID=&ID., STRING=&STRING., VALUE1=&VALUE1.,
        DATE_VAR=&DATE_VAR., REASON_1=&REASON1., REASON_2=&REASON2. ;
 quit;
%mend AUDIDAT;

%AUDIDAT(5, "Banking", 123, "01JAN70"d, "Supplement", "---");
%AUDIDAT(6, "Finance", 321, "29JAN08"d, "Supplement", "---");
%AUDIDAT(6, "Telco",   321, "15OCT94"d, "Supplement", "---");
%AUDIDAT(8, "Finance", 321, "23OCT04"d, "Supplement", "---");
```

The INSERT and the macro approach do exactly the same thing. SAS users are completely free to choose their preferred approach.

PROC PRINT now queries the intermediate status in the SAS data file DEMO2. Since all new data lines contained errors, none of these new lines were included in DEMO2.

Program 3.1-23: Retrieving the Intermediate Status of the Created SAS Data Set

```
proc print data=DEMO2 noobs ;
run ;
```

Figure 3.1-8: Feedback in SAS Output

```
ID     STRING      VALUE1     DATE_VAR
1      Banking      123        03JAN70
2      Finance      321        29JAN08
3      Banking      322        15OCT94
4      Finance      323        23OCT04
```

A second PROC PRINT now queries the created audit file DEMO2, filtered for _ATOPCODE_ "EA". The four data columns from DEMO (ID, STRING, VALUE1, DATE_VAR) and two user-defined variables (REASON_1, REASON_2) are requested. With TYPE=AUDIT, the file type is requested explicitly. The query of the audit file using PROC SQL (see Example 1) is waived at this point.

Program 3.1-24: Retrieving the Intermediate Status of an Audit Trail (PROC PRINT)

```
proc print data=DEMO2(type=audit) noobs ;
  where _atopcode_ eq "EA";
  format DATE_VAR date7. ;
  var ID STRING VALUE1 DATE_VAR REASON_1 REASON_2 ;
run ;
```

Figure 3.1-9: Feedback in SAS Output

```
ID     STRING      VALUE1      DATE_VAR      REASON_1       REASON_2

5      Banking      123        01JAN70      Supplement       ---
6      Finance      321        29JAN08      Supplement       ---
6      Telco        321        15OCT94      Supplement       ---
8      Finance      321        23OCT04      Supplement       ---
```

This result shows three differences from all previous examples:

1. SAS runs through each data row, even if it contains errors, and does not abort further processing.
2. The audit file displays *all* errors or rejected data at once.
3. The macro and INSERT approaches are more efficient than the approaches in Section 3.1.1.

Section 3.1 introduced integrity constraints, audit trails, and validation checks with integrity constraints on up to two variables. One challenge of integrity constraints is that the programming environment is currently not flexible enough to accommodate alternative approaches. For simple or one-time checks, integrity constraints might be too complex. In this case, it is possible to apply approaches from the sections that follow. With SAS syntax, programming is much more flexible than it might seem.

3.2 How to Identify and Filter Multiple Values

Duplicates are cases that have the same values in at least one key variable (for example, IDs), assuming that the entries in the key variable are correct. Duplicates are single values or complete data rows that occur more than once. Along with completeness, uniformity, and missing values, duplicates are one of the basic criteria of data quality that SAS can be used to check. The checking of all other criteria is based on these. However, checking for duplicates requires completeness and uniformity.

The conceptual opposite of duplicate data is *unique* data. Unique data are expected to occur only one single time, for example, Social Security numbers or tax identification numbers. But don't count on it. The German Federal Central Tax Office issues an 11-digit personal tax identification number (TIN), which is supposed to be as unique as a fingerprint. In more than 164,000 cases, it assigned either one taxpayer two TINs, or two taxpayers got the same TIN (Süddeutsche Zeitung, 2014).

This section deals with the handling, finding, and filtering of duplicates. When checking for duplicates, it should first be clarified conceptually whether the SAS table might only contain one case per row (situation 1), or whether there is measurement repetition and the same cases occur in several rows (situation 2). An example of situation 1 is the employee database of a company. An employee's biographical data should normally only be listed once. In this case, an occurrence of the same data more than once would be considered an error. An example of situation 2 is a purchase database where a customer orders online from the same company several times over a year or a patient visiting a doctor for treatment several times. In this case, the personal data might appear several times in the database, and an occurrence of the same data more than once would be interpreted as permissible.

There are numerous reasons for multiple occurrences of data. Possible causes of duplicate data rows could be:

- the accidental multiple appending of identical cases in SAS (for example, by INSERT or SET statements)
- the different allocation or coding of the same personal or address data
- ID variables changing over time for longitudinal data
- the automatic multiple saving of the data of a single case due to incorrect data collection (especially for online surveys)
- creating copies of identical lines or even complete tables by incorrect programs

Even with small data sets, it can happen that the same questionnaire is entered more than once.

The practical consequences of duplicate data are by no means trivial. Not so long ago, a data error inflated Wells Fargo's operational risk capital by $5.2 billion. The removal of duplicate data led to a sharp fall in Q1 operational risk-weighted assets from $403.6 billion to $338.7 billion. Naturally this also raised questions on the soundness of the bank's operations risk management (RiskNet, 2020). Other practical consequences include the following:

- The mere occurrence of duplicates affects storage capacity and computer speed.
- Depending on the type and function of duplicate data, duplicates can lead to multiple processes that consume resources and cause further damage.
- Duplicate address data leads to multiple mailings to the same address, and thus also to image damage.
- Duplicate patient data leads to incorrect billing of patients or health insurance companies.
- Duplicate product data leads to incorrect calculations of sales, excessive revenues or expenditure, and so on.
- Duplicate trigger data leads to unnecessary multiple execution of further actions such as submitting, saving, looping, clustering, and so on.
- Due to their more frequent occurrence, multiple cases are twice as likely to be sampled as cases that occur only once, and not only if duplicates are only distributed within one list, group, or cell (*within-group duplicates*).
- A more complicated situation exists if duplicates are distributed over several groups (*between-group duplicates*).
- Duplicates might already be distributed to two different groups due to one single different coding, for example, two different sampling units ("frame units").

This section presents the following options for dealing with multiple rows and values:

- Approach 1: Displaying duplicate IDs (univariate)
- Approach 2: Filtering of duplicate values (HAVING COUNT)
- Approach 3: Finding duplicate values (multivariate) (situation 2)
- Approach 4: Creating lists for duplicate rows (macro variable)
- Approach 5: Checking for duplicate entries (macro)
- Approach 6: Identifying duplicates in multiple tables

When checking for duplicates, make sure to determine whether the SAS table contains only one case per row (situation 1) or whether a measurement is repeated and the same cases occur in several rows (situation 2). The approaches described below are generally suitable for checking for intra-group and inter-group duplicates.

A data row has two different attributes related to occurrence: frequency and diversity. *Frequency* indicates whether a particular data row occurs once or more frequently. *Diversity* indicates whether a data row is unique or also occurs in copies:

- *Single* means that a row occurs only *once* in the data set. Single rows are always unique rows. (See the fourth row in the data set in Program 3.2-1: "004 b 75...".)
- *Multiple* means that a row occurs more than once in the data set. (See the third and sixth rows in the data set in Program 3.2-1: "003 b 23...".) Multiple occurring rows contain at least one unique row and one copy (or more) of that row. Single and multiple are opposites.
- *Unique* also starts with any data row; in its simplest form, a row is unique if it occurs only once. If there are several equal rows, you call *one* representative row unique. The other rows are considered duplicates of this unique row. That's why the attribute "unique" might occur more often than "single".
- *Duplicate* means that there are copies of a unique row in the data set. Duplicate rows are always also multiple rows, but not necessarily the other way round. Unique and duplicate are not necessarily opposites.

The number of unique and duplicate rows add up to the total number of multiples. See ID 002 in DUP_DEMO in Program 3.2-1. One unique (representative) and two duplicate rows (copies) make three ID 002 multiples.

Program 3.2-1: Test Data for Handling Duplicates

```
data DUP_DEMO ;
input
@ 1 ID $3. @ 5 GROUP $2. @ 7 AGE 2. @ 10 VARNUM1 1. @ 11 VARNUM2 1. @ 12 VARNUM3 1.
@ 14 VARCHAR1 $1. @ 15 VARCHAR2 $1. @ 16 VARCHAR3 $1. ;
datalines;
001 a  8 101 bca
002 a 17 010 abc
003 b 23 110 abc
004 b 75 010 bac
002 a 17 010 abc
003 b 23 110 abc
005 a 65 100 cba
002 a 17 010 abc
003 b 23 110 abc
005 a 65 100 cba
;
run;
```

3.2.1 Approach 1: Displaying Duplicate IDs (Univariate)

This first approach checks the fundamental question of whether duplicate data rows occur at all in a SAS table. If the SAS table is to contain only one case per data row (the prerequisite for situation 1), then checking the key variables can be done using several simple approaches as shown using the referenced example data set DUP_DEMO:

- In small, manageable data sets, duplicate entries in a key variable can be easily checked using a PROC FREQ or a frequency table. (See Schendera, 2004.)

Program 3.2-2: Approach 1: PROC FREQ

```
proc freq data=DUP_DEMO ;
table ID ;
run ;
```

Figure 3.2-1: Result of the FREQ Procedure

```
                          The FREQ Procedure

                                          Cumulative   Cumulative
          ID     Frequency     Percent    Frequency     Percent
          -------------------------------------------------------
          001        1         10.00          1         10.00
          002        3         30.00          4         40.00
          003        3         30.00          7         70.00
          004        1         10.00          8         80.00
          005        2         20.00         10        100.00
```

- For data that is not too large, it is also possible to query a univariate bar chart for the frequency of ID levels. Normally, all IDs should have the frequency equal to 1. Every spike beyond 1 is an indication that an ID occurs more than once. In a bar chart, multiple IDs are easy to recognize in the form of individual peaks.

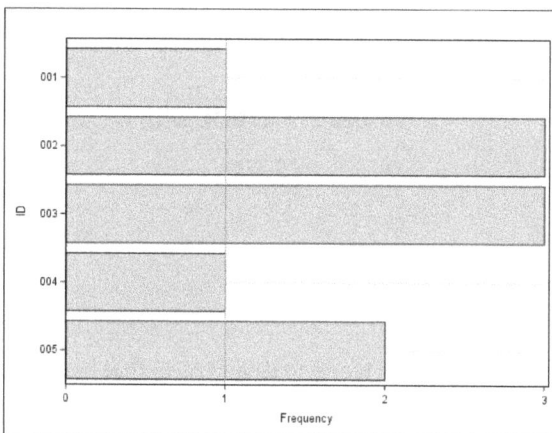

The horizontal bar chart on the left was created with PROC SGPLOT. A reference line was drawn at the value 1.

However, for voluminous tables with tens of thousands of data rows or more, there are more appropriate approaches.

- In the case of voluminous tables, an aggregating counting of the IDs using PROC SQL and COUNT is a more appropriate approach.

Program 3.2-3: Approach 1: SQL

```
proc sql ;
    select ID, count(ID) as COUNT
        from DUP_DEMO
    group by ID ;
quit;
```

Figure 3.2-2: Result of the SQL Procedure

```
ID       COUNT
--------------
001        1
002        3
003        3
004        1
005        2
```

The variable COUNT contains information about the frequency of occurrence of the IDs. If IDs are only allowed to occur once, values greater than 1 indicate that duplicate IDs occur in the data set. The value 3 in the COUNT column indicates that an ID occurs more than once. The ID 003 actually occurs three times in the data set DUP_DEMO. If levels of increments or ID variables occur more than once in an SAS table, this does not automatically mean that the rest of the row is also the same and that, apart from the ID, the data in the rows do not exist more than once in the table. It is also possible that the ID occurs more than once, but the rest of the respective data rows are correctly different. To check the correspondence of complete data rows and not just IDs, you would have to use other approaches (for example, using the NODUPRECS option).

3.2.2 Approach 2: Filtering of Duplicate Values (HAVING COUNT)

This SQL approach returns all values that occur more than once in ID. SAS outputs three values ("002", "003", "005") because these values occur more than once. This SQL program keeps all columns specified after SELECT (ID), groups all rows also by ID (GROUP BY), and keeps all ID values with a COUNT value greater than 1 (see HAVING). In this example, these three values do not represent the number of rows in DUP_DEMO.

Program 3.2-4: Approach 2: SQL (ID)	Figure 3.2-3: SQL with ID

<pre>proc sql;
   select ID
      from DUP_DEMO
         group by ID
         having count(ID) gt 1;
quit;</pre>
	<pre>ID
	--- 	
	002
	003
	005</pre>

The following SQL approach also outputs all values that occur more than once in ID. In this variant, the COUNT counting variable is explicitly created and output. In this example, the three ID values output do not represent the number of rows in DUP_DEMO.

Program 3.2-5: Approach 2: SQL (ID, COUNT)	Figure 3.2-4: SQL with ID and COUNT

<pre>proc sql;
   select ID, count(ID) as COUNT
      from DUP_DEMO
         group by ID
         having count(ID) gt 1;
quit;</pre>
	<pre>ID COUNT
	------------- 	
	002 3
	003 3
	005 2</pre>

The following SQL approach, on the other hand, outputs all rows (shortened to ID and AGE) that have the same value more than once in ID.

Program 3.2-6: Approach 2: SQL (ID, COUNT, AGE)	Figure 3.2-5: SQL with ID, AGE, and COUNT

<pre>proc sql;
   select ID, AGE
      from DUP_DEMO
         group by ID
         having count(ID) gt 1;
quit;</pre>
	<pre>ID AGE
	------------- 	
	002 17
	002 17
	002 17
	003 23
	003 23
	003 23
	005 65
	005 65</pre>

The following SQL variant is supplemented by a second grouping in the GROUP BY statement. For this reason, only one representative of each of the multiple AGE values is output in this variant. A WHERE condition > 2 is specified as the (modified) condition. Depending on the question, a positive or negative deviation from this specified (because expected) number of rows can be interpreted as an error. Alternative test options are GT, LT, NE, and so on.

Program 3.2-7: Approach 2: SQL (ID, COUNT, AGE)	Figure 3.2-6: SQL with Modified Limit Value

<pre>proc sql;
   select ID, AGE
      from DUP_DEMO
         group by ID, AGE
         having count(ID) > 2 ;
quit;</pre>
	<pre>ID AGE
	------------- 	
	002 17
	003 23</pre>

3.2.3 Approach 3: Finding Duplicate Values (Multivariate)

Before checking for duplicates, it should be clarified whether the SAS table might contain only one case per row (situation 1) or whether a measurement is repeated and the same cases occur in several rows (situation 2). This SQL approach is suitable if there should be several but the same number of data rows per group (for example, regular data deliveries or repeated measurements) (situation 2).

The following SQL program counts the occurrence of duplicate data rows. *Duplicate* is defined as rowwise equal values in the variables specified under GROUP BY (situation 2). The more variables are specified, the lower the probability of matches. Finally, all rows are kept whose COUNT value is greater than 1 (filtered for the region "Africa").

Program 3.2-8: Approach 3: Counting the Occurrence of Duplicate Data Rows

```
proc sql;
create table DUPLICATES as
select REGION, SUBSIDIARY, STORES, SALES,
            count(*) as count
    from SASHELP.SHOES
group by REGION, SUBSIDIARY, STORES
having count(*) > 1 and REGION="Africa"
order by REGION, SUBSIDIARY, STORES ;
quit;
```

Figure 3.2-7: SAS Output: Counting the Occurrence of Duplicate Data Rows

Region	Subsidiary	Stores	Sales	count
Africa	Addis Ababa	4	$67,242	2
Africa	Addis Ababa	4	$1,690	2
Africa	Addis Ababa	12	$108,942	2
Africa	Addis Ababa	12	$29,761	2
Africa	Cairo	3	$14,095	2
Africa	Cairo	3	$2,259	2
Africa	Cairo	9	$13,732	2
Africa	Cairo	9	$10,532	2
Africa	Khartoum	1	$9,244	2
Africa	Khartoum	1	$19,582	2
Africa	Kinshasa	10	$4,888	2
Africa	Kinshasa	10	$16,662	2
Africa	Luanda	2	$801	2
Africa	Luanda	2	$29,582	2

This SQL program groups all rows according to the sequence in GROUP BY, keeps all columns specified after SELECT and all rows with a COUNT value greater than 1, and filters out all rows without duplicates before storing them in the DUPLICATES table.

3.2.4 Approach 4: Creating Lists for Duplicate Rows (Macro Variable)

Another possibility to identify and filter out duplicate uni- or multivariate data is the use of PROC SORT together with the options NODUPKEY or NODUPRECS. Of all rows that have multiple entries either in the *ID* (NODUPKEY) or across the defined row (NODUPRECS), all but the first row that occur more than once can afterward be stored in a separate SAS table using DUPOUT=.

The following example is explicitly intended for duplicates of the string type and is based on a sequence of PROC SORT including DUPOUT=, PROC SQL, and the QUOTE function; another PROC SORT that accesses the macro variable &DUPLICATES created using PROC SQL; and a concluding PROC PRINT. The functionality of these steps is not explained further here. The ID variable is of type string.

Program 3.2-9: Approach 4: PROC SORT and the Options NODUPKEY or NODUPRECS

```
proc sort data=DUP_DEMO out=UNIQUES noduprecs
                       dupout=DUPLICATES;
    by _all_ ;
run ;
```

Sorting BY _ALL_ causes the entire table to be sorted through for multiple rows. Because of NODUPRECS, rows that occur once are stored in the table UNIQUES. All rows that occur more than once are stored in the table DUPLICATES.

Program 3.2-10: Approach 4: PROC SQL: Creating Lists for Duplicate Rows (Macro Variable)

```
proc sql noprint ;
      select quote(ID)
      into :DUPLICATES separated by " "
      from DUPLICATES;
quit;

proc sort data=DUP_DEMO out=DUPS_ID ;
      where ID in (&DUPLICATES.);
      by ID ;
run ;
```

Program 3.2-11: Approach 4: Printing Lists for Identical IDs and (Not) Unique Rows

```
title "Analysis for identical IDs" ;
proc print data=DUPS_ID ;
var ID ;
run ;

title "Analysis for (not) unique rows" ;
proc print data=UNIQUES  ;
run ;
```

Figure 3.2-8: SAS Output: Analysis for Identical IDs

```
Analysis for identical IDs

    Obs    ID

     1     002
     2     002
     3     002
     4     003    (abbreviated)
```

Figure 3.2-9: SAS Output: Analysis for (Not) Unique Rows

ID	GROUP	AGE	VARNUM1	VARNUM2	VARNUM3	VARCHAR1	VARCHAR2	VARCHAR3
001	a	8	1	0	1	b	c	a
002	a	17	0	1	0	a	b	c
003	b	23	1	1	0	a	b	c
004	b	75	0	1	0	b	a	c
005	a	65	1	0	0	c	b	a

The table UNIQUES contains only single rows (for example, ID 001, 002). If there are copies of these rows in table DUP_DEMO, they are stored in table DUPLICATES. If DUP_DEMO contains duplicates of every data row, especially if it is only a single copy, tables UNIQUES and DUPLICATES might contain identical data rows even in numbers. Please note: Be careful calling entries in UNIQUES unique; some might have copies in DUPLICATES.

Program 3.2-12: Approach 4: Printing Lists for Rows that Occur More than Once

```
title "Analysis for duplicate rows" ;
proc print data=DUPLICATES ;
run ;
```

Figure 3.2-10: SAS Output: Rows that Occur More than Once in DUP_DEMO

```
Analysis for duplicate rows
```

ID	GROUP	AGE	VARNUM1	VARNUM2	VARNUM3	VARCHAR1	VARCHAR2	VARCHAR3
002	a	17	0	1	0	a	b	c
002	a	17	0	1	0	a	b	c
003	b	23	1	1	0	a	b	c
003	b	23	1	1	0	a	b	c
005	a	65	1	0	0	c	b	a

The table DUPLICATES contains only rows that occur more than once in DUP_DEMO. If, for example, two identical rows occur in the output table, one is stored in the table UNIQUES and its copy in the table DUPLICATES (for example, ID 005). If no duplicate rows occur in table DUP_DEMO, the table DUPLICATES will be empty.

This example represents a transition to macro programming with PROC SQL. This check program can also be converted to a macro version.

3.2.5 Approach 5: Checking for Duplicate Entries (Macro)

The DUPLICATES macro checks a single column to see whether values occur more than once in it. The macro includes four steps. In the first step, the values in the specified column &VAR. are counted through for multiples. The result is stored in the variable DUPL_MAX. In the second step, the maximum is determined and stored in the macro variable "max_dupl". The third step repeats the first step, but this time it outputs the values that might occur more than once as column DUPLICATES directly into the SAS output. The multiple %PUTs form the fourth step. An overview of the check performed is output to the log as a summary. This macro can be extended so that it performs the checks of the program above.

Program 3.2-13: Approach 5: Checking for Duplicate Entries (Macro)

```
%macro DUPLICATES(dsname, var) ;
  proc sql ;
     create table TEMP as
    select count(*) as DUPL_MAX
        from &dsname.
          group by &var.
        having COUNT(*) > 1 ;
    select max(DUPL_MAX) into :max_dupl
        from TEMP ;
    select &var, count(*) as DUPLICATES
        from &dsname.
          group by &var.
        having COUNT(*) > 1 ;
  quit ;
%put Check performed:  Dataset: &DSNAME.. ;
%put Checked column: &VAR. ;
%put Maximum of duplicate values in &VAR.: &max_dupl. ;
%mend DUPLICATES ;


%DUPLICATES(DUP_DEMO, ID) ;
```

Figure 3.2-11: Results in SAS Log

```
Check performed:  Dataset: DUP_DEMO.
Checked column: ID
Maximum of equal values in ID:        3
```

From table DUP_DEMO, the ID column was checked for values that occur more than once. A value occurs at most three times.

Figure 3.2-12: Results in SAS Output

```
ID  DUPLICATES
--------------
002         3
003         3
005         2
```

The ID column contains three values with multiple occurrences. For example, IDs 002 and 003 each occur three times.

3.2.6 Approach 6: Identifying Duplicates in Multiple Tables

For the identification of data rows that occur multiple times in several tables, reference the following set operators:

- **INTERSECT:** Filters out the "intersection," only duplicate data rows.
- **UNION:** Combines data sets without duplicate data rows.
- **EXCEPT:** First data set without duplicate data rows in comparison with a second data set.

3.3 Identifying and Filtering Outliers

The outlier problem can massively undermine the robustness of statistical procedures. The results of data analyses can be totally distorted by a few outliers in linear regression, linear models, designed experiments, and time series analyses. The mean value should not be calculated, for example, if outliers are present because this will distort it as a location measure for the actual dispersion of the data. Even the robust t test can be distorted by outliers. With many multivariate procedures, including a cluster center or k-means analysis, outliers should be excluded from the analysis. Cluster center analysis, because of its starting value method, reacts very sensitively to outliers, which can distort the clusters. Distributions should therefore always be checked for outliers before an analysis. In regression analysis, outliers can affect regression coefficients, their standard errors, the R^2, and ultimately the validity of the conclusions reached.

Outliers have several dimensions, which can occur in combination or alone:

- Outliers can occur in univariate and multivariate (high-dimensional) data.
- Outliers can stand out semantically (qualitative) or formally (quantitative).
- Outliers can only occur in a single case, but also in certain groupings.
- Outliers can only occur sporadically, but also massively.
- Outliers can be relative to the data volume (sample size).
- Outliers can have different causes.

Outliers can occur univariate-qualitatively as, for example, a single value caused by the incorrect recording of a clinical diagnosis such as reading "hormone therapy" instead of "homeopathy." However, outliers can also occur as several multivariate-quantitative outliers caused by the simultaneous incorrect logging of several variables. Such a case could occur if data transmission via several wireless ECG probes is affected by mobile phone interference. Multivariate-qualitative outliers show their true face only in their combination, due to special semantic characteristics (for example, the phrase "old children"). Such semantic qualities only show up when you look for them.

An outlier must always be seen in relation to expectations or frames, which do not necessarily always coincide with the empirical normality. Complicating matters further is that this frame can be changed, and empirical normality can also change. Thus, a strikingly high value does not always have to be an error but can be an accurate reflection of empirical reality without fitting into a series or frame.

Outliers are therefore not necessarily wrong or inaccurately recorded. They might also be values that are correct and accurate but contrary to expectations. The former would suggest that the process of measurement should be checked, the latter the theory construction. Also, a smooth transition between real outliers and normal data cannot be excluded.

The following sections present different approaches for the identification of mainly quantitative outliers. Not every value that is formally striking is automatically false. There is no omnibus measure or procedure for the identification of outliers.

At the end of the section, there are some hints for dealing with outliers. Checking the criterion "outliers" assumes that the criteria of completeness, uniformity, duplication, and missing values have already been checked and are OK. The DQ Pyramid considers outliers as an advanced topic as it goes beyond simple data checking and might involve questioning expectations.

This section presents the following approaches for dealing with quantitative outliers:

- Approach 1: Checking for outliers using descriptive statistics such as range
- Approach 2: Checking for outliers using statistical tests (David test)
- Approach 3: Filtering outliers using conditions such as limits and intervals

Checking for qualitative outliers is usually carried out within the framework of plausibility tests, using previously defined test cases or alternative scenarios with relevant semantic properties.

3.3.1 Approach 1: Checking for Outliers Using Descriptive Statistics

Columns in SAS tables can be checked for possible outliers using descriptive statistics. Possible measures are range R (or variation width V or span R), quartile distance, and Q1 or Q3, mean absolute deviation from the median (MAD), variance, standard deviation, or the coefficient of variation.

The range R is determined by the width of the dispersion range. More precisely, it is determined by the largest and smallest value of a distribution.

$$R = x_{max} - x_{min}$$

R is based on all values of a distribution. One outlier is sufficient to significantly distort this measure of dispersion. Strikingly high R-values are indications that outliers are present, especially if several series of measured values with other dispersion ranges are available for comparison.

Program 3.3-1: Approach 1: Range R

```
proc sql;
   select max(AMOUNT) as OUTLIER_max,
   min(AMOUNT) as OUTLIER_min,
   calculated OUTLIER_max-calculated OUTLIER_min as Range_R
   from SASHELP.BUY ;
quit;
```

Figure 3.3-1: Approach 1: Range R

```
OUTLIER_max   OUTLIER_min   Range_R
-----------------------------------
      48000       -110000    158000
```

As a general rule, the larger the values achieved, the more likely there are many and/or extreme outliers. However, certain data distributions such as very large N could conceal these effects in individual statistics.

3.3.2 Approach 2: Checking for Outliers Using Statistical Tests (David Test)

Statistics also provide some so-called outlier tests. However, their careless application is extremely problematic because even non-significant outlier tests do not exclude the occurrence of outliers with certainty, but only with a certain *probability*. Here it is clearly true: "The use of outlier tests is no substitute for conscientious handling of data and plausibility checks" (Rasch et al., 1996, 571; orig.: "Die Anwendung von Ausreißertests ist kein Ersatz für gewissenhaften Umgang mit Daten und für Plausibilitätskontrollen"). Outlier tests should only be used exploratively to find initial indications of possible data or survey errors. Hartung (1999, 343-347) presents several univariate tests, including the David-Hartley-Pearson test (David test). (See also Barnett and Lewis 1994.)

$$Q = \frac{R}{s}$$

The test value Q is calculated as the quotient of the span R and the standard deviation s and compared with a tabular value Q_n. The David-Hartley-Pearson test tests the null hypothesis that the upper or lower extreme values of the distribution belong to the sample and are not outliers. If $Q < Q_n$, the null hypothesis cannot be rejected. The extreme values belong to the sample. The distribution can therefore be regarded as free of outliers.

Program 3.3-2: Approach 2: David Test	Figure 3.3-2: Approach 2: David Test
	Q=4.0048506906 N=11

```
proc summary data= SASHELP.BUY ;
    var AMOUNT ;
    output out=DVD_OUT
        range=R std=SD n=N;
    run;


data _NULL_ ;
    set DVD_OUT ;
    Q = R / SD ;
    put Q= N=;
    run;
```

The David test calculates a Q=4.00 for N=11 of the example data from SASHELP.BUY. The result is written to the SAS log.

Table 3.3-1: Q*n* Values

Since the David test is not implemented in SAS, some critical Q*n* values of the test are given here. The data are taken from Hartung (1999, 344).

N	$Q_{n, 0.95}$	$Q_{n, 0.99}$
3	2,00	2,00
4	2,43	2,45
5	2,75	2,80
6	3,01	3,10
7	3,22	3,34
8	3,40	3,54
9	3,55	3,72
10	3,69	3,88
12	3,91	4,13
15	4,17	4,43
20	4,49	4,79
30	4,89	5,25
40	5,15	5,54
50	5,35	5,77
100	5,90	6,36

The obtained Q_{11} is even *above* a provisional $Q_{12(95)}$ of 3.91 (there is no Q-value for N=11 in the table) with a confidence interval of 95% (see Table 3.3-1). The null hypothesis that there are no outliers cannot be rejected. According to the David test, the AMOUNT data do not contain any outliers. However, the David test assumes that the measurement values to be tested are realizations from a normal distribution. The AMOUNT values do not meet this requirement. For the identification of multivariate outliers by means of measures, diagrams, or statistics, please refer to Schendera (2020/2007, Section 7.3).

3.3.3 Approach 3: Filtering Outliers Using Conditions

If descriptive statistics or statistical tests have shown that columns of a SAS table contain quantitative outliers, SAS offers various options for filtering outliers using conditions. Filtering by means of a limit value and by means of calculated intervals is presented in this approach. These examples can be rewritten so that outliers are immediately set to missing or replaced by a substitute value.

Outliers Above a Limit Value (Univariate)

This SQL approach outputs AMOUNT values as outliers (>= -1000) of a metric variable (AMOUNT). In the example of a possible measure, all rows that contain a value greater or equal than -1000 in AMOUNT are filtered out. No SAS output is shown for this example.

Program 3.3-3: Approach 3: Filter by Limit

```
proc sql;
select AMOUNT
    from SASHELP.BUY
where AMOUNT ge -1000 ;
quit ;
```

Program 3.3-4: Approach 3: Measure (Variant)

```
proc sql;
create table FILTER as
 select *
    from SASHELP.BUY ;
delete from FILTER
        where AMOUNT ge -1000 ;
quit;
```

Outliers Outside a Fixed Interval (Univariate)

This SQL approach outputs AMOUNT values as outliers outside the interval -1000 to -10000 of a metric variable (AMOUNT). In the example of a possible measure, a copy of the data from AMOUNT is created in the form of the AMOUNT2 column, with the difference that all values between -1000 and -10000 are set to missing. No SAS output is shown for this example.

Program 3.3-5: Approach 3: Filter by Fixed Interval

```
proc sql;
    select AMOUNT
    from SASHELP.BUY
    where AMOUNT not between -1000 and -10000 ;
quit;
```

Program 3.3-6: Approach 3: Measure (Variant)

```
proc sql ;
create table FILTER
  as select *,
  case
    when AMOUNT not between -1000 and -10000
       then .
    else AMOUNT
  end as AMOUNT2
from SASHELP.BUY ;
quit ;
```

Outliers Outside an Interval Estimated from the Data (Univariate)

This SQL approach outputs AMOUNT values as outliers that lie outside the interval around the mean value of AMOUNT +/- 2 standard deviations. No SAS output is shown for this example.

Program 3.3-7: Approach 3: Filter by Estimated Interval

```
proc sql;
   select AMOUNT
   from SASHELP.BUY
   having AMOUNT not between MEAN(AMOUNT) - 2 * STD(AMOUNT) and
      MEAN(AMOUNT) + 2 * STD(AMOUNT)          and
      AMOUNT is not missing ;
quit ;
```

Outliers for Date Values Outside a Fixed Interval (Univariate)

This SQL approach outputs AMOUNT values as outliers for DATE values that lie outside the interval of 10/01/2005 and 10/15/2006 (MM/DD/YYYY) and are not missing. No SAS output is shown for this example.

Program 3.3-8: Approach 3: Filter Date Values by Fixed Interval

```
proc sql;
   select AMOUNT
   from SASHELP.BUY
   where DATE not between '01OCT2005'D and '15OCT2006'D and
        DATE is not missing;
quit;
```

Outliers Outside Several Fixed Intervals (Multivariate)

This SQL approach outputs AMOUNT values as outliers that lie outside the interval defined by DATE and AMOUNT. No SAS output is shown for this example.

Program 3.3-9: Approach 3: Multivariate Filter (Dates, Numerical)

```
proc sql;
   select AMOUNT
   from SASHELP.BUY
   where AMOUNT not between -1000 and -10000 and
        DATE not between '01OCT2005'D and '15OCT2006'D ;
quit;
```

Outliers Outside an Alphanumeric Interval (NOT BETWEEN)

The following SQL approach outputs all rows whose entries in the variable NAME are outside the range of "Josie" and "Philipp". The interesting thing about this condition is that "Josie" does not appear as an entry in NAME at all. Therefore, "John" is output as the next alphanumeric value. If an alphanumeric value lies exactly on the limit ("Philipp"), it is *not* output with the negative NOT BETWEEN condition. This approach also works for numeric entries (if numbers are used) and strings with mixed characters. This approach is case-sensitive.

Program 3.3-10: Approach 3: Alphanumerical Interval

```
proc sql ;
  select NAME, SEX
    from SASHELP.CLASS
    where NAME      not between
        "Josie" and "Philipp" ;
quit;
```

Figure 3.3-3: Alphanumerical Interval

```
Name        Sex
------------
... abbreviated ...
John        M
Robert      M
Ronald      M
Thomas      M
William     M
```

Handling Outliers

Now, how can you handle outliers? There is no simple answer to this seemingly simple question. The way of handling depends on how the outliers were created. (See previous examples.) I can only advise against a general or even automated procedure, especially if it is to be carried out in an unmonitored or undocumented way. The distinction between "real" outliers and outliers caused by errors requires a differentiated approach. For details and further possibilities, please refer to Schendera (2007, 198-200).

- Outliers can **remain** in the analysis if they are "real" outliers, especially if they contain vitally important or monetary information. If necessary, the expectation-guided questions with respect to selected procedures and measures must be checked to see to what extent they can deal with the particularities of the identified outliers such as robustness. For example, the ROBUSTREG procedure can be used instead of PROC REG for the calculation of a regression. A mean value as a measure of location could be replaced by a more robust measure such as an M-estimator. Another strategy could be to increase the sample size. Small, non-representative samples are more prone to outliers than large, representative samples. Depending on the data situation and the "frame," the expectations could be adapted to the data situation.
- Outliers can be **replaced** by the correct values if they are outliers caused by incorrect values. Outliers can only be replaced by other values if you are sure that these are really the correct ones via documentation, syntax, backup tables, metadata, logical derivation, and so on. Outliers can be replaced by estimated values obtained by Mean Substitution, MI, Cold or Hot Deck, or other methods. These procedures do not introduce any bias. Outliers can be replaced using specified values. For example, outliers above a certain limit value can often be set to a value below this limit in order to comply with certain requirements. Particularly for vital or monetary data, carefully consider whether this approach justifies waiving the value of the original information.
- Outliers can be **bundled** via coding when further analysis is no longer performed via the outlier value, but via the membership to a certain group. This approach is only useful if the bundling coding is possible or even appropriate to the question of the analysis.
- Outliers can be **deleted** without exception. By deleting all outliers, their influence is also completely eliminated. It is important to ensure that the outliers are distributed randomly and have a small (ideally univariate) proportion. In multivariate approaches, it is important to ensure that outliers do not concentrate on relevant predictors or the predictors for the rarer target event such as in binary logistic or Poisson regression. The price of deleting outliers is that along with the sample size the power of the procedure is reduced.
- Outliers can be **reduced**. For example, by not deleting all values, but only values outside a certain range, the influence of outliers is not completely eliminated, but only reduced. The power of procedures is reduced less than by deleting values without exception. Common ranges are the median +/- 4MAD, the mean value +/- 2Sigmas, or the α%-trimmed mean value. Depending on the definition of the rank, a different number of outliers might be identified and eliminated. The Moses test for extreme reactions checks whether extreme values in the experimental group influence the range and by default compares an observed group with a control group trimmed for outliers. Depending on the definition of the percentage or absolute number of outliers, this test might give a different result.

In all procedures, except when the data or values are not changed at all, a comparative approach should be chosen in order to estimate the effect of deleting, reducing, or even replacing outliers. These procedures are also valid for qualitative outliers, especially if they are numerically coded. The actual check for the correctness of qualitative outliers is treated in the following section on plausibility.

3.4 Uniformity: Identify, Filter, and Replace Characters

In data analysis, you need to provide uniformity on several levels: data sets, numeric variables, date variables, labels for variables and values, and unique information (usually in string form). This chapter focuses on providing alphanumeric uniformity by identifying, filtering, or replacing unwanted characters or strings. The uniformity of alphanumerical values is one of the basic criteria of data quality for which you can use SAS.

Checking all other criteria requires that uniformity, completeness, duplicates, and missing values have already been checked. However, uniformity requires at least completeness. (See Schendera, 2020/2007, Chapter 4.)

Uniformity of values and strings is important, especially in unique information such as names of persons, places, medications, dates, telephone numbers, or personnel numbers. Since names are often also used as key variables, special care must be taken when entering and checking them. The consequences of non-uniformly written strings can be serious.

- SAS is not able to automatically include upper- and lowercase strings ("SAS" versus "sas"), which are variants of the same character string, as semantically identical or uniform in an analysis. If no precautions are taken to ensure uniformity, upper- and lowercase variants are evaluated as different, although they actually belong together.
- If, for example, "identical" strings are written in upper- and lowercase, they are not sorted uniformly by PROC SORT. Instead, strings written in uppercase are sorted before strings written in lowercase, which can cause massive problems when merging tables if these strings are key variables.

Deviations from uniformity can be less obvious though. NASA lost a $125 million dollar satellite on September 23, 1999, because one team of engineers worked with metric units while another team worked with English units.

The following check and filter possibilities present four approaches:

- Approach 1: Checking for characters in longer strings. The WHERE clause is presented together with the IN condition.
- Approach 2: Complete, partially, or not at all: checking for multiple characters, strings, or substrings. The SAS functions INDEX and INDEXW are introduced for searching a character expression for *several* characters.
- Approach 3: Details: checking for single characters. For checking character expressions for single characters, the SAS functions INDEXC, FINDC, VERIFY, and COUNTC are introduced.
- Approach 4: Unifying strings, substrings, or characters. The SAS functions TRANWRD and TRANSLATE as well as the Perl regular expression PRXCHANGE are introduced for searching and replacing strings, substrings, or characters. Further approaches to search and replace strings, characters, and values are presented in the chapter on macro programming. Interested users will also find numerous SAS functions and CALL routines for the versatile handling of characters and strings (including matching) in Chapter 10.

3.4.1 Approach 1: Checking for Strings in Terms of Longer Character Strings

For a check of entries for several strings or longer character strings, the WHERE clause with the IN condition is introduced. An IN condition checks whether the value of a column specified to the left of the IN is an element of a set of predefined values or constants in brackets to the right of the expression. A constant can be a number or a string in quotation marks. The IN condition is true if the value of a column occurs in the set of specified elements. The examples presented are formulated negatively (see WHERE ... NOT IN) and search for deviations from a given correct list of strings. Depending on the perspective of the question, a WHERE ... IN can also be useful. The conditions CONTAINS, EXISTS, or LIKE can be also helpful to check for individual substrings, in addition to IN. To check the rank of substrings, the BETWEEN condition is of interest. (See the following approaches.) The macro chapter (Chapter 4) provides further approaches for processing lists of variables or values using macro variables, loops, or the PARMBUFF option.

Deviations from an Alphanumeric List (BETWEEN)

The following SQL approach outputs all rows whose entries in the variable NAME are within the range of "Josie" and "Philipp". "Josie" does not appear as an entry in NAME at all. Therefore, "Joyce" is output as the next alphanumeric value. If an alphanumeric value lies exactly on the limit ("Philipp"), it is output with the positive BETWEEN condition. A *negative* example is introduced in Program 3.3-10. This approach also works for numeric entries (if numbers are used) and strings with mixed characters. This approach is case sensitive.

Program 3.4-1: Approach 1: Deviations from an Alphanumeric List	**Figure 3.4-1: Alphanumeric List**

```
proc sql ;
  select NAME, SEX
    from SASHELP.CLASS
    where NAME between
        "Josie" and "Philipp" ;
quit;
```

```
Name      Sex
------------  ------
Joyce     F
Judy      F
Louise    F
Mary      F
Philip    M
```

Deviations in a String Variable (WHERE NOT IN)

The following SQL approach outputs all rows that are not coded as "m" or "w" in the variable SEX in SASHELP. CLASS. This approach is not case sensitive; it also works for strings with mixed characters. Every row was selected because the values of SEX in SASHELP.CLASS are all uppercase 'M' and 'F'. MYDATA contains one column (NAME) and 18 rows. No SAS output is shown in this example.

Program 3.4-2: Approach 1: Deviations in a String Variable (WHERE NOT IN) 1

```
proc sql;
create table MYDATA as
    select SEX
    from SASHELP.CLASS
    where SEX not in ("m","f") ;
quit;
```

The following SQL approach outputs all rows that have other strings in the variable NAME than specified. This approach is case sensitive. MYDATA contains two columns and 9 rows. No SAS output is shown in this example.

Program 3.4-3: Approach 1: Deviations in a String Variable (WHERE NOT IN) 2

```
proc sql ;
create table MYDATA as
 select NAME, SEX
    from SASHELP.CLASS
    where NAME not in ("Alfred","Alice","Barbara","Carol",
                       "Henry","James","Judy","Louise",
                       "Mary","Philipp","Petra","Thomas") ;
quit;
```

Deviations in Several String Variables (WHERE NOT IN)

The following SQL approach outputs all rows that have no missing values in the variables NAME and SEX in addition to the specified string entries. This approach also works for strings with mixed characters. Upper- and lowercase are taken into account. This MYDATA also contains two columns and 9 rows. No SAS output is shown in this example.

Program 3.4-4: Approach 1: Deviations in Several String Variables (WHERE NOT IN)

```
proc sql;
create table MYDATA as
    select NAME, SEX
    from SASHELP.CLASS
    where SEX not in ("m","f", " ") and
```

```
                NAME not in ("Alfred","Alice","Barbara","Carol",
                             "Henry","James","Judy","Louise","Mary",
                             "Philipp","Petra","Thomas", " ");
quit ;
```

3.4.2 Approach 2: Complete, Partially, or Not at All: Checking for Multiple Characters

SAS offers the INDEX and INDEXW functions for searching a character expression for several characters or substrings. In contrast, the INDEXC function searches a character expression for individual characters. (See Approach 3.)

INDEX searches a character expression for a character string. INDEX also triggers a response if predefined character strings only *partially* match string entries. For example, "fred" triggers a hit. INDEXW searches a character expression for a character string specified as a word. INDEXW triggers a response only if the list element matches *exactly*. For example, "Alf" does not trigger a hit. These subtle differences between INDEX and INDEXW can be desirable, depending on your requirements.

Program 3.4-5: Approach 2: Checking for Multiple Characters (INDEX, INDEXW)

```
proc sql;
select NAME as NAME_I
    from SASHELP.CLASS
    where index(NAME, "fred") or
          index(NAME, "James");
select NAME as NAME_W
    from SASHELP.CLASS
    where indexw(NAME, "Alf") or
          indexw(NAME, "James");
quit;
```

Figure 3.4-2: Characters

```
NAME_I
--------
Alfred
James


NAME_W
--------
James
```

The INDEXC function searches a character expression for certain individual characters and returns the position where these characters first appear in the scanned character expression. For example, if the names from SASHELP. CLASS were scanned for "Alfred", all entries in NAME would be output as hits because they contain a single character from "Alfred" somewhere, with two exceptions. "John" and "Thomas" would be excluded because no single character matches any character from "Alfred"; even the "a" in "Thomas" does not match the "A" in "Alfred". INDEXC is therefore less suitable for checking a character expression for exactly or partially matching character strings, but all the more suitable for excluding unwanted characters. The INDEXC function is introduced in the next section.

Using the SAS functions COUNT and FIND, a string (for example, "Barbara") could also be searched for a substring (for example, "ar"). No SAS output is shown in this example.

Program 3.4-6: Approach 2: Searching for Substrings (COUNT, FIND)

```
proc sql;
  select NAME,
    count(NAME, "ar") as NAME_ar
    from SASHELP.CLASS ;
  select NAME,
    find(NAME, "ar") as NAME_ar
    from SASHELP.CLASS ;
quit;
```

The SAS function COUNT counts the number of hits of "ar" in a string. For "B*ar*bara", for example, the value 2 is output.

The SAS function FIND returns the first *position* of "ar" in a string. For "M*ar*y" and "B*ar*bara", for example, the value 2 is therefore output.

These approaches also work for strings with mixed characters. Both functions are case sensitive.

3.4.3 Approach 3: Details: Checking for Single Characters

For checking character expressions for single characters, the SAS functions INDEXC, FINDC, VERIFY, and COUNTC are introduced.

Identifying Cases with Other than Predefined Characters (INDEXC, FINDC)

The following SQL approach outputs all rows that do not contain a single character from the character string specified after INDEXC. The deviating characters that might occur must be known in advance. This approach returns the position of the first occurrence of a character that occurs in "Jane". If no character from "Jane" occurs in the scanned character string, 0 is returned. (See VERIFY.) The INDEXC approach also works for numbers embedded in strings (strings with mixed characters). The INDEXC function is case sensitive. (See on the other hand VERIFY.)

Program 3.4-7: Approach 3: Using INDEXC	**Figure 3.4-3: Approach 3: Using INDEXC**

```
proc sql ;
   select indexc(NAME, "Jane")
        as J_CHECK, NAME
   from SASHELP.CLASS
   where calculated J_CHECK = 0 ;
quit ;
```

```
J_CHECK  Name
-----------------
   0  Philip
```

The FINDC function also checks a string to see whether the specified single characters are contained in a character expression to be scanned. In contrast to INDEXC, FINDC also enables you to specify arguments for the start position of the scan process. The FINDC function is case sensitive.

Program 3.4-8: Approach 3: Using FINDC	**Figure 3.4-4: Approach 3: Using FINDC**

```
proc sql ;
   select findc(NAME, "Jane")
        as F_CHECK, NAME
   from SASHELP.CLASS
   where calculated F_CHECK = 0 ;
quit ;
```

```
F_CHECK  Name
-----------------
   0  Philip
```

Identifying Cases with Other than Predefined Characters (VERIFY)

VERIFY returns the position where the column entries differ from any character of the string pattern specified in the parentheses for the first time. The following SQL approach outputs all rows where the NAME entries differ on the fifth position from any of the characters specified after VERIFY (in the example "Jane" plus a blank). The result returns "Janet" because the "t" differs on the fifth position from all the characters specified (specifying them backward as " enaJ" would have the same effect). With this approach, it is *not* necessary to know possibly deviating characters in advance or even their position. If no character from the specified string pattern occurs in the scanned column entries, VERIFY returns 0. (See INDEXC.) The VERIFY approach does not work for strings with mixed characters. The VERIFY function is case sensitive.

Program 3.4-9: Approach 3: Using VERIFY	**Figure 3.4-5: Approach 3: Using VERIFY**

```
proc sql ;
   select verify (NAME, "Jane")
        as V_CHECK, NAME
   from SASHELP.CLASS
   where calculated V_CHECK eq 5 ;
quit ;
```

```
V_CHECK  Name
-----------------
      5    Janet
```

Counting Individual Characters (COUNTC)

The following SQL approach outputs all rows that have the character "J" at least once in the string variable NAME. The possible deviation "J" should therefore be known in advance when using this approach. COUNTC counts the frequency of occurrence of the specified character. This approach also works for strings with mixed characters. The COUNTC function is case sensitive.

Program 3.4-10 Approach 3: Using COUNTC	Figure 3.4-6: Approach 3: Using COUNTC

```
proc sql ;
   select countc (NAME, "J")
          as C_CHECK, NAME
   from SASHELP.CLASS
   where calculated C_CHECK eq 1 ;
quit ;
```

```
C_CHECK   Name
-----------------
   1      James
   1      Jane
   1      Janet
   1      Jeffrey
   1      John
   1      Joyce
   1      Judy
```

Approach 4: Unifying Strings, Substrings, or Characters

The identification of inconsistent strings is often only half the trouble. In addition to filtering out or setting to missing, the more laborious way of standardization is often chosen. The flexibility of the DATA step for the differentiated standardization of non-uniform characters or strings is unsurpassed. However, even with PROC SQL and the skillful use of SAS functions it is possible to achieve a lot. The following SQL approaches illustrate interesting applications of the SAS functions TRANWRD, TRANSLATE, and PRXCHANGE. If you simply strive for uniform upper- or lowercase strings (for example, "SAS" versus "sas"), you can use the UPCASE or LOWCASE functions. In contrast, PROPCASE returns a string in lowercase except the first character, which is returned in uppercase.

Searching and Replacing a String with a Constant

Using TRANWRD, the following SQL approach searches for the substring "Ja" from the column NAME of the SASHELP file CLASS and replaces it with the constant "XXX". The TRANWRD function is case sensitive. "Ja" is interpreted as a substring.

Program 3.4-11: Approach 3: Using TRANWRD

```
proc sql outobs=5;
select NAME as NAME_old,
   tranwrd(NAME, "Ja", "XXX") as NAME_new
   from SASHELP.CLASS
   where calculated NAME_new contains ("XX");
quit;
```

To compare, we request the original entries in NAME and the modified entries in NAME_new. OUTOBS= is used to limit the number of rows in the output below. For the various differences to the TRANSLATE function, see the example.

Figure 3.4-7: Approach 3: Using TRANWRD

```
NAME_old   NAME_new
-----------------------------------
James      XXXmes
Jane       XXXne
Janet      XXXnet
```

Searching and Replacing Several Strings with a Constant

Using PRXCHANGE, the following SQL approach searches for any combination of concatenated uppercase (here "B", "C", "M" or "J") character values with "a" ("Ba" or "Ca") in the variable NAME, replaces it with "XXX" if that pattern was found, and stores as it as NAME_new. "−1" repeats the pattern-matching replacement until the end of NAME. The PRXCHANGE function is case sensitive.

Program 3.4-12: Approach 3: Using PRXCHANGE (Constant)

```
proc sql outobs=5;
select NAME as NAME_old,
       prxchange("s/[BCMJ]a/XXX/", -1, NAME)
             as NAME_new
   from SASHELP.CLASS
   where calculated NAME_new contains ("XX");
quit;
```

For illustration, the original entries are requested and for comparison the modified string entries.

Figure 3.4-8: Approach 3: Using PRXCHANGE (Constant)

```
NAME_old   NAME_new
----------------------------------
Barbara    XXXrbara
Carol      XXXrol
James      XXXmes
Jane       XXXne
Janet      XXXnet
```

Changing Comma-separated Characters

Using PRXCHANGE, the following SQL approach searches for comma-separated strings in the column NAME of the SAS file FREQUENTFLYERS, swaps their sequence, and removes the comma. (See further notes on this file in Chapter 5, Section 5.2.)

Program 3.4-13: Approach 3: Using PRXCHANGE (Comma-separated Characters)

```
proc sql outobs=5 ;
select NAME as NAME_old,
       prxchange("s/(\w+), (\w+)/$2 $1/", -1, NAME)
             as NAME_new
   from MY_PATH.FREQUENTFLYERS ;
quit;
```

For illustration, the original and modified string entries are requested.

Figure 3.4-9: Approach 3: Using PRXCHANGE (Comma-separated Characters)

```
NAME_old                    NAME_new
-------------------------------------------
COOPER, LESLIE              LESLIE COOPER
LONG, RUSSELL               RUSSELL LONG
BRYANT, ALTON               ALTON BRYANT
NORRIS, DIANE               DIANE NORRIS
PEARSON, BRYAN              BRYAN PEARSON
```

Searching and Replacing Several Characters with Other Characters

Using TRANSLATE, the following SQL approach searches for the characters "J" and "a" from the column NAME of the SASHELP file CLASS and replaces them with the corresponding characters "XXX". A "J" is replaced by an "X", as is an "a". "Ja" is not interpreted as a substring. There are no values to replace for the third (target) X; it is ignored accordingly. The TRANSLATE function is case sensitive.

Program 3.4-14: Approach 3: Using TRANSLATE

```
proc sql outobs=5 ;
   select NAME as NAME_old,
       translate(NAME, "XXX", "Ja") as NAME_new
       from SASHELP.CLASS
   where calculated NAME_new contains ("X");
quit;
```

For illustration, the original entries are requested and for comparison the modified string entries.

Figure 3.4-10: Approach 3: Using TRANSLATE

```
NAME_old   NAME_new
-----------------
Barbara    BXrbXrX
Carol      CXrol
James      XXmes
Jane       XXne
Janet      XXnet
```

There are two differences to the TRANWRD function:

- TRANSLATE replaces character by character; TRANWRD replaces substring by substring.
- With TRANSLATE, the characters to be inserted are specified first, with TRANWRD the substrings to be replaced.

For further possibilities of filtering, scanning, and also standardization, see Chapter 10 on SAS functions and routines. There you will find numerous Perl regular expression functions and routines for the matching of characters and strings. For example, PRXPAREN returns the last parenthesis match for which there is a pattern match. PRXSUBSTR returns the position and length of a substring that matches a pattern. PRXMATCH searches for a pattern match and returns the position where the pattern was found. SAS also offers many other SAS functions and routines for handling characters and strings, from ANYALNUM (searches a string for an alphanumeric character and returns the first position where it was found), to NOTPRINT (searches a string for a non-printable character and also returns the first position where it was found), to the already introduced VERIFY.

Chapter 4: Macro Programming with PROC SQL

PROC SQL supports the use of the SAS macro language. The focus of this chapter is less on the introduction to programming SAS macros, but rather on their application. This chapter presents numerous examples of SAS macros in application-oriented sections, which will extend the possibilities of working with PROC SQL, simplify it through automation, and accelerate it.

Technically speaking, macros are the automation of recurring processes within applications. Applied to SAS, they are the automated processing of the same application for many different variables, the efficient processing of different applications for the same variable, or a combination of both.

Because macros are based on SAS syntax, including PROC SQL, macros also embody all the advantages of syntax programming, including validation, automation and reusability, speed, openness, clarity, and systematization. Because macros in turn systematize and automate the SAS syntax itself, it is safe to say that macros embody syntax efficiency. By means of macros, SAS syntax programming can be automated and used repeatedly without the need to write or adapt new syntax commands. The following are additional advantages of SAS macros (Schendera, 2005, 147ff.):

- Increased efficiency: SAS macros increase the efficiency of syntax programming.
- Speed: Macros generally run faster than repeatedly executed normal instructions and are often pro-grammed faster.
- Brevity: SAS macros reduce long syntax programs with uniform steps to one step that is repeated.
- Clarity: The compression to a few essential steps makes programs more manageable, systematic, and comprehensible.
- Error-free: If SAS macro code and complex logic are correctly organized and adapted to the data in a macro program, the execution is also error-free. Conversely, writing the same number of names for data sets and variables, instructions, and conditions "by hand" not only involves a great deal of effort, but also increases the probability of carelessness and errors if the program is long enough.
- Combinability: Macros can be used with SAS syntax. This means that macros can be inserted into already existing programs. One can imagine this in such a way that first "normal" SAS syntax is processed before the macro, then a macro, then again sections of "normal" SAS syntax.
- Enhanced performance: Performance enhancement means that you can access and use hundreds of SAS provided macros, including macros for Base SAS, SAS/ETS, SAS/IML, SAS/OR, SAS/STAT, and many more. So, you can not only increase your performance toward efficiency, but also your scope of performance.

A key difference between programming with PROC SQL and SAS macros is that PROC SQL processes values of numeric and character variables. Programming with the SAS Macro Facility, on the other hand, processes the text of SQL programs. Another central difference is that in SAS macros, the SAS macro code overrides SAS code and is executed before the SAS code. This difference is particularly relevant when working with SAS macro programs. (See the introductory notes in Section 4.2.)

This chapter is divided into four sections.

- Section 4.1 introduces working with SAS macro variables.
- Section 4.2 introduces programming SAS macro programs.

- Section 4.3 introduces the most important elements of the SAS macro language, including SAS macro functions, SAS macro statements, and interfaces.
- Sections 4.4–4.10 introduce interesting SAS macro programs for special applications, for example, for working with multiple SAS files, restructuring and transposing SAS data sets, or retrieving SAS information.

This chapter begins with SAS macro variables. As the simplest variant of macros, SAS macro variables are useful placeholders for texts. (See Section 4.1.) Macro variables can be automatic SQL macro variables provided by SAS or macro variables defined by the user. Automatic macro variables provide global information such as the current operating system, the SAS version, or the concrete number of rows processed by any PROC SQL statement. Macro variables can also be used outside macro programs directly in normal SQL or SAS syntax. With INTO, it is possible with PROC SQL to store alphanumeric or missing values queried or calculated in macro variables and to use the stored values in further steps. From these macro variables, these values can be passed back to titles, log, output, other macros, and so on.

More complex operations such as condition-based or repeated execution of tasks can be performed with SAS macro programs. (See Section 4.2.) Macro programs are embedded between %MACRO and %MEND and generally contain macro variables, SQL or SAS syntax, and other elements of the SAS Macro Facility. When macros are passed to SAS, these macros in turn generate SAS code that SAS executes. The most important elements of the SAS macro language are introduced in Section 4.3. Starting in Section 4.4, exemplary SAS macro programs for special applications are presented.

4.1 Macro Variables

Macro variables are:

- Macro variables are "slots", that is, interfaces or placeholders for texts. The entries in macro variables are text values. Even values that appear to be numbers are text values. Macro variables are case-sensitive.
- Macro variables can contain the names of variables, paths, libraries, SAS files, texts for titles, and much, much more. Automatic SAS macro variables (see Section 4.1.1) contain various return codes, date and time of the start of the execution of a SAS job or session, name of the current graphics device driver or the most current SAS data set, the name of the currently executing macro, batch job or SAS procedure or SAS process, and much more. Automatic SQL macro variables (see Section 4.1.2) feedback the number of rows processed by a particular PROC SQL statement, the number of iterations executed by the inner loop of PROC SQL, and also various status and return codes.
- For beginners, macro variables are initially easier to handle than macro programs. Therefore, this section treats macro variables before macro programs.
- SAS offers three types of macro variables: automatic SAS macro variables (see Section 4.1.1), automatic SQL macro variables (see Section 4.1.2) and user-defined macro variables (see Section 4.1.3).
- The maximum length for the name of a macro variable in SAS 9.4 is 32 characters. The length of a macro variable is automatically determined by the assigned text entries and does not need to be explicitly defined via a LENGTH statement. The maximum length of macro variable values is 65,534 characters.
- The contents of entries in macro variables can be displayed using the %PUT statement or the SAS option SYMBOLGEN. The %PUT statement writes the contents of system- or user-defined macro variables to the SAS log. The call to the %PUT statement can be enriched with text comments. The %PUT statement displays only the entries in the specified macro variables. SYMBOLGEN (or SGEN) shows which concrete values replace the placeholders. SYMBOLGEN displays the entries in all macro variables of a SAS macro program. SYMBOLGEN is therefore considered more convenient. Depending on the number of macro variables in the final result, it might also be less clear than the %PUT statement. SYMBOLGEN can be overridden with NOSYMBOLGEN (or NOSGEN). SQL also offers the possibility of storing entries in macro variables in normal SAS variables or passing them from a local macro variable to a global macro variable.
- At first, the function of macro variables is not to include one or more observations, such as fields of a SAS table. However, the contents of macro variables can be stored in already existing fields of SAS tables or in new fields created explicitly for this purpose. The function of macro variables is to accelerate the writing

and execution of SAS programs. Macro variables are tools and placeholders for text. An obvious application is the processing of the same query for many different variables. Before the user executes the SAS program for each of the variables individually, a macro program executes the SAS program for all variables one after the other. This is possible because macro variables are able to dynamically include one or more texts in a SAS program as placeholders (for example, texts of variable names).

- Macro variables can be referenced inside and outside macro programs. Macro variables can be specified and referenced anywhere in a SAS program (except in data rows). If macro variables are referenced outside a macro program, this is called open SAS code (open code).

- Macro variables can be defined by the user or by SAS. Automatic macro variables are available throughout the SAS session. User-defined macro variables are available either globally or locally, depending on their definition. (See below.)

- Macro variables can be stored either in the global or local symbol table. Macro variables created in open code using PROC SQL and also system-defined macro variables are always stored in the global symbol table. Macro variables created by macro programs can be stored either in the local or the global symbol table.

- In open SAS code, macro variables are defined by a %LET statement, in macro programs by defining the macro. A macro variable is always preceded by at least one "&" (ampersand). Two ampersands are then required if a macro variable consists of a macro variable and preceding text. Three ampersands are required if a macro variable consists of *two* macro variables (without preceding text). A double ampersand ensures that macro variables composed in this way are read and interpreted correctly by the SAS macro processor.
Two "&&" or more will then resolve into one "&". Ideally, a macro variable should end with a period. A period at the end is essential if the macro variable in question is to be followed by further text. In PROC SQL, the definition of a macro variable is preceded by a colon, for example, after INTO.

Section 4.1.1 introduces SAS automatic macro variables. For automatic SAS SQL macro variables, the name and functionality are already defined by SAS. Section 4.1.2 introduces special SAS SQL macro variables. Section 4.1.3 introduces how users can define their own macro variables. For user-defined SAS macro variables, the name and functionality can be defined by the user.

All of these variants of SAS macro variables are extremely useful for simple text replacement, but they cannot perform complex operations such as conditioned (conditional) or repeated execution of tasks. SAS macro programs are required for this purpose. (See Section 4.2.)

4.1.1 Automatic SAS Macro Variables

With automatic SAS macro variables, the name and functionality of the variables are already defined by SAS. For example, to automatically include date, time, or user in a SAS output, the automatic SAS macro variables SYSDATE9, SYSDAY, and SYSUSERID need to be specified after the TITLE statement.

Titles are used to provide readers of a report with the most important information at a glance. Instead of updating this information manually, which is often prone to errors, it is helpful to have SAS update it automatically. The following examples show how macro variables can form an interface or a "bridge" between analysis and report.

Program 4.1-1: Working with Automatic SAS Macro Variables

```
title1 "Management Summary (Reporting: &sysuserID)" ;
title2 "Date: &sysdate9, &systime (&sysday)" ;
proc freq data=SASHELP.CLASS ;
run ;
```

For example, if SAS is started on March 1, 2009, at 9:40 and this program is executed immediately after the call, SAS will add the following titles to the SAS output. (The PROC FREQ result is not shown.)

Figure 4.1-1: Working with Automatic SAS Macro Variables: SAS Output

```
Management Summary (Reporting: CFG Schendera)
Date: 01MAR2009, 09:40 (Sunday)
```

If the displayed program is executed again at 10:15 (without SAS having been started again beforehand), the displayed time is not updated to 10:15 in the new SAS output. The displayed time will be updated to 10:15 in the next output of PROQ FREQ if SAS was restarted at 10:15 before the next execution of the program. Alternatively, use the SAS functions DATE and TIME. (See Section 4.3.)

In open SAS code, macro variables such as &SYSTIME must be specified within double quotation marks. Macro variables within single quotation marks are not resolved. All automatic SAS macro variables are global (except SYSPBUFF) and are created when SAS is started. SYSPBUFF is a local macro variable. SAS offers many other SAS automatic macro variables. The interested user is referred to the extensive SAS documentation of several dozens of SAS automatic macro variables (SAS Institute, 2016e). Other interesting ways to work with SAS SQL automatic macro variables are presented in the next section.

Table 4.1-1: Categories and Descriptions of Automatic SAS Macro Variables (Selection)

SAS Macro Function	Description (Source: *SAS® 9.4 Macro Language*, 5th ed.)
Access: Read and write	
SYSBUFFER	Text that is entered in response to a %INPUT statement when there is no corresponding macro variable.
SYSCC	Current condition code that SAS returns to the operating environment ("OE").
SYSCMD	Last unrecognized command from the command line of a macro window.
SYSDEVIC	Name of the current graphics device.
SYSDMG	Return code that reflects an action taken on a damaged data set.
SYSDSN	Libref and name of the most recently created SAS data set.
SYSFILRC	Return code from the last FILENAME statement.
SYSLAST	Name of the SAS data set created most recently.
SYSLCKRC	Return code from the most recent LOCK statements.
SYSLIBRC	Return code from the last LIBNAME statement.
SYSLOGAPPLNAME	Value of the LOGAPPLNAME system option.
SYSMSG	Text to display in the message area of a macro window.
SYSPARM	Character string that can be passed from the OE to SAS program steps.
SYSPBUFF	Contains text supplied as macro parameter values.
SYSRC	Last return code generated by the operating system.
Access: Read only	
SYSCHARWIDTH	Contains the character width value (1 or 2).
SYSDATE	Character string that contains the date on which a SAS job or session began executing (with 2-digit year).
SYSDATE9	Character string that contains the date on which a SAS job or session began executing (with 4-digit year).
SYSDAY	Day of the week on which a SAS job or session began executing.
SYSENCODING	Contains the name of the SAS session encoding.
SYSENV	Reports whether SAS is running interactively (front, back).
SYSERR	Return code status by some SAS procedures and DATA step.
SYSERRORTEXT	Text of the last error message for display in the SAS log.

(Continued)

Table 4.1-1: (*Continued*)

SAS Macro Function	Description (Source: *SAS® 9.4 Macro Language*, 5th ed.)
SYSHOSTNAME	Host name of the computer that is running the SAS process.
SYSINDEX	Number of macros that have started execution in the current SAS job or session.
SYSINFO	Return codes provided by some SAS procedures.
SYSJOBID	Name of the current batch job or user ID (depending on OE).
SYSMACRONAME	Name of the currently executing macro.
SYSMENV	Invocation status of the macro that is currently executing.
SYSNCPU	Current number of CPUs available to SAS for computations.
SYSODSPATH	Current Output Delivery System (ODS) pathname.
SYSPROCESSID	Process ID of the current SAS process.
SYSPROCESSNAME	Process name of the current SAS process.
SYSPROCNAME	Name of the procedure (or DATASTEP for DATA steps) currently being processed.
SYSSCP	Identifier for the OE, for example, "WIN".
SYSSCPL	Often more specific identifier for the OE, for example, "X64_10PRO".
SYSSITE	Number of SAS Site.
SYSSTARTID	ID generated from the last STARTSAS statement.
SYSSTARTNAME	Process name generated from the last STARTSAS statement.
SYSTCPIPHOSTNAME	Host names of the local and remote computers when multiple TCP/IP stacks are supported.
SYSTIME	Character string that contains the time at which a SAS job or session began executing.
SYSUSERID	User ID or login of the current SAS process.
SYSVER	Release number of SAS software that is running.
SYSVLONG	Release number and maintenance level of SAS software that is running (with 2-digit year).
SYSVLONG4	Release number and maintenance level of SAS software that is running (with 4-digit year).
SYSWARNINGTEXT	Text of the last warning message formatted for display in the SAS log.

Note: *%put _automatic_ ;* enables the complete output of the currently provided automatic macro variables in the SAS log.

4.1.2 Automatic SQL Macro Variables

Even with automatic SAS SQL macro variables, the name and functionality of the variables are already defined by SAS. (For automatic SAS macro variables, see Section 4.1.1.) PROC SQL updates the values in three automatically created macro variables after executing certain SQL statements:

- **SQLOBS** returns the number of rows processed by a particular PROC SQL statement, for example, SELECT or DELETE. If NOPRINT is specified, the concrete SQLOBS value depends on whether an output table or one or more macro variables were specified. If no output table and/or only one macro variable was created, the value of SQLOBS is equal to 1. Whereas if an output table and/or list of macro variables was specified, SQLOBS corresponds to the number of rows in the output table or the number of processed rows.
- **SQLOOPS** represents the number of iterations executed by the inner loop of PROC SQL. The number of iterations usually increases proportionally with the complexity of a query.
- **SQLRC** is a status variable and returns codes for the success or failure of a PROC SQL statement. "0" stands for a successful execution of an SQL statement without errors. The code "4" stands for a "warning"; the SQL statement continues to be executed. All other codes ("8", "12", "16", "24" and "28") stand for errors, mainly caused by the user, system, or SAS.

Other, more database-specific SQL macro variables are SQLXRC (DBMS-specific return code) and SQLXMSG (DBMS-specific error return code). SQLXMSG can contain special characters.

For example, if the call of the automatic macro variable (from %PUT) is executed after the actual SELECT clause (which in this example requests STORES > 20), the user receives the requested information in the SAS output.

Program 4.1-2: Working with Automatic SQL Macro Variables

```
proc sql ;
  select REGION, PRODUCT, SUBSIDIARY
    from  SASHELP.SHOES
    where REGION= "Africa" and STORES  > 20 ;
  %put SQL_N_ROWS=|| &sqlobs ||
       SQL_ITER=|| &sqloops ||
       SQL_STATUS=|| &sqlrc || ;
quit ;
```

Figure 4.1-2: Working with Automatic SQL Macro Variables: SAS Output

```
Region                        Product         Subsidiary
------------------------------------------------------
Africa                        Boot            Algiers
Africa                        Sandal          Algiers
Africa                        Men's Casual    Cairo
Africa                        Boot            Khartoum
Africa                        Boot            Nairobi
```

SAS, however, also writes the following output generated by the automatic macro variables in the log.

Figure 4.1-3: Working with Automatic SQL Macro Variables: Output in SAS Log

```
SQL_N_ROWS=|| 5 ||      SQL_ITER=|| 16 ||      SQL_STATUS=|| 0 ||
```

SQL_N_ROWS contains the number of rows found (N=5). SQL_ITER contains the number of iterations of the inner loop of PROC SQL (N=16). As a status variable, SQL_STATUS returns the success of the SELECT statement using the code "0". (See also the next example.)

A practical application of SQL macro variables is the formatting of "empty" SQL queries without hits or results. If a query is unsuccessful, SAS only displays a message in the log that no rows were selected, for example, if the REGION "Africa" does not contain any STORES with values above 50. The following unformatted query does not yet produce a hit.

Program 4.1-3: "Empty" Query: Unformatted

```
proc sql ;
  select REGION, PRODUCT, SUBSIDIARY
    from  SASHELP.SHOES
    where REGION= "Africa" and STORES  > 50 ;
 quit ;
```

SAS outputs the following note to the log by default.

SAS Log 4.1-1: "Empty" Query: Unformatted

```
NOTE: No rows were selected.
```

The following application takes advantage of SQLOBS' ability to count the number of rows that have been processed by a particular PROC SQL statement. For example, if no rows have been processed, then SQLOBS gets a 0, otherwise a positive number.

Program 4.1-4: "Empty" Query: Formatted

```
proc sql ;
   select REGION, PRODUCT, SUBSIDIARY
      from  SASHELP.SHOES
      where REGION= "Africa" and STORES  > 50 ;
         select " " length = 45
         label = "The last query did not produce a hit!"
         from SASHELP.SHOES (obs=1)
         where &SQLOBS. = 0 ;
quit ;
%put &sqlobs. ;
```

The second WHERE clause creates a text string for the situation (caused by the first WHERE clause) that no rows were processed. This text string is output to the SAS log instead of the default note. Since no output table has been created and no rows have been processed, the value in SQLOBS is 0. The string created after the SELECT clause is not queried as a value, but newly created. This trick causes the following row to be displayed in the log.

SAS Log 4.1-2: "Empty" Query: Formatted

```
The last query did not produce a hit!
---------------------------------------------
```

Another possible application of SQL macro variables is the automatic creation of macro variables. In the following example, the macro variables SUBS*1* to SUBS*max* are created for each unique instance of SUBSIDIARY (from the SASHELP file SHOES) and displayed using the macro LISTSUBS.

Program 4.1-5: Creating a Series of Macro Variables

```
proc sql ;
select unique(SUBSIDIARY)
   from SASHELP.SHOES
   order by SUBSIDIARY ;
select unique(SUBSIDIARY) into :SUBS1 - :SUBS&sqlobs
   from SASHELP.SHOES
   order by SUBSIDIARY ;
quit ;
```

In the SQL step, the macro variables SUBS*1* to SUBS*max* are created for the unique levels of SUBSIDIARY from the SASHELP file SHOES. The first SELECT statement is necessary for determining the SQLOBS value for the second SELECT statement. The specific value is determined automatically by *SQLOBS*.

Program 4.1-6: Example: Calling a Series of Macro Variables Including SQL Macro Variables

```
%macro LISTSUBS ;
%put Number of all subsidiaries: &sqlobs. ;
      %do i=1 %to &sqlobs ;
             %put Subsidiary No.&i: &&SUBS&i ;
       %end ;
%put SQL_N_ROWS=|| &sqlobs || SQL_ITER=|| &sqloops ||
      SQL_STATUS=|| &sqlrc || ;
%mend ;

%LISTSUBS ;
```

The LISTSUBS macro calls all created macro variables one after another using a DO-TO loop, from *1* to *sqlobs(=max)*. The SAS output contains all unique levels of SUBSIDIARY from SASHELP.SHOES. The log also displays all macro variables from SUBS1 to SUBS53 with additional text and numbering. This time, the parameters sqlobs, sqloops and sqlrc, which were also queried, return the number of rows passed through (N=53), the number of iterations of the inner loop (N=522), and the code "0" for the status.

Figure 4.1-4: Output (Abbreviated)
```
Number of all subsidiaries: 53

Subsidiary No.1: Addis Ababa
Subsidiary No.2: Al-Khobar
Subsidiary No.3: Algiers
Subsidiary No.4: Auckland
Subsidiary No.5: Bangkok
. . .
Subsidiary No.49: Tel Aviv
Subsidiary No.50: Tokyo
Subsidiary No.51: Toronto
Subsidiary No.52: Vancouver
Subsidiary No.53: Warsaw


SQL_N_ROWS=|| 53 ||        SQL_ITER=|| 522 ||        SQL_STATUS=|| 0 ||
```

SAS offers additional automatic SQL macro variables. The interested user is referred to the SAS documentation for further automatic SQL macro variables.

Table 4.1-2: Categories and Descriptions of Automatic SQL Macro Variables (Selection)

SAS Macro function	Description
Regarding the execution of SAS job	
SQLEXITCODE	Highest return code that occurred from some types of SQL insert failures (written to the SYSERR macro variable).
SQLOBS	Number of rows that were processed by an SQL procedure statement, for example, SELECT or DELETE.
SQLOOPS	Number of iterations that the inner loop of PROC SQL processes.
SQLRC	Status values that indicate the success or failure of an SQL procedure statement.
DBMS-specific Return Code	
SQLXMSG	Descriptive information and the DBMS-specific return code for the error that is returned by the Pass-Through Facility. May contain special characters, so use %SUPERQ when querying.
SQLXRC	DBMS-specific return code that is returned by the Pass-Through Facility.

Digression: Macro Variables in Local and Remote Environments

For processes that run between local and remote hosts, the possibilities with the macros %SYSRPUT and %SYSLPUT might be interesting. The **%SYSRPUT** macro enables you to take a value from a macro variable stored on the remote host. %SYSRPUT assigns this value to a macro variable on the local host. %SYSRPUT is similar to the %LET statement because it assigns a value to a macro variable. The difference to %LET is that %SYSRPUT transfers the value of a variable to the local host, but processing takes place on the remote host, whereas %LET processes on the local host. In contrast, the **%SYSLPUT** macro (along with the /REMOTE= option) enables you to create a macro variable in a certain server session. This is necessary because asynchronous remote executions can generally result in multiple server sessions. %SYSRPUT and %SYSLPUT are introduced in the SAS macro statements.

4.1.3 User-defined SAS Macro Variables (INTO)

With user-defined SAS macro variables, the name and functionality can be defined by the user. Using PROC SQL, individual values or entire lists of values can be easily determined, transferred to macro variables as text, and then taken up again and processed further elsewhere in the SAS program.

PROC SQL offers a very convenient way of creating user-defined macro variables by using INTO. INTO is always specified after the SELECT statement in PROC SQL. In PROC SQL, the user-defined macro variable is preceded by a colon after INTO instead of a "&". From another perspective, PROC SQL's INTO is always accompanied by user-defined macro variables. The default function of INTO is to store rows of a table into one or more macro variables. If values cannot be calculated with PROC SQL or can only be calculated with difficulty, you can use the two routines SYMPUT or SYMPUTX. (See also the example below for CALL SYMPUTX.) Note that the macro variables specified after INTO are preceded by a ":" and not by an "&". The variable names after INTO can be specified individually explicitly (see Section 4.1) or in the form of lists. (See Sections 4.2, 4.3, and 4.4.) Values that are passed to the macro variables are considered by SAS as text.

Although it is irrelevant for macro programming whether you are working with SQL or with the DATA step, this section describes how to work with macro variables in SQL and how the values that they contain are passed to titles, SAS output, or further SQL steps. However, because certain values or analyses can sometimes not be calculated with PROC SQL or can only be calculated with great difficulty, Section 4.3.4 describes how to work with macro variables in the DATA step.

Passing Values to Titles, Log, or Output

This section introduces you to working with macro variables in PROC SQL. This section uses very simple programs to show how easy it is to work with SAS macros, and how convenient it is to pass values to titles, SAS output, the DATA step, or other SQL steps.

In the following example, the average value of SALES is calculated from the file SASHELP.SHOES. The macro variable &MEANSALE is used to pass the obtained mean value as a string to the title. Note that if macro variables are specified after INTO, they are preceded by a ":" instead of a "&". Only one value is written to this macro variable.

Program 4.1-7: Passing a Calculated Value to a Title

```
proc sql ;
select mean(SALES) into :meansale
from SASHELP.SHOES ;
quit;
title " All stores with above average sales (>=&meansale., $)";
```

Figure 4.1-5: Passing a Calculated Value to a Title: Output

```
     All stores with above average sales (>=85700.17, $)
```

Note: The mean of SALES is calculated and displayed in the title instead of the "placeholder." If the query "mean(SALES)" does not yield a value, no value can be passed to the macro variable. At this point, the headline contains an empty space or a period. This example is extended in the following section.

Tip: To check whether and which value was passed to a macro variable, you only need to pass the following statement to SAS:

Program 4.1-8: Using %PUT: Checking Values Passed to a Macro Variable

```
%put &MEANSALE. ;
```

After %PUT, only the macro variable to be checked is specified. If the macro variable contains one or more values, %PUT writes them all to the SAS log. An output can look like this, for example:

Figure 4.1-6: Checking Values Passed to a Macro Variable: Output

```
nn    %put &MEANSALE. ;
85700.17
```

The macro variable &MEANSALE. contains the value 85700.17. The number *nn* before %PUT (in the log) reflects the row number of the SAS log and is completely irrelevant for the content of the result.

Passing to a Title and a Condition of a Subsequent SQL Query I

Macro variables can pass values also to subsequent SQL statements. Macro variables are particularly useful when it is foreseeable that the values to be passed must be specified or updated more than once, for example, in the case of repetitive data deliveries, changing specifications for analysis, and so on. Automatic updating by means of macro variables is much more effective than by hand.

In the following example, the mean of SALES is written to the macro variable (:MEANSALE) and from there passed to a title and into the WHERE condition of a subsequent SQL query. NOPRINT is used to suppress the MEANSALE value in the SAS output. Only one value is written to this macro variable.

Program 4.1-9: Passing to a Title and a Condition of a Subsequent SQL Query I

```
proc sql noprint ;
    select mean(SALES) into :meansale
        from SASHELP.SHOES ;
title "All stores with sales at least 5x higher
            than the average ($ &meansale.):";
        create table MYDATA as
    select Region, Subsidiary, Product, Sales
        from SASHELP.SHOES
            where SALES >= 5*&meansale. ;
quit;
```

Note: The average of SALES is calculated and displayed in the title instead of the "placeholder." If the query "mean(SALES)" returns no value or a missing value, no value can be passed to the macro variable. Depending on the programming variant, the title either contains a period or is not output at all. For the WHERE condition, SAS outputs the following error message to the log.

SAS Log 4.1-3: Error Message in the SAS Log

```
NOTE: There were 0 observations read from the data set SASHELP.SHOES
      WHERE 0 ;         /*    an obviously wrong WHERE-condition (FALSE)    */
```

NOTE: Table WORK.MYDATA created, with 0 rows and 4 columns.

Figure 4.1-7: Passing to a Title and a Condition of a Subsequent SQL Query: Output (Abbreviated) I

```
All stores with sales at least 5x higher than the average ($ 85700.17):
```

Region	Subsidiary	Product	Sales
Canada	Vancouver	Men's Dress	$757,798
Canada	Vancouver	Slipper	$700,513
Canada	Vancouver	Women's Dress	$756,347
Central America/Caribbean	Kingston	Men's Casual	$576,112

Passing to a Title and a Condition of a Subsequent SQL Query II

The average of SALES is written to the macro variable (:MEANSALE) and from there passed to a title and another SQL query. NOPRINT is used to suppress the MEANSALE value in the SAS output. Only one value is written to this macro variable.

Program 4.1-10: Passing to a Title and a Condition of a Subsequent SQL Query II

```
proc sql noprint ;
select avg(SALES) into :meansale
from SASHELP.SHOES ;
title " Above-average sales in the 'Asia' region (Limit: &meansale.)" ;
create table MYDATA as
select Region, Subsidiary, Product, Sales
from SASHELP.SHOES
      where REGION  = "Asia" and SALES > &meansale. ;
quit ;
```

Figure 4.1-8: Passing to a Title and a Condition of a Subsequent SQL Query: Output (Abbreviated) II

```
Above-average sales in the 'Asia' (Limit: 85700.17)

    Region      Subsidiary    Product          Sales
    Asia        Seoul         Men's Dress      $116,333
    Asia        Seoul         Slipper          $149,013
```

The INTO presented in this section represents an important interface to the SAS Macro Facility. Sections 4.3.3 and 4.3.4 contain further examples and notes on the interaction between macro variables, SQL, and the DATA step.

4.1.4 Macro Variables, INTO and Possible Loss of Precision

In the case of user-defined macro variables in combination with numeric values, especially when using INTO, you should be aware of the *possible* truncation when passing floating point numbers. Truncation might occur when truncating too many decimal places in floating point numbers, causing an undesirable loss of precision. The more precise a value is, the more decimal places it has. 0.66666666666666, for example, is more precise than 0.666667 or 0.67. Monetary data such as currency rates require the greatest possible accuracy for analysis, filtering, and further processing.

In the following SQL example, two variants of the result of dividing 2 by 3 are generated. The variant *without* loss of precision is created directly in column FRACTION1 using SET. A variant *with loss* of precision can occur by passing the FRACTION1 value to the macro variable FRACTION2 using INTO.

Program 4.1-11: Loss of Precision (DATA step and SQL)

```
proc sql noprint feedback ;
   create table SAMPLE (FRACTION1 num);
   insert into SAMPLE
      set FRACTION1 = 2/3;
      select FRACTION1 into :FRACTION2
   from SAMPLE ;
quit ;


data _null_ ;
   set  SAMPLE;
      FRACTION2 = &FRACTION2;
   put (FRACTION:)(/= 18.17);
run ;
```

A DATA step requested both FRACTION values. You should now expect that the values FRACTION1 and FRACTION2 match perfectly. In fact, the FRACTION2 value (in the macro variable) has considerably fewer decimal places than the directly created FRACTION1 value.

Note: In the PUT function, the decimal place specification (second value, 17) must be smaller than the width specification (first value, 18). Decimal places and width were intentionally set higher than 16 to demonstrate how SAS would proceed.

Figure 4.1-9: Loss of Precision (DATA step and SQL)

```
FRACTION1=.66666666666666600
FRACTION2=.66666700000000000
```

In order to ensure the required number of decimal places for values in a macro variable, the precise FRACTION1 value can be passed to the macro variable FRACTION2 in the format of a hexadecimal string at the moment of transfer. The hexadecimal string (not displayed) for 0.666666666666 is "3FE55555555555".

Program 4.1-12: Avoiding Loss of Precision (SQL)

```
proc sql noprint feedback ;
   create table SAMPLE2 (FRACTION1 num);
   insert into SAMPLE2
      set FRACTION1 = 2/3;
      select put (FRACTION1, hex16.) into :FRACTION2
   from SAMPLE2 ;
quit ;
```

The PUT function passes the exact value of FRACTION1 as a hexadecimal string to the macro variable FRACTION2.

Program 4.1-13: Avoiding Loss of Precision (DATA step)

```
data _null_ ;
   set SAMPLE2 ;
      FRACTION2 = input("&FRACTION2", hex16.) ;
      put (FRACTION :) (/= best32.) ;
run ;
```

If both FRACTION values are now requested using the DATA step, the FRACTION2 value in the macro variable matches the directly created FRACTION1 value perfectly.

Figure 4.1-10: Avoiding Loss of Precision (DATA step and SQL)

```
FRACTION1=0.66666666666666
FRACTION2=0.66666666666666
```

An alternative solution can be the SYSEVALF function. The following statement results in 0.666666666666.

Program 4.1-14: Avoiding Loss of Precision (%SYSEVALF)

```
%put %sysevalf(2/3);
```

If the desired precision cannot be achieved in spite of the format assignment or by means of the SYSEVALF function, the loss of precision could be caused by suboptimal hardware or differences between operating and database management systems.

4.1.5 The Many Roads Leading to a Macro Variable

SAS macro variables are versatile and ubiquitous. This also means that macro variables (and macros!) can be accessed or referenced in different ways. The following example illustrates several possibilities using a compact example:
- creating text: via a macro (the macro "INPUT1"), via a local macro variable (INPUT2), a reference to macro or macro variable (i., ii.), a reference to created variables (iii., iv.), via concatenating created variables (v. and vi.) or a direct reference to macro or macro variable (cvii.)
- referencing a macro by calling *"%Macro"* (INPUT1) or as a reference to a macro variable using RESOLVE (ii., INPUT2)
- the creation of calculated variables via CALCULATED (iv.) or via concatenation (v. to vii.)
- concatenating calculated variables without (v.) or with blanks (vi., vii.)
- concatenation via direct references to macro and macro variable. (See vii.)

Program 4.1-15: Options to Access Macro Variables

```
/* Macro INPUT1 (create) */
%macro INPUT1;
San
%mend INPUT1;

/* Macro variable INPUT2 (create) */
%let INPUT2=Francisco ;

proc sql outobs=2 ;
create table MY_CALLS as
select
/* i. Reference to macro INPUT1 (call) */
"%INPUT1"               as NEW_VAR1,
/* ii. Reference to macro variable INPUT2 */
resolve("&INPUT2")  as NEW_VAR2,
/* iii. Reference to macro INPUT1 (call) */
resolve("%INPUT1")  as NEW_VAR3,
/* iv. Passing variable NEW_VAR2 (CALCULATED) */
calculated NEW_VAR2 as NEW_VAR4,
/* v. Reference to NEW_VAR1 and NEW_VAR2 (CALCULATED) */
trimn (calculated NEW_var1)||calculated NEW_var2
                    as NEW_VAR5,
/* vi. Reference to NEW_VAR1 and NEW_VAR2 (CALCULATED) */
calculated NEW_var1||trim(" ")||calculated NEW_var2
                    as NEW_VAR6,
/* vii. Reference to macro INPUT1 and macro variable INPUT2 */
"%INPUT1"||trim(" ")||resolve("&INPUT2")
                    as NEW_VAR7
from SASHELP.CLASS ;
quit ;
proc print data=MY_CALLS noobs ;
run ;
```

Figure 4.1-11: Results of Options to Access Macro Variables

NEW_VAR1	NEW_VAR2	NEW_VAR3	NEW_VAR4	NEW_VAR5	NEW_VAR6	NEW_VAR7
San	Francisco	San	Francisco	SanFrancisco	San Francisco	San Francisco
San	Francisco	San	Francisco	SanFrancisco	San Francisco	San Francisco

Example vii. does not require the creation of variables as an intermediate step. The direct reference to macro "INPUT1" or macro variable "INPUT2" during concatenation might therefore be more efficient than example vi.

4.2 Macro Programs with PROC SQL

Before answering the obvious question of what macro programs actually are, another question should be asked and answered first: what are the basic differences between SAS macro code and SAS code? There are two main differences between SAS macro code and SAS code:

- Different syntax. The similarity of some commands and functionalities sometimes obscures this difference. SAS macro programs can be recognized by the % symbol and the & symbol.
- SAS macro code is processed *before* SAS code. This means that SAS macro code is "translated" into SAS code in the compiler. After successful translation, only the SAS code is executed. The underlying SAS macro code is stored in the SAS catalog SASMACR (default) and in principle no longer plays an active role after the execution of its translation.

These two differences taken together are central to understanding the interaction between SAS macro code and SAS code. "Macro" is an abbreviation of "macroinstruction" and can best be translated as "higher-level command." The "higher-level" consists of a SAS macro program organizing both the SAS syntax and the logic of its execution. Thus, as a higher-level command, it specifies which slots are to be replaced by strings and which further loops and iterations are to be run through. The specific execution is then basically a SAS program. The SAS macro program precedes in time and is also conceptually superior.

4.2.1 What Are Macro Programs?

What are macro programs?
- Macro programs are "slots," that is, interfaces or placeholders for texts. Macro variables start with an "&". Macro programs, on the other hand, start with a **"%"**. An "&" or "%" at the beginning of a character string initiate the SAS macro processor.
- Macro programs are limited by **%macro** and **%mend**. The macro is defined between %macro and %mend. (See also the concluding notes on the macros %STPBEGIN and %STPEND for stored processes.) The %macro *macro name* is at the beginning, and %mend is at the end of the macro program definition. It is not necessary to specify *macro name* after %mend, but it is recommended for complex macros for reasons of clarity. A macro is called using *%macro name*. The % character in *%macro name* causes the macro processor to search for a compiled SAS macro *macro name* in the SAS catalog SASMACR (default) and to execute it, if available.
- The default SAS catalog **SASMACR** is located in the temporary SAS directory WORK and will be deleted after the SAS session is terminated. To re-execute SAS macros, SAS offers several possibilities:

  - SAS macros can be executed again in a new SAS session as SAS macro programs.
  - SAS macro programs can be included and executed using %INCLUDE.
  - SAS macros can be stored and executed in a permanent user-defined SAS catalog using the Stored Compiled Macro Facility. (See the MSTORED and SASMSTORE= system options.)
  - Another possibility is the Autocall Facility, initiated with the system options MAUTOSOURCE and SASAUTOS=. For the last two approaches mentioned, additional host-dependent settings might also be required.

- A macro call can, but does not necessarily have to, end with a **semicolon**. Between definition and call, the macro is compiled. As soon as SAS compiles the text of a macro, the interfaces or placeholders are filled with the "actual" text. Only after this step has been completed, the SAS macro is fully compiled, stored in the SAS catalog SASMACR (default) and executed as SAS code.
- Macro programs can be embedded in **open** SAS code or created **stand-alone**.
- Macro programs can contain **comments**. The simplest way to place comments in SAS macro code is between " /* " and " */ " (block comment). The variant of the simple line comment " * ; " only works for SAS code or SAS code in macros, but not for SAS macro code. The variant " % * ; " only works for SAS macro code, but not for SAS code. The " /* */" variant works for SAS code and SAS macro code. A tip to deactivate larger passages of SAS macro code is to define this code as an autonomous SAS macro with MACRO and MEND, but not to call it.
- Macro programs can themselves contain macro variables. (See Section 4.1.)
- Macro programs can contain parameters. SAS offers various possibilities for passing macro parameters (styles). The **Name Style** is the most efficient. The reason is that only here the macro name begins with a "%". This directly causes the word scanner to pass the token to the macro processor. With the other two styles (Command Style, Statement Style), the word scanner must first check this, which takes up unnecessary time. Command Style and Statement Style will not be discussed further here. For the Name Style, the macro parameters are defined after *%macro name* in brackets. Two variants can be distinguished. The definition is the same for both variants: In the parentheses after *%macro name*, the macro parameters are specified as they are used in the macro program.

Variants of Name Style: Definitions and Calls

Starting point:
PROC SQL program
	Program 4.2-1: Name Style Macro Variant of a PROC SQL Program
	<pre>proc sql ;
	   create table OUTPUT
	      as select AGE
	      from SASHELP.CLASS ;
	quit ;</pre>

Macro variant:
Converting the SQL program into a macro (INTRO1)
	<pre>%macro INTRO1(OUTDATA,VAR,INDATA);
	proc sql ;
	   create table &OUTDATA.
	      as select &VAR.
	      from &INDATA.;
	quit ;
	%mend INTRO1 ;</pre>

Calling INTRO1 using position parameters	**Program 4.2-2: Variations of Calling a Name Style macro** right: <pre>%*INTRO1*(OUTPUT, AGE, SASHELP.CLASS);</pre> wrong (altered order!): <pre>%*INTRO1*(AGE, SASHELP.CLASS, OUTPUT);</pre>
Calling INTRO1 using keyword parameters	correct call: <pre>%*INTRO1*(OUTDATA=OUTPUT,VAR=AGE,INDATA=SASHELP.CLASS) ;</pre> correct call: <pre>%*INTRO1*(VAR=AGE,INDATA=SASHELP.CLASS,OUTDATA=OUTPUT) ;</pre>

Converting the SQL program in a macro (INTRO2) with defaults.	**Program 4.2-3: Defining and Calling a Macro with Defaults**
	<pre>%macro INTRO2(OUTDATA=OUTPUT,
	              VAR=AGE, INDATA=SASHELP.CLASS);
	proc sql ;
	   create table &OUTDATA.
	      as select &VAR.
	      from &INDATA.;
	quit ;
	%mend INTRO2 ;</pre>
	Calling the macro including the defaults (cf. *%macro* line):
	<pre>%INTRO2;</pre>
	Overwriting all the defaults (cf. call):
	<pre>%*INTRO2*(OUTDATA=OUTPUT2,VAR=HEIGHT,
	                    INDATA=SASHELP.CLASS);</pre>
	Partial keeping of defaults:
	<pre>%*INTRO2*(VAR=WEIGHT,INDATA=SASHELP.CLASS);</pre>

Converting the SQL program in a macro (INTRO3) for a mixed call (see definition and call).	**Program 4.2-4: Defining and Calling a Mixed Macro (Defaults)**
	<pre>%macro INTRO3(OUTDATA,VAR=AGE, INDATA=SASHELP.CLASS);
	proc sql ;
	   create table &OUTDATA.
	      as select &VAR.
	      from &INDATA.;
	quit ;
	%mend INTRO3 ;</pre>
	Mixed call:
	<pre>%*INTRO3*(OUTPUT3, VAR=HEIGHT,INDATA=SASHELP.CLASS);</pre>

Passing the entries of a local macro variable to a global macro variable. (See %LET and %GLOBAL.) HERE denotes the global macro variable.

Program 4.2-5: Passing from Local to Global Macro Variables

```
%macro INTRO3(OUTDATA,VAR=AGE, INDATA=SASHELP.CLASS);
%global HERE ;
%let HERE=&VAR. ;
proc sql ;
   create table &OUTDATA.
      as select &VAR.
   from &INDATA.;
quit ;
%mend INTRO3 ;

%INTRO3(OUTPUT3, VAR=HEIGHT, INDATA=SASHELP.CLASS);

proc sql ;
select &HERE.
from OUTPUT3 ;
```

- The **call** is different. (See the different calls of macro %INTRO1.)

  ○ For **keyword** parameters, the desired entries are specified after an equal sign after the name of the respective parameter; in the call, the sequence of the parameters is irrelevant.
  ○ With **position** parameters, the desired entries need only be separated by a comma; however, it can be a disadvantage that the sequence of the macro parameters must correspond exactly to the sequence in the macro definition.

  A simultaneous ("mixed") use of **both** types of parameters is also possible, in which case the position parameters must come before the keyword parameters in the definition and call. (See macro INTRO3.) The keyword parameters have the advantage that they allow certain preset values to be passed to the program already when the macro is defined. These values can be taken over or even overwritten as desired when the macro is called. (See the different calls of macro INTRO2.)
- Macro programs can contain further elements of the SAS macro language. Elements of the SAS macro language include

  ○ **SAS Macro Functions** (for example, %INDEX, %LENGTH, %SUBSTR, %SYSEVALF)
  ○ **SAS Macro Statements:** "Macro versions" of SAS functions (%if, %where, %do %to, %do %while, %do %until, and so on)
  ○ Various **Interfaces** from the SAS Macro Facility to the DATA step, especially the SAS macro CALL routines (for example, CALL EXECUTE, CALL SYMPUTX, and so on)
  ○ **Automatic SAS and SQL Macro Variables** (see Sections 4.1.1 and 4.1.2)

- In addition to macro variables, a SAS macro can contain numerous other functions, routines, and statements. The **functionality** and **flexibility** of macro programs can be much more complex than that of macro variables. A SAS macro program is a program written in SAS syntax (including Base SAS, SAS Macro Facility, and procedures such as PROC SQL). If a user is familiar with DATA step programs and functions, there is no major difference to writing a SAS macro.
- Macro variables can be stored either in local or global symbol tables. The central difference is that if macro variables are stored in a **global symbol table** using %GLOBAL or CALL SYMPUTX or in open code with INTO from PROC SQL, they are available as "global" macro variables at any point of a SAS session. This means that global macro variables can also be further processed in PROC SQL, DATA steps, or other SAS macros. If a macro variable is "only" stored in a **local symbol table** (for example, explicitly using %LOCAL), it is available only as a "local" macro variable within the macro in which it was created. Local macro variables cannot be processed in PROC SQL, DATA steps, or other SAS macros. A typical example of local macro variables are the listed examples of macros with position and keyword parameters. These parameters are at first local and are only valid for the respective macro in which they were defined. Only an explicit assignment to a global macro variable makes it possible to make the values of a local macro variable available to PROC SQL, DATA steps, or other SAS macros. The last example for %INTRO3 illustrates

how entries can be passed from a local to a global macro variable using %LET and %GLOBAL. The subsequent PROC SQL uses the global macro variable &HIER to take the values from the local macro variable &VAR. and process them further. Incidentally, if a variable is defined outside a macro using %LET=, it is a global macro variable. If it is defined inside using %LET, it is a local macro variable.

- For the SAS Macro Facility, SAS provides numerous SAS macro functions and SAS macro statements, as well as various interfaces between the SAS Macro Facility to PROC SQL and the DATA Step. (See Section 4.3.)

Note: The macros **%STPBEGIN** and **%STPEND** are similar to %MACRO and %MEND in functionality and structure. They also appear as a pair and enclose a process to be executed. However, %STPBEGIN and %STPEND are only used for SAS Stored Processes (new in SAS 9). SAS Stored Processes are nearly any SAS programs that are hosted on servers and are described by metadata.

Program 4.2-6: %STPBEGIN and %STPEND for SAS Stored Processes

```
%global _ODSSTYLE ;
%let _ODSSTYLE = MyODSStyle ;

%stpbegin ;
 title "Shoe Sales By Region and Product" ;
 proc report data = sashelp.shoes nowindows ;
    column region product sales ;
    define region / group ;
    define product / group ;
    define sales / analysis sum ;
    break after region / ol summarize suppress skip ;
 run ;
 %stpend ;
```

%STPBEGIN and %STPEND are set before and after the end of the stored process in order to initialize or terminate its processing using ODS (Output Delivery System). SAS Stored Processes require the BI Manager or the Stored Process Manager. SAS recommends not using %STPBEGIN or %STPEND if the stored process does not use ODS. The call of %STPBEGIN or %STPEND must be terminated with a semicolon. Whoever can define SAS macros can also program SAS Stored Processes. The first steps in programming SAS macros are shown in the next sections.

4.2.2 Let's Do It: SAS Macros with the %LET Statement

Let's Do It: Simple SAS Macros with the %LET statement (One Parameter)

With the %LET statement, a macro variable can be defined and a value can be assigned at the same time. The repeated passing of this one value using the %LET statement to a SAS macro can already save a lot of effort. How does something like this look like in practice?

In the first macro %LETSDOIT1, the macro variable ANA_VAR is a placeholder. The subsequent %LET statement passes the desired text to this placeholder, in this case the name of a variable. The concluding %LETSDOIT1 calls the originally defined macro, whereby the name of the desired variable is inserted instead of the placeholder (for example, SEX or AGE). Each time %LETSDOIT1 is called, the macro is executed for the variable specified under %LET. For space reasons, no results are specified in these first two examples.

Program 4.2-7: Defining and Calling %LETSDOIT1

Defining the macro

```
%macro letsdoit1;
    title "Analysis of Column &ana_var." ;
       proc freq data=SASHELP.CLASS ;
           table &ana_var. ;
        run ;
%mend letsdoit1;
```

Macro Calls:

```
%let ana_var=SEX ;
%letsdoit1 ;

%let ana_var=AGE ;
%letsdoit1 ;
```

In the %LETSDOIT2 macro, the macro variable LIMIT is a placeholder. The subsequent %LET statement passes the desired text to this placeholder, in this case texts for values. The concluding %LETSDOIT2 calls the originally defined macro, whereby the text for the desired values is inserted instead of the placeholder (for example, 1.2 or -2.05). After each macro call, the feedback from the SAS log is shown.

Program 4.2-8: Defining and Calling %LETSDOIT2

```
%macro letsdoit2;
    title "Analysis of S0666 by Limit &limit." ;
        proc print data=SASHELP.SYR1001 ;
            where S0666 < &limit. ;
        run ;
%mend letsdoit2;
```

Macro Call I:

```
%let limit=1.2 ;
%letsdoit2 ;
```

Macro Call II:
```
%let limit=%str(-2.05) ;
%letsdoit2 ;
```

SAS Log 4.2-1: %LETSDOIT2: Calls I and II

Call I:

```
NOTE: There were 52 observations read from the data set SASHELP.SYR1001.
      WHERE S0666<1.2;
```

Call II:

```
NOTE: There were 51 observations read from the data set SASHELP.SYR1001.
      WHERE S0666<-2.05;
```

Note: As soon as a %LET statement is called again, the value in the macro variable is updated. In order to be able to pass signs, leading and trailing blanks to the macro variable, the relevant text ("-2.05") must be specified in a %STR function (see last example). The macros with the %LET statement can alternatively be rewritten into different macros with the same functionality. In the macro LETSDOIT3, the macro variables are specified in the bracket after the macro name. The number "-2.05" after the macro call is set instead of the macro variable LIMIT and passed to the execution.

Program 4.2-9: Defining and Calling %LETSDOIT3

```
%macro letsdoit3 (LIMIT);
    title "Analysis of S0666 by Limit &limit." ;
        proc print data=SASHELP.SYR1001 ;
            where S0666 < &limit. ;
        run ;
%mend letsdoit3;
%letsdoit3 (-2.05) ;
```

In the macro LETSDOIT4, the macro variables are also specified in the brackets after the macro name. However, in this variant, the number "-2.05" is first passed to the macro call via the %LET statement and then to LIMIT and execution.

Program 4.2-10: Defining and Calling %LETSDOIT4

```
%macro letsdoit4 (LIMIT);
   title "Analysis of S0666 by Limit &limit." ;
      proc print data=SASHELP.SYR1001 ;
         where S0666 < &limit. ;
      run ;
%mend letsdoit4;
%let limit=%str(-2.05) ;
%letsdoit4 (&LIMIT.) ;
```

The macros LETSDOIT2 to LETSDOIT4 have the same functionality and basically achieve the same result. The repeated calling of the %LET statement can already save a lot of effort; the clear disadvantage of the %LET statement is that only one macro variable can be defined at a time. By the way, the %LET statement is regarded as a SAS macro statement and described there. The next section shows how several parameters can be passed to a macro at once. The presented macros are more sophisticated and also contain numerous elements of the SAS macro language. This section is a preview of the possibilities that will be further explored in Section 4.3.

4.2.3 Listwise Execution of Commands

This section illustrates how SAS macros can be used to easily repeat the same actions. The following will be introduced:

- The %REPEATED macro: repeatedly calling a macro by hand.
- The macros LISTING1 through LISTING6: automatically repeated listwise macro calling using the PARMBUFF option.

The first macro, %REPEATED, shows how several parameters can be passed to a SAS macro. The listwise calling of this macro can also save a lot of effort. In principle, the approach is simple. At first, after %MACRO and %MEND, an arbitrary name for the macro is defined. In the brackets after the name after %MACRO, the required placeholders are put together. If the placeholders called later do not occur in these brackets, it causes problems when executing the macro. Placeholders can be named arbitrarily; however, it is recommended to assign placeholders with meaningful names. Between %MACRO and %MEND, the SAS code is inserted. The placeholders needed at the beginning of the code now appear again (often several times, see VAR), only their spelling is different. In the (listwise) macro calls (see %REPEATED), the texts to be inserted are now specified instead of the placeholders (from the parentheses after %MACRO). Each macro call now executes the SQL program with the desired parameters instead of the placeholders. Finally, the feedback from the SAS log is listed.

Program 4.2-11: Defining and Calling %REPEATED

```
%macro REPEATED(dsname, var, compare, value);
   proc sql ;
         create table &var. as
         select &var.
         from &dsname.
         where &var. &compare. &value.;
      quit ;
%mend REPEATED ;
%REPEATED(SASHELP.CLASS, SEX, =, "M")  ;
%REPEATED(SASHELP.CLASS, AGE, >, 12)  ;
%REPEATED(SASHELP.CLASS, NAME, EQ, "John")  ;
%REPEATED(SASHELP.AIR, AIR, GE, 200)  ;
```

SAS Log 4.2-2: %REPEATED

```
10  %REPEATED(SASHELP.CLASS, SEX, =, "M")  ;
NOTE: Table WORK.SEX created, with 10 rows and 1 columns.

11  %REPEATED(SASHELP.CLASS, AGE, >, 12)  ;
NOTE: Table WORK.SEX created, with 12 rows and 1 columns.
```

```
12  %REPEATED(SASHELP.CLASS, NAME, EQ, "John")  ;
NOTE: Table WORK.SEX created, with 1 rows and 1 columns.

13  %REPEATED(SASHELP.AIR, AIR, GE, 200)  ;
NOTE: Table WORK.SEX created, with 96 rows and 1 columns.
```

Note: Specifications in the macro call are separated by commas. Every character specified in the macro call, except the separators, is interpreted as text, including values and apostrophes. In the REPEATED example, the variant of position parameters was used. In contrast to keyword parameters, the sequence of the macro variables in the macro call must correspond to the sequence after %MACRO. Repeatedly calling a macro with several placeholders can already save a lot of effort; however, it would be even more efficient if you could also avoid the listwise calling "by hand." The next section shows how a macro can be automatically executed several times.

Examples for the Automatic Listwise Execution of Macros (One Parameter as List)

The previous examples triggered SAS commands or applications by repeatedly calling a macro, either with one or more placeholders. The following SQL examples illustrate how a macro can be executed automatically multiple times. All these different applications work with the PARMBUFF option. The **PARMBUFF option** (or PBUFF) assigns an entire list of values to a called macro. The values are passed to the placeholders (for example, &LISTMEM.) and also to the automatic macro variable SYSPBUFF. SYSPBUFF, in turn, resolves the text provided by PARMBUFF and passes it to the already known %LET statement (but this time within a macro).

The basic functionality of all application examples is the same. The macro call (LISTING1) contains a list of two or more elements (AGE and SEX). The PARMBUFF option assigns the elements to the macro (LISTING1), as well as to the automatic macro variable SYSPBUFF. Using &LISTMEM, all elements of the list are passed to PROC SQL iteratively one after the other in a %DO %WHILE loop and processed further as long as elements from the list are present. The report of the listwise requested output is waived; instead, the feedback from the SAS log is compiled.

Macro LISTING1: Listwise Creation of Tables (Processing of Texts)

Program 4.2-12: Defining and Calling %LISTING1: Text (Tables)

```
%macro LISTING1/parmbuff ;
   %let num=1 ;
   %let listmem=%scan(&syspbuff.,&num.) ;
   %do %while(&listmem. ne) ;
      proc sql ;
             create table SUBSET_&listmem. as
             select &listmem.
             from SASHELP.CLASS ;
         quit ;
      %let num=%eval(&num. +1 ) ;
      %let listmem=%scan(&syspbuff., &num.) ;
   %end ;
%mend LISTING1 ;
%LISTING1(AGE, SEX)  ;
```

SAS Log 4.2-3: %LISTING1

```
NOTE: Table WORK.SUBSET_AGE created, with 19 rows and 1 columns.
NOTE: Table WORK.SUBSET_SEX created, with 19 rows and 1 columns.
```

Note: The two texts from the macro call each pass through the SQL program. The table SASHELP.CLASS is therefore processed twice by this single call. In the tables SUBSET_AGE and SUBSET_SEX, subsets are created in the shape of the columns AGE and SEX.

Macro LISTING2: Listwise Printing of Columns (Processing of Texts)

Program 4.2-13: Defining and Calling %LISTING2: Text (Columns)

```
%macro LISTING2/parmbuff ;
   %let num=1 ;
   %let listmem=%scan(&syspbuff.,&num.) ;
   %do %while(&listmem. ne) ;
      proc print data=SASHELP.CLASS ;
         var &listmem. ;
      run ;
      %let num=%eval(&num. + 1) ;
      %let listmem=%scan(&syspbuff., &num.) ;
   %end ;
%mend LISTING2 ;

%LISTING2(AGE, SEX)  ;
```

SAS Log 4.2-4: %LISTING2

```
NOTE: There were 19 observations read from the data set SASHELP.CLASS.
NOTE: There were 19 observations read from the data set SASHELP.CLASS.
```

Note: The two texts from the macro call pass through the SQL program twice. As a result, this single call creates two tables with the names SUBSET_AGE and SUBSET_SEX.

Macro LISTING3: Listwise Creation of Tables (Processing of "Numbers")

Program 4.2-14: Defining and calling %LISTING3: Numbers

```
%macro LISTING3/parmbuff ;
   %let num=1 ;
   %let listmem=%scan(&syspbuff.,&num.) ;
   %do %while(&listmem. ne) ;
      proc sql ;
            create table AGE_&listmem. as
            select *
            from SASHELP.CLASS
            where AGE >= &listmem. ;
         quit ;
      %let num=%eval(&num. +1 ) ;
      %let listmem=%scan(&syspbuff., &num.) ;
   %end ;
%mend LISTING3 ;

%LISTING3(11, 13, 15);
```

SAS Log 4.2-5: %LISTING3

```
NOTE: Table WORK.AGE_11 created, with 19 rows and 5 columns.
NOTE: Table WORK.AGE_13 created, with 12 rows and 5 columns.
NOTE: Table WORK.AGE_15 created, with 5 rows and 5 columns.
```

Note: The three values from the macro call cause SAS to pass through the SQL program three times. As a result, this single call creates three tables named AGE_11, AGE_13, and AGE_15, respectively. The values in the macro call are unsigned, so they are interpreted as positive without exception.

Macro LISTING4: Flexible Handling of Negative Values (Processing of "Numbers")

The LISTING4 macro has two special features. The first special feature concerns the mode of the executed program. While all the macros presented so far ran through one standard program iteratively, this macro contains a %IF %DO condition that causes the execution of two program variants (depending on the requirements). This macro works differently depending on whether the values are positive or negative, and it executes correspondingly different program variants. The second special feature concerns the handling of negative values. Negative values often cannot be processed directly because of their sign. For example, a minus sign cannot be passed directly to a table name because this character would violate the conventions for names for SAS data sets. A simple variant is to pass unsigned values to SAS, but to explicitly inform SAS via the %LET statement that the lists contain positive or negative values so that the passed values can be handled differently after %IF %DO. In this example, the distinction between positive and negative values is made by the user. The following macro LISTING5 gives a first hint how this "manual" step could be solved automatically with SAS.

Program 4.2-15: Defining and Calling %LISTING4: Negative Values

```
%macro LISTING4/parmbuff;
%if &listval.=NEG %then %do ;
   %let num=1 ;
   %let listmem=%scan(&syspbuff.,&num.) ;
   %do %while(&listmem. ne) ;
      proc sql ;
              create table AMOUNT_NEG_&listmem. as
              select *
              from SASHELP.BUY
              where AMOUNT <= &listmem.*-1 ;
         quit ;
         proc print data=AMOUNT_NEG_&listmem. ;
         run ;
         %let num=%eval(&num. +1 ) ;
       %let listmem=%scan(&syspbuff., &num.) ;
         %end ;
   %end ;
%if &listval.=POS %then %do ;
   %let num=1 ;
   %let listmem=%scan(&syspbuff.,&num.) ;
   %do %while(&listmem. ne) ;
      proc sql ;
              create table AMOUNT_POS_&listmem. as
              select *
              from SASHELP.BUY
              where AMOUNT <= &listmem. ;
         quit ;
         proc print data=AMOUNT_POS_&listmem. ;
         run ;
         %let num=%eval(&num. +1 ) ;
       %let listmem=%scan(&syspbuff., &num.) ;
          %end ;
    %end ;
%mend LISTING4 ;
%let listval=POS ;
%LISTING4(1000, 2000, 50000);
%let listval=NEG ;
%LISTING4(1000, 2000, 50000);
```

SAS Log 4.2-6: %LISTING4

```
NOTE: Table WORK.AMOUNT_POS_1000 created, with 10 rows and 2 columns.
NOTE: Table WORK.AMOUNT_POS_2000 created, with 10 rows and 2 columns.
NOTE: Table WORK.AMOUNT_POS_50000 created, with 11 rows and 2 columns.
NOTE: Table WORK.AMOUNT_NEG_1000 created, with 10 rows and 2 columns.
NOTE: Table WORK.AMOUNT_NEG_2000 created, with 8 rows and 2 columns.
NOTE: Table WORK.AMOUNT_NEG_50000 created, with 2 rows and 2 columns.
```

Note: The user makes a distinction between positive and negative values. The preceding %LET statement explicitly tells SAS whether the lists contain only positive or negative values. The respective values given cause SAS to pass through the SQL program correspondingly frequently.

Macro LISTING5: Listwise Creation of Tables (One Parameter as Macro Variable)

The previous examples triggered a macro iteratively by passing lists of values to the macro in the macro call. The following example demonstrates how SAS can take over the compilation and pass it to the macro.

As a first step, the desired values are stored in a macro variable separated by commas using PROC SQL. Then, in the second step, the actual macro is defined. In the third step, in the macro call, the macro variable is now passed to SAS instead of the individual values as before. This macro also indicates how value lists with exclusively positive or negative values could be created: Namely with preceding SQL steps that store the value lists filtered in macro variables accordingly.

Program 4.2-16: Defining and Calling %LISTING5: Listwise Calling

```
proc sql noprint ;
   select unique(NAME)
      into :NAME_LST separated by " , "
         from SASHELP.CLASS
order by NAME ;

  %macro LISTING5/parmbuff ;
     %let num=1;
     %let memname=%scan(&syspbuff,&num) ;
     %do %while(&memname. ne) ;
        proc sql ;
          create table DATA_&memname.
          as select *
          from SASHELP.CLASS
          where NAME eq "&memname." ;
        quit ;
        %let num=%eval(&num+1) ;
        %let memname=%scan(&syspbuff,&num) ;
     %end;
  %mend LISTING5 ;

 %LISTING5(&NAME_LST.) ;

%put &NAME_LST. ;
```

SAS Log 4.2-7: %LISTING5 (Excerpt)

```
NOTE: Table WORK.DATA_ALFRED created, with 1 rows and 5 columns.
NOTE: Table WORK.DATA_ALICE created, with 1 rows and 5 columns.
NOTE: Table WORK.DATA_BARBARA created, with 1 rows and 5 columns.
    (abbreviated)
NOTE: Table WORK.DATA_RONALD created, with 1 rows and 5 columns.
NOTE: Table WORK.DATA_THOMAS created, with 1 rows and 5 columns.
NOTE: Table WORK.DATA_WILLIAM created, with 1 rows and 5 columns.
```

Figure 4.2-1: %LISTING5: Output in Log

```
%put &NAME_LST. ;
Alfred , Alice , Barbara , Carol , Henry , James , Jane , Janet , Jeffrey , John , Joyce ,
Judy , Louise , Mary , Philip , Robert , Ronald , Thomas , William
```

Note: The values from the macro variable cause SAS to pass through the SQL program correspondingly frequently. This single call creates 19 tables with the names DATA_ALFRED, DATA_ALICE, and so on. The values in the macro call are then displayed in the SAS output using %PUT for checking purposes. LISTING5 contains no processing variants.

Macro LISTING6: Converting from SAS to EXCEL (Processing of Texts)

Macro LISTING6 is a special application for exporting SAS tables into an EXCEL format. You might have been confronted with two problems in the past: either the SAS formats were not adopted by EXCEL or the EXCEL file was many times larger than the original SAS table. Differences by a factor of 10 are not uncommon, which is unacceptable for SAS data volumes above the two-digit MB range. The example is based on the SAS sample data FATCOMP. However, to illustrate the power of macros, the names of the formats have been aligned with those of the variables (see PROC FORMAT), so columns and formats have the same names.

Program 4.2-17: Demo Data and Formats

```
data FatComp ;
        input EXPOSE RESPONS Count;
        datalines;
    0  0   6
    0  1   2
    1  0   4
    1  .  11
    ;
proc format;
    value EXPOSE   1='High Cholesterol Diet'
                   0='Low Cholesterol Diet';
    value RESPONS 1='Yes'
                   0='No';
    run;


data _TEMP ;
set FATCOMP;
run ;
```

Program 4.2-18: Defining and Calling %LISTING6: Converting SAS Formats to EXCEL Values

```
%macro LISTING6/parmbuff ;
    %let num=1 ;
    %let listmem=%scan(&syspbuff.,&num.) ;
        %do %while(&listmem. ne) ;
            proc sql undo_policy=none;
                create table _TEMP as select *,
                    put(&listmem., &listmem..) as &listmem._FT
        %let num=%eval(&num. + 1) ;
        %let listmem=%scan(&syspbuff., &num.) ;
                from _TEMP ;
            quit;
        %end ;
%mend LISTING6 ;

%LISTING6(EXPOSE,RESPONS)   ;

proc print data=_TEMP ;
run ;

proc export data=_TEMP
    outfile="C:\SAS_formatted.xls" replace;
run;
```

Note: Since this approach overwrites the original entries, a copy of the original file is created with the name _TEMP for security reasons, which is then used exclusively for further work. The two texts from the macro call pass through the SQL program twice. Alternatively, an approach with a macro variable could have been used.

The PUT function can be used to store the format of a variable (RESPONSE or RESPONSE.) as values of a new column (RESPONS_FT). All variables are kept in the example. Via PROC EXPORT, the SAS table is stored in the EXCEL file "SAS_formatted.xsl" and there under the tab "_TEMP". As can be seen in the screenshot, the created table now stores the original SAS formats as Excel values (columns D and E).

Figure 4.2-2: %LISTING6: Screenshot of EXCEL File "SAS_formatted.xsl"

	A	B	C	D	E	F
1	EXPOSE	RESPONS	Count	EXPOSE_FT	RESPONS_FT	
2	0	0	6	Low Cholesterol Diet	No	
3	0	1	2	Low Cholesterol Diet	Yes	
4	1	0	4	High Cholesterol Diet	No	
5	1		11	High Cholesterol Diet	.	
6						

The second goal was also achieved: the bloating of the created EXCEL table was avoided. This macro can be extended by further functionalities, including deleting the original variables using ALTER TABLE and DROP or DELETE in PROC DATASETS, removing the automatically created punctuation for empty cells (see cell E5, highlighted by a black frame) using COMPBL or COMPRESS, and so on.

4.2.4 Condition-based Execution of Commands

SAS offers numerous possibilities for the condition-based execution of commands. Conditions are generally defined as if-conditions, supplemented by a predefined action that is executed when a condition is met. If an if-condition is met, the defined action is executed. Actions can be defined as PROC SQL analyses, execution of other SAS procedures such as PROC CONTENTS, jumping to certain positions in the SAS code, execution of other SAS macros and many more. If an if-condition does not apply, you should specify an alternative action to be executed. If no if-condition is met, but an alternative action was not specified, undesired results might be the result.

In SAS, *values*, *columns*, or *SAS data sets* can fulfill an if-condition (see examples 1 to 4), or *users* can specify the fulfillment of a certain condition (see examples 5 and 6). SAS also offers the CASE expression. The SAS macro statements %DO %TO %BY, %DO %UNTIL, %DO %WHILE, and %IF %THEN are presented in Section 4.3.2. This section illustrates the following variants:

- **Example 1:** If entries in a column fulfill a certain condition, then execute either the specified action or the specified alternative.
- **Example 2:** If a data set fulfills a certain condition, then execute either the specified action or the specified alternative.
- **Example 3:** If the attributes of one or more values fulfill a certain condition, then execute either the specified action or the specified alternative.
- **Example 4:** If a DATA step process fulfills a certain condition, then execute a macro.
- **Example 5:** If users specify a certain condition, then execute either the specified action or the specified alternative.
- **Example 6:** If users specify a certain condition, then go to a specific position in the SAS code.

Example 1: If Entries in a Column Fulfill a Certain Condition, Then Execute Either the Specified Action or the Specified Alternative

The if-condition in the IF_THENDO_ELSE macro is as follows if entries in capital letters occur in COUNTRY (possibly due to %UPCASE). If capitalized "CHINA" entries in the COUNTRY column of the SAS file SASHELP.PRDSAL2 occur in the macro IF_THENDO_ELSE, they trigger a specific PROC SQL analysis filtered by entries for "China". On the other hand, if capitalized "CANADA" entries occur in the IF_THENDO_ELSE macro, they trigger a filtered SQL analysis for "Canada".

Program 4.2-19: Defining and Calling Macro %IF_THENDO_ELSE

```
* Example for UPCASE (SAS Macro Function);
%macro IF_THENDO_ELSE(COUNTRY) ;
%if %upcase(&COUNTRY.)= CHINA %then %do ;
        title "Analysis for China";
                proc sql ;
                        select sum(ACTUAL) as ACTUAL_sum,
                        sum(PREDICT) as PREDICT_sum
                        from SASHELP.PRDSAL2
                        where COUNTRY = "China" ;
        quit;
%end ;
%else %if %upcase(&COUNTRY.)= CANADA %then %do ;
        title "Analysis for Canada";
                proc sql ;
                        select sum(ACTUAL) as ACTUAL_sum,
                        sum(PREDICT) as PREDICT_sum
                        from SASHELP.PRDSAL2
                        where COUNTRY = "Canada" ;
        quit;
%end ;
%mend IF_THENDO_ELSE ;

%IF_THENDO_ELSE(Canada) ;
```

Figure 4.2-3: %IF_THENDO_ELSE: Output

```
Analysis for Canada
ACTUAL_sum PREDICT_sum
---------------------
3197833 3431016
```

Example 2: If a Data Set Fulfills a Certain Condition, Then Execute Either the Specified Action or the Specified Alternative

The if-condition in the FILE_EXIST macro is if the specified SAS data set exists. If the SAS data set SASHELP.CLASS exists in the macro FILE_EXIST, it triggers a PROC CONTENTS query. If the specified SAS data set should not exist, the prepared error message is output to the SAS log.

Program 4.2-20: Defining and Calling Macro %FILE_EXIST

```
* Example for SYSFUNC+EXIST (SAS Macro Function);
%macro FILE_EXIST(IN_DATA);
        %if %sysfunc(exist(&IN_DATA)) %then
                %do;
                proc contents data=&IN_DATA. ;
                run;
        %end;
        %else
                %put The dataset &IN_DATA. does not exist.;
%mend FILE_EXIST;

%FILE_EXIST(sashelp.claass);
```

Figure 4.2-4: %FILE_EXIST: Feedback in SAS Log

```
The dataset sashelp.claass does not exist.
```

Example 3: If the Attributes of One or More Values Fulfill a Certain Condition, Then Execute Either the Specified Action or the Specified Alternative

The if-condition in the macro VAR_TYP is if the specified values are of type numeric. If in the macro VAR_TYP the values specified under &A. *and* &B. are of type numeric, the sum of both values is calculated. If one of the two specified values is not numeric, the prepared error message is output to the log.

Program 4.2-21: Defining and Calling macro %VAR_TYP

```
* Example for DATATYP (Autocall Macro);
%macro VAR_TYP(A,B);
        %if (%datatyp(&A.)=NUMERIC and %datatyp(&B.)=NUMERIC)
                %then %do;
                        %put The result is: %sysevalf(&A.+&B.).;
        %end;
        %else %do;
%put NOTE: Mathematical operation requires numeric input! ;
        %end;
%mend VAR_TYP ;


%VAR_TYP(A=10, B=20);


%VAR_TYP(A=10, B=XYZ);
```

Figure 4.2-5: %VAR_TYP: Output in SAS Log

```
%VAR_TYP(A=10, B=20);
The result is: 30.
%VAR_TYP(A=10, B=XYZ);
NOTE: Mathematical operation requires numeric input!
```

The next example is different from the previous examples. In contrast to a macro in which a condition is met and corresponding alternative actions are triggered, DATA step processes trigger the execution of a previously defined macro.

Example 4: If DATA Step Processes Fulfill a Certain Condition, Then Execute a Macro

The example consists of a DATA step and a macro. In the end, the DATA step triggers the macro. First, a macro TRIG_MAC is created. The macro TRIG_MAC triggers the analysis of filtered data. However, the macro is not executed yet. The DATA step creates a temporary SAS file TO_CHECK including a new variable ID. In the DATA step, a temporary variable FILE_END is created after END=. As soon as the last observation (row) is processed, the value of FILE_END is 1, otherwise 0. FILE_END is not stored in the SAS data set. As soon as the last observation is processed and a value in ID exists, the macro TRIG_MAC is triggered. In summary, the DATA step creates a temporary SAS file TO_CHECK. As soon as the last observation of this procedure has been processed, the macro created previously is triggered and the data in TO_CHECK is processed further.

Program 4.2-22: Defining and Calling Macro %TRIG_MAC

```
* Example for CALL EXECUTE (SAS Macro Call Routine);
%macro TRIG_MAC ;
        proc sql;
        select *
                from TO_CHECK ;
        quit;
%mend TRIG_MAC ;
data TO_CHECK ;
```

```
        retain ID ;
set SASHELP.BUY end=file_end ;
        if DATE >= "01JAN2004"d
                then
        do;
        ID + 1;
                output;
        end;
if file_end and ID then call execute("%TRIG_MAC");
        run;
```

Examples 5 and 6 differ from examples 1 to 4 in that *users* now specify that a certain condition applies and not, for example, values, columns, or SAS data sets.

Example 5: If Users Specify a Certain Condition, Then Execute Either the Specified Action or the Specified Alternative

The if-condition in the macro DOIF_YESNO is: If a user specifies the parameters "yes" or "no" in the call of DOIF_YESNO, then execute a PROC SQL step (if "yes") or call meta information of the specified SAS file using PROC CONTENTS. It is important that the parameters in YES_NO are not information within SAS, but parameters that are passed to SAS "from outside" by the user.

Program 4.2-23: Defining and Calling Macro %DOIF_YESNO

```
* Example for UPCASE (SAS Macro Function);
%macro DOIF_YESNO(YES_NO=,IN_DATA=) ;
        %if %upcase(&YES_NO)=YES %then
        %do ;
                proc sql ;
                create table DEMO as select * from &IN_DATA. ;
                quit ;
        %end ;
        %else %if %upcase(&YES_NO)=NO %then
        %do ;
                proc contents data=&IN_DATA. ;
                run ;
        %end ;
%mend DOIF_YESNO ;

%DOIF_YESNO(YES_NO=no,IN_DATA=sashelp.class) ;
```

Example 6: If Users Specify a Certain Condition, Then Go to a Specific Position in the SAS Code

The if-condition in the macro IF_GOTO is: If a user specifies the parameters "yes" or "no" in the call of IF_GOTO, then jump to a specific position in the SAS code. SAS statements are not executed after %GOTO, only when the "landing" position specified after %GOTO ("HERE", without percentage sign and colon) is reached. In the example, the PROC CONTENTS section and possibly all other SAS statements are skipped and only the PROC SQL section specified after %HERE: ("HERE" with percentage sign and colon) is executed again. In example 6, the parameters in YES_NO are not information within SAS, but are passed to SAS "from outside" on the part of the user.

Program 4.2-24: Defining and Calling Macro %IF_GOTO

```
* Example for %UPCASE and %GOTO (SAS Macro Statement);
%macro IF_GOTO(YES_NO=,IN_DATA=) ;
        %if %upcase(&YES_NO)=YES %then
        %goto HERE
/* HERE after %GOTO without percentage sign and colon */;
                proc contents data=&IN_DATA. ;
run ;
```

```
              /* ... many other SAS Statements ... */
              /* ... many other SAS Statements ... */
              /* ... many other SAS Statements ... */
/* now %HERE with percentage sign and colon */
    %HERE: proc sql ;
              create table DEMO as select * from &IN_DATA. ;
              quit ;
%mend IF_GOTO ;

%IF_GOTO(YES_NO=no,IN_DATA=sashelp.class) ;
```

The elements of the SAS macro language will be introduced in more detail in the following sections. (See 4.3.1 for SAS macro functions and 4.3.2 for SAS macro statements.)

4.2.5 Tips for the Use of Macros

Through automation, SAS macros and macro variables can dramatically increase the efficiency of working with SAS SQL. SAS macros can themselves be optimized for efficiency and performance. To that end, this section provides several recommendations. This section is not intended to be a short introduction to programming SAS macros. Refer to Burlew (2001, 1998) and SAS Institute (2016e, 1994).

Use SAS (SQL) Macro Variables

SAS macro variables such as SYSDATE9 and SYSDAY and SAS SQL macro variables such as SQLOBS and SQLOOPS provide you with information that you can use immediately and conveniently. Check the SAS Online Help to see what other macro variables are offered by SAS.

Use SAS Macro Programs

Check your work to see whether it can be simplified by macro programs. Repeated, identical retrieval, or analysis sequences are ideal candidates for macro programming. You save yourself the time for programming and the system the time for execution. SAS macros are processed faster than normal SAS code. This is because, unlike repetitive SAS programs, a macro is compiled only once during a job, regardless of how often it is called.

Use SAS Features to Program and Execute SAS Macros

You can use the Stored Compiled Macro Facility for production jobs, reducing execution time by accessing previously compiled SAS macros. However, using the Stored Compiled Macro Facility during application development is not considered efficient. The SAS Autocall Facility might also be of interest to users. With the SAS Autocall Facility, SAS macros once written and validated can be stored and used in any other application. By centrally storing numerous autocall macros in a single directory, search time can be reduced and efficiency can be increased. With PROC SQL you can also retrieve information of last used macros.

Program 4.2-25: SAS Dictionaries: Retrieve Information of Last Used Macros

```
proc sql;
create table MYDATA as
   select libname, memname, memtype, objname, objtype, objdesc,
         created, modified, alias, level from dictionary.catalogs
       where memname in ('SASMACR') ;
quit ;
proc print ;
run ;
```

SAS Won't Leave You Alone: Let SAS Help You

The SAS system options MLOGIC, MLOGICNEST, MPRINT, MPRINTNEST, and SYMBOLS support troubleshooting and problem solving when programming SAS macros. Especially for SAS macro variables, various options of the %PUT statement are of interest including _ALL_, _AUTOMATIC_, _GLOBAL_, _LOCAL_ or _USER_. The DATA step Debugger can also be helpful. In checked or productive macros, however, this option should be switched off (for example, using NOMLOGIC, NOMPRINT, and NOSYMBOLGEN), because otherwise they delay the execution unnecessarily. Use the SAS Online Help.

Check and Document Your Macros: Help Others

If you can assume that your macros will be used by colleagues or successors, check whether you can easily change your macro from a special macro to a macro for general use. Check your macros with test cases that contain ideal data that should not lead to errors as well as incorrect data that triggers errors. Document your macros. Explain what your macro does, and also point out clearly what it does not do. Remember that many users are irresponsible enough to just read through the macro calls. Also point out possible risks there.

Gain Routine in Programming the Basics of SAS Macros

A MACRO statement starts with the definition of the macro, assigns the macro name and, if necessary, further parameters and options. A macro is concluded with MEND. Do not use reserved terms or the prefixes AF, DMS, SQL, and SYS for macro names and variables. A macro cannot contain the statements CARDS, DATALINES, or PARMCARDS. In SAS code, a macro call must always be preceded by the definition of a macro.

Differentiate between Local and Global Macro Variables

SAS offers two types of macro variables: global and local. Global macro variables can be used at any point in a SAS session and can also be processed in DATA steps. Local macro variables can only be used within a macro or its further nesting. Global and local macro variables can be created in different ways using SAS (for example, using %GLOBAL or %LOCAL, but also %LET). If a variable is defined outside a macro using **%LET**=, it is a global macro variable; if it is defined inside, it is a local macro variable.

Learn to Assess the Advantages and Disadvantages of SAS Statements

For example: local macro variables. Local macro variables can only be used within a macro or its further nesting. However, they can be created with the same name as already existing variables, which protects them from being accidentally overwritten.

For example: %INCLUDE. The advantage of %INCLUDE is that it does not need to compile the code beforehand, which means less execution time. The disadvantage of %INCLUDE is that for proper execution, the exact physical storage location must be known and correctly passed to the SAS program, which means more complex programming and validation.

Create SAS Macros in Name Style

Of the styles to call a macro, the name style is considered the most efficient. The reason is that only here the macro name starts with a "%". This directly causes the word scanner to pass the token to the macro processor. With the other two styles (Command Style, Statement Style), the word scanner must first check this, which takes time unnecessarily.

Do Not Nest Definitions of SAS Macros

Nesting macros within other macros is considered unnecessary and inefficient. Instead, macros should be defined separately, so that only the call to the macro within another macro is nested, but not its definition.

Experience has shown that defining macros separately has the added advantage of making them easier to understand and validate. In the following macro MYREPORT, for example, only the call for the macro TITLES for the headings is nested, but not its definition. The definition takes place outside, even before MYREPORT.

Program 4.2-26: Define SAS Macros Separately (Do Not Nest): Example
```
%macro TITLES ;
title1 "Management Summary for &REGION" ;
   %if &STORES. <= 5 and &STORES. > 0 %then
      %do ;
        title2 "- Only small number of stores (N=&STORES.)-";
      %end ;
%mend TITLES ;

%macro MYREPORT(REGION, STORES);
   proc means data=SASHELP.SHOES ;
      where REGION="&REGION" and STORES > &STORES. ;
   class PRODUCT ;
        var SALES ;
      %TITLES
   run ;
%mend MYREPORT ;

%MYREPORT(Africa, 10) ;
```

The only nested "macro definitions" allowed are comments of larger passages of SAS macro code to disable them.

More Tips

Set macro variables that are no longer needed to zero, for example, using *%let macro variable=*. Otherwise, the previously stored values might be output again the next time the program is called.

Use multiple ampersands when working with very long strings. SAS resolves multiple ampersands successively, for example, from && to && to &. This causes SAS to re-scan macro variables with very long values.

Finally, take your time with the extensive SAS documentation. SAS can do a lot, and much more than you might first assume. Take the time to find the answer to your specific question in this vast sea of information.

4.3 Elements of the SAS Macro Language

In addition to the macro variables (see Section 4.1), SAS provides numerous SAS macro functions and SAS macro statements for the Macro Facility. The elements of the SAS macro language include:

- SAS Macro **Functions** (see Section 4.3.1)
- SAS Macro **Statements** (see Section 4.3.2)
- **Interfaces** between the SAS Macro Facility and PROC SQL and the DATA step, especially SAS Macro CALL Routines (see Sections 4.3.3 and 4.3.4)
- **Automatic** SAS macro variables (see Sections 4.1.1 and 4.1.2)

4.3.1 SAS Macro Functions

A SAS macro function is similar to a SAS function. A SAS macro function processes an argument and outputs a result. One difference is that a macro function starts with a "%". Unlike SAS macro functions, values returned by SAS functions might be truncated. Please refer to the current SAS documentation for details (SAS Institute, 2016e).

SAS macro functions include the handling of characters/strings, the evaluation of logical and numerical expressions, and quoting (masking). Macro functions for handling characters or strings modify character strings or return information about them. With %SUBSTR, for example, it is possible to extract a string from character strings.

Since the SAS macro processor is text-based, numbers are interpreted as letters by default. Evaluation functions make it possible to evaluate arithmetic and logical expressions on text as numbers. Evaluation functions (%EVAL) perform several operations. They convert the operands in the argument from text format into numerical values, perform the required arithmetic or logical (Boolean) operation on the numerical values, and convert the result back into one or more letters.

As elements of the SAS macro language, certain special characters (for example, %, &, ; or "") have a different meaning for the macro processor than for "normal" SAS. Macro functions for quoting mask special characters and mnemonic operators so that the macro processor interprets them as normal text and not as elements of the SAS macro language (for example, %QUOTE). SAS does not regard %QSCAN, %QSUBSTR, and %QUPCASE as belonging to the quoting functions because they not only mask a character, but also process it further.

Of the other SAS macro functions, %SYSFUNC is particularly noteworthy. %SYSFUNC allows the execution of normal SAS functions (for example, ATTRN, DATE, EXIST, PUTN, TRANSLATE). For the few exceptions, please refer to the current SAS documentation (for example, INPUT, LAG, MISSING, PUT, RESOLVE, SYMGET). %SYSFUNC is not intended for SAS macro functions. All SAS macro functions can be used when defining a macro as well as in open SAS code. (See the following overview in Table 4.3-1.)

Table 4.3-1: Categories and Descriptions of SAS Macro Functions (Selection)

SAS Macro Function	Description
Characters/Strings	
%INDEX	Returns the position of the first character of a string.
%LENGTH	Returns the length of a string.
%SCAN, %QSCAN	Search for a word that is specified by its position in a string. %QSCAN masks special characters and mnemonic operators in its result.
%SUBSTR, %QSUBSTR	Produce a substring of a character string. %QSUBSTR masks special characters and mnemonic operators in its result.
%UPCASE, %QUPCASE	Convert values to uppercase. %QUPCASE masks special characters and mnemonic operators in its result.
Evaluation	
%EVAL	Evaluates arithmetic and logical expressions using integer arithmetic.
%SYSEVALF	Evaluates arithmetic and logical expressions using floating-point arithmetic.
Quoting/Masking	
%BQUOTE, %NRBQUOTE	Mask special characters and mnemonic operators in a resolved value at macro execution. %BQUOTE and %NRBQUOTE do not require to mark quotation marks.
%QUOTE, %NRQUOTE	Mask special characters and mnemonic operators in a resolved value at macro execution. %QUOTE and %NRQUOTE require to mark quotation marks without a match with a preceding %.
%STR, %NRSTR	Mask special characters and mnemonic operators in constant text at macro compilation. %STR and %NRSTR require to mark quotation marks and brackets without a match with a preceding %.
%SUPERQ	Masks all special characters and mnemonic operators at macro execution but prevents further resolution of the value.
%UNQUOTE	Unmasks all special characters and mnemonic operators for a value during macro execution.

(Continued)

Table 4.3-1: (*Continued*)

SAS Macro Function	Description
Other Functions	
%SYMEXIST	Returns an indication of the existence of a macro variable.
%SYMGLOBL	Returns an indication as to whether a macro variable is global in scope.
%SYMLOCAL	Returns an indication as to whether a macro variable is local in scope.
%SYSFUNC, %QSYSFUNC	Execute SAS functions or user-written functions.
%SYSGET	Returns the value of the specified operating-environment variable.
%SYSPROD	Reports whether a SAS software product is licensed at the site.

Note: For the quoting/masking functions, NR means "Not Resolved" (% and & are also masked). "B", on the other hand, stands for "By itself" (for characters usually occurring in pairs, such as quotation marks or brackets).

Digression

In addition to the SAS macro functions, SAS also provides autocall macros. For reasons of simplification, useful autocall macros are also presented in the introduction to the SAS macro functions (with a corresponding reference). Similar syntax, programming rules, and functionality lead to the fact that autocall macros are often mistaken for SAS macro functions. SAS autocall macros include: %CMPRES, %DATATYP, %LEFT, %QCMPRES, %QLEFT, %QTRIM, and %TRIM, and %VERIFY. Autocall macros are provided in the SASMACR directory.

Examples of SAS Macro Functions in Alphabetical Order

With the exception of the autocall macros after %EVAL and %SYSEVALF, selected SAS macro functions are presented in alphabetical order.

%EVAL and %SYSEVALF Functions

The SAS macro functions evaluate arithmetic and logical expressions using integer arithmetic (%EVAL) or floating-point arithmetic (%SYSEVALF).

Program 4.3-1: Examples for %EVAL/%SYSEVALF Functions

```
* Inputs * ;
%let VAR_A=12 ;
%let VAR_B=34 ;
%let VAR_C=12.3 ;
%let VAR_D=34.5 ;

* Right and wrong operations * ;

/* %EVAL */
/* %EVAL for integers (right) */
%put %eval(12 + 34);
/* %EVAL for integers (wrong; see decimal places) */
%put %eval(12 / 34);
/* %EVAL for integers (right) */
%put %eval(&VAR_A. + &VAR_B.);
/* %EVAL for integers (wrong; see decimal places) */
%put %eval(12.3 / 34.5);
/* %SYSEVALF */
/* %SYSEVALF for floating point values (right) */
%put %sysevalf(12.3 / 34.5);
/* %SYSEVALF for floating point values (right) */
%put %sysevalf(&VAR_C. / &VAR_D.);
```

```
/* %SYSEVALF for floating point values (right) */
%put %sysevalf(&VAR_A. / &VAR_B.);
```

Figure 4.3-1: %EVAL/%SYSEVALF Output in SAS Log

```
* Right and wrong operations * ;

/* %EVAL */
/* %EVAL for integers (right) */
%put %eval(12 + 34);
46
/* %EVAL for Integers (wrong; see decimal places) */
%put %eval(12 / 34);
0
/* %EVAL for integers (right) */
%put %eval(&VAR_A. + &VAR_B.);
46
/* %EVAL for Integers (wrong; see decimal places) */
%put %eval(12.3 / 34.5);
ERROR: A character operand was found in the %EVAL function or %IF condition where a numeric
operand is required. The condition was: 12.3 / 34.5
/* %SYSEVALF */
/* %SYSEVALF for floating point values (right) */
%put %sysevalf(12.3 / 34.5);
0.35652173913043
/* %SYSEVALF for floating point values (right) */
%put %sysevalf(&VAR_C. / &VAR_D.);
0.35652173913043
/* %SYSEVALF for floating point values (right) */
%put %sysevalf(&VAR_A. / &VAR_B.);
0.35294117647058
```

Tip: If SAS returns an error in connection with %EVAL although this function does not occur in the program, it is recommended to check the program for expressions with %DO %TO %BY, %DO %UNTIL, %DO %WHILE, or %IF %THEN.

%CMPRES and %QCMPRES Autocall Macros

%CMPRES and %QCMPRES are often used together with %EVAL/%SYSEVALF. %CMPRES and %QCMPRES remove multiple blanks as well as leading and trailing blanks. For example, %CMPRES returns a nonquoted result even if the argument itself is quoted; arguments are resolved and processed as elements of the macro language. %QCMPRES, on the other hand, returns the quoted argument. Arguments are not processed as elements of the macro language but as text. As mentioned above, %CMPRES and %QCMPRES look like SAS macro functions but are actually autocall macros.

Program 4.3-2: Examples for %CMPRES/%QCMPRES Functions

```
%let X=%nrstr(%eval(&VAR_A. + &VAR_B.));
%put Result for CMPRES ('Result'): %cmpres(&X.);
%put Result for QCMPRES ('Expression'): %qcmpres(&X.);
```

Figure 4.3-2: %CMPRES/%QCMPRES Output in SAS Log

```
Result for CMPRES ('Result'): 46
Result for QCMPRES ('Expression'): %eval(&VAR_A. + &VAR_B.)
```

%INDEX Function

The %INDEX macro function returns the position of the first character of a string. The TMPERM macro checks whether a dot (character) exists in a name of a SAS data set (argument). If there is a dot in a SAS data set name (in the example: "SASHELP.CLASS"), the macro concludes that this is a permanent data set and returns a corresponding response and the position of the character in the SAS data set name. The logic of the example assumes that the name for a temporary SAS record is *not* preceded by WORK.

Program 4.3-3: Example for %INDEX Function

```
%macro TMPERM (IN_DATA) ;
        %let POS =%index(&IN_DATA.,.);
        %if %index(&IN_DATA.,.) > 0
        %then %put The dataset &IN_DATA. is permanent
                (there is a dot at position &POS.). ;
        %else %put The dataset &IN_DATA. is temporary
                (there is not a dot). ;
%mend TMPERM ;

%TMPERM (SASHELP.CLASS) ;
```

Figure 4.3-3: %INDEX Output in SAS Log

```
The dataset SASHELP.CLASS is permanent
(there is a dot at position 8).
```

%SCAN and %QSCAN Functions

%SCAN and %QSCAN macro functions search for a word that is specified by its position in a string. %QSCAN masks special characters and mnemonic operators in its result.

Example 1: %SCAN Function

The macro variable NO_VAR passes the value 2 as parameter to the SAS macro function %SCAN and also as suffix to the created SAS file DATEN_2. The macro variable VAR_LIST passes the words "age", "height", and "weight" as arguments to the SAS macro function %SCAN. As a result, %SCAN selects the second word from the list of strings and stores it in the temporary SAS file DATA_2.

Program 4.3-4: Example 1: %SCAN Function

```
* Inputs * ;
%let VAR_LIST= age height weight ;
%let NO_VAR= 2 ;

proc sql ;
title "SELECT for the &no_var.. variable of the list";
create table DATEN_&NO_VAR.
as select %scan(&VAR_LIST., &NO_VAR.)
from SASHELP.CLASS ;
quit ;
proc print noobs ;
run ;
```

Figure 4.3-4: %SCAN Output in SAS Log

```
SELECT for the 2. variable of the list

Height
 69.0
 56.5
 65.3
(Output abbreviated)
```

Example 2: Differences between %SCAN and %QSCAN

Using the %SCAN and the %QSCAN macro function, the same %NRSTR expression is searched for a word at the 3rd position. However, the results of %SCAN and %QSCAN are different. The %NRSTR macro function, which is also used, is explained in the following section.

Program 4.3-5: Example 2: Differences between %SCAN and %QSCAN

```
* Inputs * ;
%let w=wwww ; %let x=xxxx ; %let y=yyyy ; %let z=zzzz ;
* %NRSTR expression * ;
%let TOTAL=%nrstr(&w*&x*&y*&z);

* Outputs (without, %SCAN, %QSCAN) * ;
%put Output of expression: &TOTAL. ;
%put Result of scan using SCAN: %scan(&TOTAL.,3,*);
%put Result of scan using QSCAN: %qscan(&TOTAL.,3,*);
```

Figure 4.3-5: %SCAN/%QSCAN Output in SAS Log

```
Output of expression: &w*&x*&y*&z
Result of scan using SCAN: yyyy
Result of scan using QSCAN: &y
```

The SAS macro function %SCAN returns the value "yyyy" of the third macro variable. In contrast, the SAS macro function %QSCAN returns the *name* of the third macro variable "&y".

%STR and %NRSTR Functions

%STR and %NRSTR macro functions mask special characters and mnemonic operators in constant text while compiling the macro. For comparison, %QUOTE and %NRQUOTE mask while executing a macro. Paired special characters such as " " and ( ) are automatically masked. Non-paired special characters such as quotation marks " , " and brackets (,) must be marked by a preceding %. In %BQUOTE, %NRBQUOTE, and %SUPERQ, non-paired symbols need not be marked. The function of %STR and %NRSTR can best be illustrated in conjunction with the %PUT statement. %PUT writes texts or values of macro variables to the SAS log.

%STR should not be used in functions or calls with a list of parameter values because the macro processor does not recognize the arguments of a function or the parameter values due to the masking of non-paired brackets. If an argument in %STR, %NRSTR, %QUOTE, or %NRQUOTE contains a % character that precedes a question mark or parenthesis, a second % character must be specified. The second % prevents the first % from masking the question mark or bracket.

Program 4.3-6: Examples for %STR

```
%put One line of text: Without special characters and so on. ;

%put %str(proc sql ; select * from SASHELP.CLASS ; quit ; ) ;

%put %str(title "Performance 100%%");

%put %str(Text with the special character 'semicolon' (;), e.g. 'run;'.) ;

%put %str(Text with the special character 'apostrophe' (%'), e.g. 'isn%'t it'.) ;
```

Program 4.3-7 Examples for %NRSTR

```
%put How the %nrstr(%nrstr) function works! ;

%put %nrstr(Text with the special character 'percentage' (%%), e.g. 100%%.) ;
```

```
%put %nrstr(footnote "Song by: 'Peter&Paul&Mary'";);

%put %nrstr(Text with the name of a macro variable, e.g. &PCT_STRG.) ;

%macro fnotes (singer=%nrstr(Peter&Paul&Mary));
footnote "Sung by: &singer.";
%mend fnotes;

%fnotes();
```

For the sake of clarity, the outputs of %STR and %NRSTR are divided into two different figures.

Figure 4.3-6: Output in SAS Log: %STR Function

```
One line of text: Without special characters and so on.
proc sql ; select * from SASHELP.CLASS ; quit ;
title "Performance 100%";
Text with the special character 'semicolon' (;), e.g. 'run;'.
Text with the special character 'apostrophe' ('), e.g. 'isn't it'.
```

The first two %STR examples contain neither paired nor non-paired special characters and are not explained further. In the third %STR example, to output the string "Performance 100%" with a percentage sign in the SAS log, the percentage sign must be marked by a preceding %. Omitting the preceding % would prevent the string from being correctly demarked. In the fourth %STR example, there is a pair of brackets and quotation marks; both are automatically masked. In the last %STR example, there are two non-paired apostrophes in addition to a pair of brackets, each of which must be explicitly marked with a preceding %.

Figure 4.3-7: Output in SAS Log: %NRSTR Function

```
How the %nrstr function works!
Text with the special character 'percentage' (%), e.g. 100%.
Text with the name of a macro variable, e.g. &PCT_STRG.
footnote "Song by: 'Peter&Paul&Mary'";
```

Figure 4.3-8: %NRSTR Output

```
                     Sung by: Peter&Paul&Mary
```

The first %NRSTR example illustrates the basic function of the %NRSTR macro function. A character string for a SAS macro function is masked and processed as text. In the second %NRSTR example, percentage signs are marked by a preceding %. In the third and fifth examples, the %NRSTR macro function masks the ampersands ("&"), preventing them from being misinterpreted as macro variables "&Paul" or "&Mary". The fourth %NRSTR example masks a macro variable, thus preventing it from being resolved, and processes it as text.

%SUBSTR Function

The SAS macro function %SUBSTR extracts a substring from a string. Up to three parameters can be specified in the brackets after %SUBSTR. The first parameter, the argument, denotes a string or a text expression. If the argument might contain certain special characters, using %QSUBSTR instead of %SUBSTR is recommended. The second parameter, the position, specifies the position of the first character in the string or text expression from which one or more characters are to be read. The value for the position must not be greater than the total number of characters in the string. The third parameter, the length, is optional and denotes the length of the string that is to be read from position. The value for the length must not be greater than the number of characters from the specified position to the end of the string.

Program 4.3-8: Example for %SUBSTR function

```
proc sql ;
title "Analysis of &SYSDATE.";
create table DATA_%substr(&SYSDATE.,6)
as select *
```

```
from SASHELP.CLASS ;
quit;
proc print ;
run ;
```

The example references the macro variable SYSDATE twice. The complete string of SYSDATE is stored in the title. (See TITLE.) However, %SUBSTR only reads a substring from the macro variable SYSDATE and uses this text part to name the temporary SAS data set to be created. The value 6 for the position means that the five characters in front of it, "28NOV", are ignored and only all values from the 6th position onwards are used to name the SAS file. This example does not use a value for the length.

Figure 4.3-9: %SUBSTR Output in SAS Log

```
Analysis of 17DEC20
      (output abbreviated)
```

SAS Log 4.3-1: Output in SAS Log

```
NOTE: Table WORK.DATA_20 created, with 19 rows and 5 columns.
      (output abbreviated)
```

%SYSFUNC Function

The %SYSFUNC SAS macro function executes functions of the SAS language or user-written functions in the SAS Macro Facility. %SYSFUNC's scope of performance is extremely powerful, so several examples are presented below.

- Example 1: %SYSFUNC + EXIST
- Example 2: %SYSFUNC + INDEX / INDEXW
- Example 3: %SYSFUNC + OPEN, ATTRN, and CLOSE
- Example 4: %SYSFUNC + PUTN

Example 1: %SYSFUNC + EXIST

Using EXIST, %SYSFUNC checks whether a certain element occurs in a SAS directory.

Program 4.3-9: Example 1: %SYSFUNC + EXIST

```
%macro FILE_EXIST(IN_DATA);
      %if %sysfunc(exist(&IN_DATA)) %then
      %do;
            proc contents data=&IN_DATA. ;
            run;
      %end;
      %else
            %put The dataset &IN_DATA. does not exist.;
%mend FILE_EXIST;
%FILE_EXIST(sashelp.claass);
```

Figure 4.3-10: %SYSFUNC+EXIST Output in SAS Log

```
The dataset sashelp.claass does not exist.
```

The macro FILE_EXIST checks whether the SAS file CLAASS exists in the SASHELP directory. If the element that you are looking for does not occur in the SAS directory, a message is output to the log.

Example 2: %SYSFUNC + INDEX or INDEXW

Using INDEX and INDEXW, %SYSFUNC searches a character expression for a specified character string. The difference between the two functions is that INDEXW only triggers a response if the list element matches exactly. However, the INDEX function also triggers a response if specified character strings only *partially* match string entries. Depending on the requirements, this can be desirable. A possible undesired effect is illustrated by the following two examples, which differ only in the use of the two functions and are otherwise identical. First, we will discuss the accidental triggering by nonmatching or truncated list elements with the INDEX function. Then we will look at an example of INDEXW function triggering by exactly matching list elements.

INDEX triggers an action if predefined strings match a character expression in whole or in part.

Program 4.3-10: Example 2: %SYSFUNC + INDEX: Triggering by Nonmatching List Elements

```
%macro M_INDEX(IN_VAR) ;
%if %sysfunc(index(Alfred Henry James
                Alice Barbara Carol, &IN_VAR.)) %then
      %do ;
      data DATA_&IN_VAR. ;
            set SASHELP.CLASS ;
                  where index(NAME, "&IN_VAR.");
            run ;
      proc print data=DATA_&IN_VAR. ;
            run ;
      %end ;
%mend M_INDEX ;


/* Calls for INDEX: Alfred / Alf */


%M_INDEX(Alfred) ;


%M_INDEX(Alf) ;
```

After these two calls, the M_INDEX macro creates a file with the name "DATA_Alfred" and "DATA_Alf", since the specified strings "Alfred" or "Alf" appear in whole or in part in the NAME column of the SAS file SASHELP.CLASS.

SAS Log 4.3-2: Output in SAS Log

```
NOTE: There were 1 observations read from the data set SASHELP.CLASS.
NOTE: There were 1 observations read from the data set WORK.DATA_ALF.
```

The *content* of "DATA_Alfred" in the NAME column is "Alfred". The content of "DATA_Alf" in the NAME column is not "Alf", but *also* "Alfred". The action to create a SAS file with a given string is also triggered if specified strings match string entries only partially ("Alf").

INDEXW triggers an action only if specified strings *exactly* match a character expression.

Program 4.3-11: Example 2: %SYSFUNC + INDEXW: Triggering by Exactly Matching List Elements

```
%macro M_INDEX_W(IN_VAR) ;
%if %sysfunc(indexw(Alfred Henry James
                Alice Barbara Carol, &IN_VAR.)) %then
      %do ;
      data DATA_&IN_VAR. ;
            set SASHELP.CLASS ;
                  where indexw(NAME, "&IN_VAR.");
            run ;
      proc print data=DATA_&IN_VAR. ;
            run ;
      %end ;
%mend M_INDEX_W;


/* Calls for INDEXW: Alfred / Alf */


%M_INDEX_W(Alfred) ;


%M_INDEX_W(Alf) ;
```

Only the file with the name "DATA_Alfred" is created because the element "Alfred" occurs exactly in the NAME column of the SASHELP.CLASS file.

SAS Log 4.3-3: Output in SAS Log

```
NOTE: There were 1 observations read from the data set WORK.DATA_ALFRED.
NOTE: There were 1 observations read from the data set WORK.DATA_ALF.
```

The file "DATA_Alf" was not created because the string "Alf" does not exactly match the entries in NAME. INDEXW triggers actions only for exactly matching list elements.

Example 3: %SYSFUNC + OPEN, ATTRN, and CLOSE

%SYSFUNC opens a SAS data set with OPEN and closes a SAS data set with CLOSE and queries the value of a numeric attribute for the specified SAS data set with ATTRN between these steps.

Program 4.3-12: Example 3: %SYSFUNC + OPEN, ATTRN and CLOSE

```
%let DSNAME=%sysfunc(open(SASHELP.CLASS)) ;
%let N_OBS=%sysfunc(attrn(&DSNAME., nobs)) ;
%let CLOSE=%sysfunc(close(&DSNAME.)) ;
title %upcase("&N_OBS. Dance Students in %sysfunc(date(), monname.) of Year
                                    %sysfunc(date(),year.)";);

proc sql ;
select n(NAME) as N,
 mean(age) as Mean_Age format=5.2,
 mean(height) as Mean_Height format=5.2
from SASHELP.CLASS;
quit ;
```

Figure 4.3-11: %SYSFUNC+OPEN, ATTRN, and CLOSE Output in SAS Output

```
19 DANCE STUDENTS IN  DECEMBER OF YEAR 2020

     N  Mean_Age  Mean_Height
    ----------------------------
    19    13.32       62.34
```

Using the three %LET statements, the number of rows from SASHELP.CLASS is queried and stored in the macro variable N_OBS and referenced in the title. (See TITLE.) Using DATE, %SYSFUNC queries the current date. The %UPCASE macro function also converts all characters in the title to uppercase.

Example 4: %SYSFUNC + PUTN

PUTN enables you to specify a numerical format in runtime. PUTN, PUTC, INPUTN or INPUTC can be used together with %SYSFUNC, instead of PUT and INPUT. In the first macro example, PUTN assigns the format DOLLARX10.2 to numeric values. In the second macro example, PUTN assigns condition-based value labels to numeric values.

In the FMT macro, %SYSFUNC with PUTN assigns the format DOLLARX10.2 to the AMOUNT values.

Program 4.3-13: Example 4: %FMT: %SYSFUNC + PUTN

```
%macro FMT(AMOUNT) ;
    %put %sysfunc(putn(&AMOUNT.,dollarx10.2)) ;
%mend FMT ;

%FMT(12345.67) ;
```

Figure 4.3-12: %SYSFUNC+PUTN Output in SAS Log

```
$12.345,67
```

In the FMTCAT macro, %SYSFUNC with PUTN assigns condition-based value labels to the MYVALUE values. If the value of MYVALUE is 0, it receives the label "exact". If a value is above or below 0, it receives the labels "too high" or "too low".

Program 4.3-14: Example 4: %FMTCAT: %SYSFUNC + PUTN

```
proc format ;
  value GROUPING
  0<-high = "Too high "
      0 = "   Exact "
  low-<0  = "Too low "
      ;
run ;

%macro FMTCAT(MYVALUE);
  %put The value &MYVALUE. is: %sysfunc(putn(&MYVALUE.,GROUPING.)) ;
%mend FMTCAT ;
%FMTCAT(123.45) ;
%FMTCAT(0)      ;
%FMTCAT(-123.45);
```

Figure 4.3-13: %FMTCAT: %SYSFUNC+PUTN Output in SAS Log

```
  %FMTCAT(123.45) ;
The value 123.45 is: Too high
  %FMTCAT(0)      ;
The value 0 is: Exact
  %FMTCAT(-123.45);
The value -123.45 is: Too low
```

Table 4.3-2 lists selected functions and arguments for the %SYSFUNC SAS macro function. For details and more SAS functions and arguments, please refer to the current SAS documentation.

Table 4.3-2: Functions and Arguments for %SYSFUNC

Name	Description (Source: *SAS® 9.4 Macro Language*, 5th ed.)
ATTRC	Returns the value of a character attribute for the specified SAS data set.
ATTRN	Returns the value of a numeric attribute for specified SAS data set.
CEXIST	Verifies the existence of a SAS catalog or SAS catalog entry.
CLOSE	Closes a SAS data set.
CUROBS	Returns the number of the current observation.
DCLOSE	Closes a directory.
DINFO	Returns specified information items for a directory.
DNUM	Returns the number of members in a directory.
DOPEN	Opens a directory.
DOPTNAME	Returns a specified directory attribute.
DOPTNUM	Returns the number of information items available for a directory.
DREAD	Returns the name of a directory member.
DROPNOTE	Deletes a note marker from a SAS data set or an external file.
DSNAME	Returns the data set name associated with a data set identifier.
EXIST	Verifies the existence of a SAS library member.
FAPPEND	Appends a record to the end of an external file.

(Continued)

Table 4.3-2: (*Continued*)

Name	Description (Source: *SAS® 9.4 Macro Language,* 5th ed.)
FCLOSE	Closes an external file, directory, or directory member.
FCOL	Returns the current column position in the File Data Buffer (FDB).
FDELETE	Deletes an external file.
FETCH	Reads the next nondeleted observation from a SAS data set into the Data Set Data Vector (DDV).
FETCHOBS	Reads a specified observation from a SAS data set into the DDV.
FEXIST	Verifies the existence of an external file associated with a fileref.
FGET	Copies data from the File Data Buffer (FDB).
FILEEXIST	Verifies the existence of an external file by its physical name.
FILENAME	Assigns or deassigns a fileref for an external file, directory, or output device.
FILEREF	Verifies that a fileref has been assigned for the current SAS session.
FINFO	Returns a specified information item for a file.
FNOTE	Identifies the last record that was read.
FOPEN	Opens an external file.
FOPTNAME	Returns the name of an information item for an external file.
FOPTNUM	Returns the number of information items available for an external file.
FPOINT	Positions the read pointer on the next record to be read.
FPOS	Sets the position of the column pointer in the FDB.
FPUT	Moves data to the FDB of an external file starting at the current column position.
FREAD	Reads a record from an external file into the FDB.
FREWIND	Positions the file pointer at the first record.
FRLEN	Returns the size of the last record read or the current record size for a file opened for output.
FSEP	Sets the token delimiters for the FGET function.
FWRITE	Writes a record to an external file.
GETOPTION	Returns the value of SAS or graphics option.
GETVARC	Assigns the value of a SAS data set variable to a character DATA step or macro variable.
GETVARN	Assigns the value of a SAS data set variable to a numeric DATA step or macro variable.
LIBNAME	Assigns or deassigns a libref for a SAS library.
LIBREF	Verifies that a libref has been assigned.
MOPEN	Opens a directory member file.
NOTE	Returns an observation ID for current observation of a SAS data set.
OPEN	Opens a SAS data file.
PATHNAME	Returns the physical name of a SAS library or an external file.
POINT	Locates an observation identified by the NOTE function.
REWIND	Positions the data set pointer to the beginning of a SAS data set.
SPEDIS	Returns a number for the operation required to change an incorrect keyword in a WHERE clause to a correct keyword.
SYSGET	Returns the value of the specified host environment variable.
SYSMSG	Returns the error or warning message produced by the last function that attempted to access a data set or external file.

(*Continued*)

Table 4.3-2: (*Continued*)

Name	Description (Source: *SAS® 9.4 Macro Language,* 5th ed.)
SYSRC	Returns the system error number or exit status of the entry most recently called.
VARFMT	Returns the format assigned to a data set variable.
VARINFMT	Returns the informat assigned to a data set variable.
VARLABEL	Returns the label assigned to a data set variable.
VARLEN	Returns the length of a data set variable.
VARNAME	Returns the name of a data set variable.
VARNUM	Returns the number of a data set variable.
VARTYPE	Returns the data type of a data set variable.

Note: SAS functions that are not available with %SYSFUNC or %QSYSFUNC: DIF, DIM, HBOUND, IORCMSG, INPUT, LAG, LBOUND, MISSING, PUT, RESOLVE, SYMGET; variable information functions such as VNAME; and SAS File I/O Functions. INPUTN, INPUTC, PUTN, or PUTC can be used with %SYSFUNC or %QSYSFUNC instead of INPUT and PUT. Values returned by SAS functions might be truncated.

%UPCASE Function

The %UPCASE macro function converts all values in an argument to uppercase.

Program 4.3-15: Example for %UPCASE function

```
%let DSNAME=%sysfunc(open(SASHELP.CLASS)) ;
%let N_OBS=%sysfunc(attrn(&DSNAME., nobs)) ;
%let CLOSE=%sysfunc(close(&DSNAME.)) ;
title %upcase("&N_OBS. Dance Students in %sysfunc(date(),
              monname.) of Year %sysfunc(date(),year.)";);

proc sql ;
select n(NAME) as N,
mean(age) as Mean_Age format=5.2,
mean(height) as Mean_Height format=5.2
from SASHELP.CLASS;
quit ;
```

Figure 4.3-14: %UPCASE Output

```
19 DANCE STUDENTS IN  DECEMBER OF YEAR 2020

N Mean_Age Mean_Height
-------------------------------
19 13.32 62.34
```

The %UPCASE macro function converts all characters except numbers in the title to uppercase. The %SYSFUNC options OPEN, ATTRN, and CLOSE have already been explained above.

4.3.2 SAS Macro Statements

The function of a SAS macro statement is comparable to that of a SAS statement. Like SAS macro functions, they always start with a %. A SAS macro statement instructs the macro processor to perform an operation. A macro statement usually consists of a series of keywords, SAS names, special characters, and operands and ends with a semicolon. Macro statements can be used to assign values (%LET), branch (%GOTO), or even execute conditional (%IF %THEN) and iterative processing (%DO with %TO, %UNTIL, or %WHILE) of SAS statements. Some SAS macro statements can only be used when defining macros (%ABORT), while others can be used when defining macros and in open SAS code (%LET). See Table 4.3-3 for an overview of macro statements.

Table 4.3-3: Categories and Descriptions of SAS Macro Statements (Selection)

SAS Macro Statements	Description (Source: *SAS® 9.4 Macro Language,* 5th ed.)
Valid in macro definitions and open SAS code	
%* Comment	Macro comment statement: designates comment text.
%COPY	Copies specified items from a SAS macro library.
%DISPLAY	Displays a macro window.
%GLOBAL	Creates macro variables that are available during the execution of an entire SAS session.
%INPUT	Supplies values to macro variables during macro execution.
%LET	Creates a macro variable and assigns it a value.
%MACRO	Begins a macro definition.
%PUT	Writes text or macro variable information to the SAS log.
%SYMDEL	Deletes the specified variable or variables from the macro global symbol table.
%SYSCALL	Invokes a SAS call routine.
%SYSEXEC	Issues operating-environment commands.
%SYSLPUT	Creates a new macro variable or modifies the value of an existing macro variable on a remote host or server.
%SYSRPUT	Assigns the value of a macro variable on a remote host to a macro variable on the local host.
%WINDOW	Defines customized windows.
Valid in macro definitions only	
%ABORT	Stops the macro that is executing along with the current DATA step, SAS job, or SAS session.
%DO	Begins a %DO group.
%DO, iterative	Executes a section of a macro repetitively based on the value of an index variable.
%DO %UNTIL	Executes a section of a macro repetitively until a condition is true.
%DO %WHILE	Executes a section of a macro repetitively while a condition is true.
%END	Ends a %DO group.
%GOTO	Branches macro processing to the specified label. (See %label.)
%IF %THEN %ELSE	Conditionally process a portion of a macro.
%LABEL	Identifies the destination of a %GOTO statement.
%LOCAL	Creates macro variables that are available only during the execution of the macro where they are defined.
%MEND	Ends a macro definition.
%RETURN	Execution causes normal termination of the currently executing macro.

Examples of SAS Macro Statements in Alphabetical Order

Using the SASHELP data set PRDSAL2, the following SAS language statements are illustrated:

- %DO %TO, %DO %WHILE, and %DO %UNTIL for iterative processing of PROC SQL statements. These statements are introduced using the same example and then supplemented by further examples if necessary.
- %IF %THEN %DO %ELSE for the condition-based execution of PROC SQL statements.

%DO %TO Function

The DO_TO macro starts executing with the specified value in &START. ("1995"). With each execution, the counter increment in &START. is increased incrementally by one unit. The DO_TO macro is repeated until the &STOP. value specified after %TO. value ("1998") is reached. More precisely, the DO_TO macro is stopped as soon as a value outside the range of &START. and &STOPP. would be reached. The values under &START. and STOPP. must be integers. The macro variable &RP_VALUE. contains the respective value; their number determines how often the loop is run through. (See example 4.)

Program 4.3-16: %DO %TO Function (%DO_TO with &START. to &STOPP.)

```
%macro DO_TO(START,STOPP) ;
%do RP_VALUE =&START. %to &STOPP. ;
title "Analysis for the year &RP_VALUE.";
proc sql ;
select COUNTRY, sum(ACTUAL) as ACTUAL_sum,
                sum(PREDICT) as PREDICT_sum
from SASHELP.PRDSAL2
where YEAR=&RP_VALUE.
group by COUNTRY ;
quit;
%end ;
%mend DO_TO ;


%DO_TO(1995, 1998) ;
```

The macro DO_TO executes the analysis from TITLE to QUIT for the years 1995, 1996, 1997, and 1998. &START. (see "1995") defines the first analysis and &STOPP. (see "1998") defines the last analysis. In addition, &START/&STOP values are transferred to the macro variable RP_VALUE. and from there to the title and the WHERE clause.

Figure 4.3-15: %DO_TO Output

```
Analysis for the year 1995              Analysis for the year 1996

Country ACTUAL_sum PREDICT_sum          Country ACTUAL_sum PREDICT_sum
-----------------------------------     -----------------------------------
Canada 712440.8 755785.6                Canada 708818.4 769110.4
Mexico 465915.2 518104                  Mexico 472192.8 528722.4
U.S.A. 2112784 2240967                  U.S.A. 2198152 2264191


Analysis for the year 1997              Analysis for the year 1998

Country ACTUAL_sum PREDICT_sum          Country ACTUAL_sum PREDICT_sum
-----------------------------------     -----------------------------------
Canada 890551 944732                    Canada 886023 961388
Mexico 582394 647630                    Mexico 590241 660903
U.S.A. 2640980 2801209                  U.S.A. 2747690 2830238
```

Note: For space reasons, SAS outputs, which normally follow each other vertically, are placed next to each other on this page.

The DO_TO2 macro creates five copies of the same data set. The only difference between the individual copies is the suffix of the file name. The macro DO_TO2 starts to execute the same analysis incrementally according to a counter. The macro DO_TO2 is repeated until the value for &RP_VALUE. (5) is reached. The macro DO_TO2 is no longer executed after a value outside the range of &i. and &RP_VALUE.

Program 4.3-17: %DO %TO Function (%DO_TO2, 1 to &i.)

```
%macro DO_TO2(RP_VALUE) ;
 %do i=1 %to &RP_VALUE. ;
     proc sql ;
     create table TABLE_&i. as
     select COUNTRY, sum(ACTUAL) as ACTUAL_sum,
                     sum(PREDICT) as PREDICT_sum
   from SASHELP.PRDSAL2   ;
```

```
       quit;
   %end ;
%mend DO_TO2 ;


%DO_TO2(5) ;
```

SAS Log 4.3-4: Output in SAS Log

```
NOTE: Table WORK.TABLE_1 created, with 23040 rows and 3 columns.
NOTE: Table WORK.TABLE_2 created, with 23040 rows and 3 columns.
NOTE: Table WORK.TABLE_3 created, with 23040 rows and 3 columns.
NOTE: Table WORK.TABLE_4 created, with 23040 rows and 3 columns.
NOTE: Table WORK.TABLE_5 created, with 23040 rows and 3 columns.
     (abbreviated)
```

It is also possible to nest several %DO %TO loops. In the DO_TO3 macro, all entries of the QUARTER loop are first processed (1, 2, and so on) for the first value of the YEAR loop (1995), then the second value of the YEAR loop, and so on. The DO_TO3 macro runs through a total of 16 loops, creating 16 copies of the same data set. The only difference between the individual copies is the suffix of the file name.

Program 4.3-18: Nested %DO %TO Function (%DO_TO3)

```
%macro DO_TO3 ;
 %do YEAR=1995 %to 1998 ;
       %do QUARTER=1 %to 4 ;
       proc sql ;
       create table TABLE_&YEAR._&QUARTER. as
       select COUNTRY, sum(ACTUAL) as ACTUAL_sum,
                       sum(PREDICT) as PREDICT_sum
       from SASHELP.PRDSAL2    ;
       quit;
        %end ;
 %end ;
%mend DO_TO3 ;

%DO_TO3 ;
```

No further parameters need to be passed to the DO_TO3 macro in the call.

SAS Log 4.3-5: Output in SAS Log

```
NOTE: Table WORK.TABLE_1998_1 created, with 23040 rows and 3 columns.
NOTE: Table WORK.TABLE_1998_2 created, with 23040 rows and 3 columns.
NOTE: Table WORK.TABLE_1998_3 created, with 23040 rows and 3 columns.
     (abbreviated)
```

%DO %WHILE Function

The DO_WHILE macro starts executing with the specified value in RP_VALUE (1995). With each execution, the counter (increment) is increased by one unit. The DO_WHILE macro is repeated as long as the specified &STOPP. value (1997) is less than or equal to the incrementing &RP_VALUE.

Program 4.3-19: %DO %WHILE Function (%DO_WHILE with &STOPP.)

```
%macro DO_WHILE(STOPP) ;
%let RP_VALUE=1995 ;
%do %while (&RP_VALUE. <= &STOPP.) ;
title "Analysis for the year &RP_VALUE.";
       proc sql ;
               select COUNTRY, sum(ACTUAL) as ACTUAL_sum,
                               sum(PREDICT) as PREDICT_sum
```

```
                from SASHELP.PRDSAL2
                where YEAR=&RP_VALUE.
                group by COUNTRY ;
        quit;
%let RP_VALUE=%eval(&RP_VALUE+1);
%end ;
%mend DO_WHILE ;


%DO_WHILE(1997) ;
```

The macro DO_WHILE executes the same analysis for the years 1995, 1996, and 1997. &RP_VALUE. (see "1995") defines the first analysis, and &STOPP. (see "1997") defines the last analysis. Entries in &RP_VALUE. are passed to the title and the WHERE clause.

Figure 4.3-16: %DO_WHILE Output

```
Analysis for the year 1995
Analysis for the year 1996
Analysis for the year 1997
     (abbreviated)
```

The macro DO_WHILE2 is a variant for strings. The DO_WHILE2 macro starts with an execution for the first predefined value (Canada) and then moves on to the next entry (Mexico). If there were four entries in the DO_WHILE2 call, they would be processed in the specified order. If "Mexico" was specified first in the call to DO_WHILE2, then the analysis would be executed first for "Mexico" and only then for "Canada" and so on.

Program 4.3-20: %DO %WHILE Function (%DO_WHILE2 with &i.)

```
%macro DO_WHILE2(IN_COUNTRY) ;
%let i=1 ;
%do %while (%scan(&IN_COUNTRY.,&i.) ne) ;
%let LAND=%scan(&IN_COUNTRY.,&i.) ;
        proc sql ;
                create table SUMMARY_&LAND. as
                        select sum(ACTUAL) as ACTUAL_sum,
                        sum(PREDICT) as PREDICT_sum
                from SASHELP.PRDSAL2
                where COUNTRY="&LAND.";
        quit;
%let i=%eval(&i.+1) ;
%end ;
%mend DO_WHILE2 ;


%DO_WHILE2(Canada Mexico) ;
```

SAS Log 4.3-6: Output in SAS Log

```
NOTE: Table WORK.SUMMARY_CANADA created, with 1 rows and 2 columns.
NOTE: Table WORK.SUMMARY_MEXICO created, with 1 rows and 2 columns.
     (abbreviated)
```

%DO %UNTIL Function

The DO_UNTIL macro starts executing with the specified value in RP_VALUE (1995). With each execution, the counter is increased incrementally by one unit. The DO_UNTIL macro is therefore repeated until the specified &STOPP. value (1996) is equal to the incrementing &RP_VALUE.

Program 4.3-21: %DO %UNTIL Function (%DO_UNTIL with &STOPP.)

```
%macro DO_UNTIL(STOPP) ;
%let RP_VALUE=1995 ;
%do %until (&RP_VALUE. = &STOPP.+1) ;
```

```
title "Analysis for the year &RP_VALUE.";
      proc sql ;
            select COUNTRY, sum(ACTUAL) as ACTUAL_sum,
                              sum(PREDICT) as PREDICT_sum
            from SASHELP.PRDSAL2
            where YEAR=&RP_VALUE.
            group by COUNTRY ;
      quit;
%let RP_VALUE=%eval(&RP_VALUE+1);
%end ;
%mend DO_UNTIL ;

%DO_UNTIL(1996) ;
```

The macro DO_UNTIL executes the same analysis for the years 1995 and 1996. &STOP. (see "1996") defines the last analysis. &STOPP. values are also transferred to the macro variable RP_VALUE and from there passed to the title and WHERE clause. The correction by "+1" in %DO %UNTIL leads to the condition that the repeated execution of the macro stops at "STOP" *equal* to "1996". If this correction were omitted, DO_UNTIL would stop one unit earlier at "1995".

Figure 4.3-17: %DO_UNTIL Output in SAS Output

```
Analysis for the year 1995
Analysis for the year 1996
      (abbreviated)
```

%IF %THEN %DO %ELSE Function

The main difference between the %IF %THEN %DO %ELSE statement as an element of the SAS macro language and the IF THEN ELSE statement as an element of the Base SAS language is the following: the %IF %THEN %DO %ELSE macro statement generates condition-based predefined SAS code and executes it. The IF THEN ELSE statement just executes SAS statements condition-based during the DATA step. The macro IF_THEN2 illustrates how the %IF %THEN %DO %ELSE statement generates and executes condition-based predefined SAS code. For space reasons, the SAS output will not be shown and instead only the relevant output will be presented and explained in the SAS log.

In principle, there are two functionalities in the macro IF_THEN2. If a valid entry occurs under &IN_VAR., an SQL analysis grouped by &IN_VAR. is triggered. The operator "NE [Blank]" means "as long as a parameter value is not missing" or as long as a parameter value exists. If there is no or no valid entry under &IN_VAR., a non-grouped SQL analysis is triggered. The operator "EQ [Blank]" means "as soon as a parameter value is missing".

Program 4.3-22: %IF %THEN %DO %ELSE Function (%IF_THEN2)

```
%macro IF_THEN2(IN_VAR) ;
proc sql ;
      create table SUMMARY_&IN_VAR. as
      select sum(ACTUAL) as ACTUAL_sum,
      sum(PREDICT) as PREDICT_sum
%* If IN_VAR is not missing in the call: *;
%if &IN_VAR. ne %then
    %do ;
    , &IN_VAR.
      from SASHELP.PRDSAL2
      group by &IN_VAR. ;
    %end ;
%* If IN_VAR is missing in the call: *;
%else %if &IN_VAR. eq %then
    %do ;
    from SASHELP.PRDSAL2 ;
    %end ;
      quit;
%mend IF_THEN2 ;
```

```
%IF_THEN2(country) ;
%IF_THEN2() ;
%IF_THEN2(year) ;
```

If the value of the macro variable IN_VAR is COUNTRY, the IF_THEN2 macro generates and executes the following predefined SAS code.

Program 4.3-23: %IF_THEN2 with COUNTRY in IN_VAR

```
proc sql ;
create table SUMMARY_country as select sum(ACTUAL) as ACTUAL_sum,
sum(PREDICT) as PREDICT_sum , country from SASHELP.PRDSAL2 group by country ;
quit;
```

If the value of the macro variable IN_VAR is a blank, the IF_THEN2 macro generates the following predefined SAS code.

Program 4.3-24: %IF_THEN2 with Blank in IN_VAR

```
proc sql ;
create table SUMMARY_ as select sum(ACTUAL) as ACTUAL_sum, sum(PREDICT) as
PREDICT_sum from SASHELP.PRDSAL2 ;
quit;
```

If the value of the macro variable IN_VAR is YEAR, the IF_THEN2 macro generates the following SAS code.

Program 4.3-25: %IF_THEN2 with YEAR in IN_VAR

```
proc sql ;
create table SUMMARY_year as select sum(ACTUAL) as ACTUAL_sum, sum(PREDICT)
as PREDICT_sum , year from SASHELP.PRDSAL2 group by year ;
quit;
```

Users will find further examples for %IF %THEN %DO in Sections 4.2.3 and 4.2.4 on listwise and condition-based execution of commands.

Tip: The expressions %DO %TO %BY, %DO %UNTIL, %DO %WHILE, and %IF %THEN perform an evaluation by automatically calling the %EVAL function. If SAS returns an error in connection with %EVAL, although no %EVAL occurs in the program, it is recommended to check the program for expressions such as %DO %TO %BY, %DO %UNTIL, %DO %WHILE, or %IF %THEN.

4.3.3 Interfaces I: From the SAS Macro Facility to the DATA Step

SAS offers eight interfaces for the interaction of the SAS Macro Facility with PROC SQL and the DATA step. The execution of the SAS Macro Facility happens before PROC SQL or a DATA step. This means that at the moment of executing PROC SQL or a DATA step, all information is already available from the macro processing. The interfaces between the SAS Macro Facility and PROC SQL or a DATA step include SAS macro CALL routines, SAS macro functions, and INTO. This section covers the SAS macro CALL routines and their function as an interface from the SAS Macro Facility to PROC SQL or a DATA step (for INTO, see the next section). The SAS macro CALL routines allow:

- **transferring** information from a macro variable to PROC SQL or a DATA step (see CALL SYMPUT or SYMPUTX)
- **executing** SAS macros depending on values from PROC SQL or a DATA step (see CALL EXECUTE)
- **resolving** (RESOLVE) or deleting (CALL SYMDEL) macro variables
- **querying** information about whether a specified variable exists or is global or local (SYMEXIST, SYMGLOBL, SYMLOCAL)

Table 4.3-4: Categories and Descriptions of DATA Step Interfaces (Selection)

SAS Macro Function	Description
SAS Macro CALL Routine	
CALL EXECUTE	*Execution:* Resolves the argument and executes the resolved value immediately (if the value is an element of the SAS macro language) or only at the next step (if the value is a SAS statement).
CALL SYMDEL	*Delete:* Deletes the specified macro variable in the argument.
CALL SYMPUT, CALL SYMPUTX	*Read or write:* Passes a value from a DATA step to a macro variable.
SAS Macro Function	
RESOLVE	*Resolution:* Resolves the value of a text expression during the execution of a DATA step.
SYMEXIST	*Information:* Returns an indication about whether the specified variable exists.
SYMGET	*Read or write:* Returns the value of a macro variable during the execution of a DATA step.
SYMGLOBL	*Information:* Returns an indication about whether the specified variable is global.
SYMLOCAL	*Information:* Returns an indication about whether the specified variable is local.

Examples of SAS Macro CALL Routines in Alphabetical Order

The following SAS macro CALL routines are illustrated using the SASHELP data set SHOES:

- CALL EXECUTE
- CALL SYMPUTX
- CALL SYMPUT

The SAS macro function SYMGET is explained using the SAS file SASHELP.CLASS.

CALL EXECUTE

The two examples illustrate how CALL EXECUTE resolves an argument and executes the resolved value. For both examples, it is necessary to create a scaled down, temporary subset SHOES from the SAS file SASHELP.SHOES in a preparatory step. A further example of CALL EXECUTE can be found in the chapter on condition-based executing of macros. (See macro TRIG_MAC.)

Program 4.3-26: Preparing a Temporary Subset SHOES

```
proc sql ;
create table SHOES as select * from SASHELP.SHOES
where REGION in ("Asia", "Africa", "Canada", "Pacific")
and PRODUCT in ("Men's Casual", "Women's Casual") ;
quit ;
```

Example 1: CALL EXECUTE without IF Conditions

Apart from the above preparatory step, this example consists of three steps:

1. A temporary partial data set REG_SALES is created from the SASHELP file SHOES.
2. A PCTABLE macro is created (but not yet executed).
3. CALL EXECUTE calls the PCTABLE macro and uses PROC TABULATE and the temporary table *SHOES* referenced therein to create a result table for each existing region in the *REG_SALES* file. In other words, each valid REGION level in REG_SALES triggers the execution of the PCTABLE macro and thus the processing of the SHOES table.

This example illustrates two further aspects in addition to the CALL EXECUTE function: (a) by calling macros, standardized tables can be created automatically and (b) separate data sets, for example, SHOES and REG_SALES, can be related to each other without any complications.

Program 4.3-27: Example 1: CALL EXECUTE without IF Conditions

```
proc sql ;
create table REG_SALES as
select unique REGION as REGION_N
from SASHELP.SHOES ;
quit ;

%macro PCTABLE (REGION_N) ;
proc tabulate data = SHOES ;
where REGION="&REGION_N.";
title "Table for Region &REGION_N." ;
class SUBSIDIARY PRODUCT ;
var SALES ;
tables SUBSIDIARY="" All="Total",
     (PRODUCT="Product Line"
      All="Total")*SALES=""*(sum="Sum")
      /rts=18 box=[label="Sales in the region -&REGION_N.- and
                          the respective subsidiaries:"];
run ;
%mend PCTABLE ;

data _null_ ;
set REG_SALES ;
call execute('%PCTABLE('||REGION_N||')') ;
run ;
```

In the CALL EXECUTE example in Program 4.3-27, the TABULATE tables for the regions "Africa", "Asia", and so on are automatically generated for each existing level in the REGION field without having to specify a single IF condition or value list.

SAS Log 4.3-7: Output in SAS Log

```
NOTE: There were 9 observations read from the data set WORK.SHOES.
      WHERE REGION='Africa';
NOTE: There were 3 observations read from the data set WORK.SHOES.
      WHERE REGION='Asia';
NOTE: There were 8 observations read from the data set WORK.SHOES.
      WHERE REGION='Canada';
NOTE: There were 10 observations read from the data set WORK.SHOES.
      WHERE REGION='Pacific';
Log output abbreviated.
```

Figure 4.3-18: CALL EXECUTE without IF Output in SAS Output (Sample)

Table for Region Asia

Sales in the region -Asia- and the respective subsidiaries:	Product Line		Total
	Men's Casual	Women's Casual	
	Sum	Sum	Sum
Bangkok	.	5389.00	5389.00
Seoul	11754.00	20448.00	32202.00
Total	11754.00	25837.00	37591.00

Example 2: CALL EXECUTE with IF Conditions

The second example differs from the first example only in that additional IF conditions are formulated in the third step. In the first step, a temporary partial data set REG_SALES2 is created from the temporary SHOES. In the second step, a macro PCTABLE2 is created but not yet executed. In the third step, CALL EXECUTE calls the macro PCTABLE2 if the IF conditions are met and triggers the standardized creation of tables using PROC TABULATE for each existing region in the REG_SALES2 file.

Program 4.3-28: Example 2: CALL EXECUTE with IF Conditions

```
proc sql ;
create table REG_SALES2 as
select REGION, SUBSIDIARY, PRODUCT,
sum(SALES) as SALE_SUM
from SHOES group by REGION, SUBSIDIARY, PRODUCT ;
quit;


%macro PCTABLE2 (REGION) ;
proc tabulate data = SHOES ;
where REGION="&REGION." ;
class SUBSIDIARY PRODUCT ;
var SALES ;
tables SUBSIDIARY="" All="Total",
       (PRODUCT="Product Line"
        All="Total")*SALES=""*(sum="Sum")
       /rts=18 box=[label="Sales in the region -&REGION.- and
                          the respective subsidiaries:"];
run ;
%mend PCTABLE2 ;


data _null_ ;
set REG_SALES2 ;
if PRODUCT in ("Men's Casual") and SALE_SUM > 200000
then call execute('%PCTABLE2('||REGION||')');
run ;
```

SAS Log 4.3-8: Output in SAS Log (Selection)

```
NOTE: There were 9 observations read from the data set WORK.SHOES.
      WHERE REGION='Africa';
NOTE: There were 8 observations read from the data set WORK.SHOES.
      WHERE REGION='Canada';
NOTE: There were 10 observations read from the data set WORK.SHOES.
      WHERE REGION='Pacific';
```

Figure 4.3-19: CALL EXECUTE with IF Output in SAS Output (Sample)

Sales in the region -Africa- and the respective subsidiaries:	Product Line Men's Casual Sum	Women's Casual Sum	Total Sum
Addis Ababa	67242.00	51541.00	118783.00
Algiers	63206.00	.	63206.00
Cairo	360209.00	328474.00	688683.00
Khartoum	9244.00	19582.00	28826.00
Kinshasa	.	17919.00	17919.00
Luanda	62893.00	.	62893.00
Total	562794.00	417516.00	980310.00

The table for REGION="Asia" is not created because the IF condition for SALE_SUM is not met. Interested users can find further information about the construction of multidimensional tables using PROC TABULATE in Schendera (2004, Chapter 11.4.5).

CALL SYMPUTX and SYMPUT

CALL SYMPUTX and CALL SYMPUT pass a value from a DATA step to a macro variable. The difference between CALL SYMPUTX and CALL SYMPUT is that CALL SYMPUTX also automatically removes leading and trailing blanks. The following example is based on the calculation of location and dispersion measures. PROC SQL does not contain an analysis function for calculating the median; however, a median can be obtained easily using the SAS function MEDIAN or a routine. In the following example, the median is obtained with the routine SYMPUTX and stored in a macro variable. CALL SYMPUTX has the same functionality as CALL SYMPUT with the additional appealing advantage that this routine automatically removes leading and trailing blanks. In the following example, PROC MEANS is used to obtain the median for the SALES column of the SAS file SASHELP.SHOES, CALL SYMPUTX passes it to a macro variable, and SQL includes it in a WHERE clause finally.

Program 4.3-29: Examples for CALL SYMPUTX

```
proc means data = SASHELP.SHOES noprint ;
var SALES ;
output out = MED_DATA (drop=_TYPE_ _FREQ_)
      median(SALES) = SALES_MED  ;
run ;


data MED_DATA ;
set MED_DATA ;
      call symputx ("MEDIAN", SALES_MED) ;
run ;


proc sql ;
create table MYDATA as
select *
from SASHELP.SHOES
      where SALES ge 0 and SALES le &MEDIAN. ;
quit;
```

Note: PROC MEANS obtains the median for the SALES column of SASHELP.SHOES and stores it as value SALES_MED in the data set MED_DATA. The parameters _TYPE_ and _FREQ_, which are created automatically, can but do not have to be removed from MED_DATA. Using CALL SYMPUTX, the subsequent DATA step transfers the value in the SALES_MED column to the macro variable MEDIAN. The notation for the macro variable in CALL SYMPUTX differs from the macro variable specified after INTO in two ways. The variable name is not preceded by a colon, and the name is also enclosed in quotation marks. The concluding PROC SQL includes the macro variable &MEDIAN in a WHERE clause and filters the data from SASHELP.SHOES according to this condition. If CALL SYMPUT is used for the same example, the only difference is to remove the leading and trailing blanks "by hand," for example, using TRIM and LEFT.

Program 4.3-30: Examples for CALL SYMPUT

```
proc means data = SASHELP.SHOES noprint ;
var SALES ;
output out = MED_DATA (drop=_TYPE_ _FREQ_)
      median(SALES) = SALES_MED  ;
run ;


data _NULL_ ;
set MED_DATA ;
    call symput("MEDIAN", trim(left(SALES_MED))) ;
run ;
```

```
proc sql ;
create table MYDATA as
select *
from SASHELP.SHOES
        where SALES ge 0 and SALES le &MEDIAN. ;
quit ;
```

Note: "data _NULL_" with CALL SYMPUT is chosen instead of the DATA step with CALL SYMPUTX and overwriting the same file with itself. The keyword _NULL_ enables you to program in the DATA step environment without having to create a data set. This approach is generally more efficient than overwriting a file since no data set is created and therefore no data has to be read or written.

SYMGET

SYMGET returns the value of a macro variable during the execution of a DATA step. SYMGET allows at least three different types of arguments:

- Text in quotation marks
- A column from a SAS file whose values correspond to names of macro variable
- An expression that resolves to a macro variable name

The latter two SYMGET functions are illustrated in the following approaches. In all three approaches, boys and girls are assigned a numeric code from the SASHELP file CLASS and stored in the newly created column CODE. Boys receive CODE=0 and girls CODE=1.

Program 4.3-31: Approach without SYMGET: Macro Variables in CASE/WHEN/THEN

```
%let M=0 ; %let F=1 ;

proc sql ;
select *,
        case
                when SEX = "M"
                then  &M.
                else &F.
        end as CODE
from SASHELP.CLASS ;
quit;
```

The first example works without SYMGET. This approach uses a condition linked to the SEX column to assign the two macro variables to the newly created CODE column.

Program 4.3-32: Approach with SYMGET I: Levels of Variable SEX

```
proc sql ;
select *,
        case
                when SEX = "M"
                then symget('M')
                else symget('F')
        end as CODE
from SASHELP.CLASS ;
quit;
```

The second example works with SYMGET. The two SYMGET expressions after THEN and ELSE resolve to the names of the macro variables &M. and &F. defined under %LET. This approach uses a condition linked to the SEX column to assign the entries of the two macro variables &M. and &F. to the newly created CODE column.

Program 4.3-33: Approach with SYMGET II: Variable SEX (Same Result)

```
proc sql ;
      select *, symget(SEX) as CODE
      from SASHELP.CLASS ;
quit;
```

The third example also works with SYMGET. This approach assumes that the *names* of the macro variables &M. and &F. (defined under %LET above) correspond to the *levels* of the SEX column. SYMGET uses the values of the macro variables &M. to assign the value "0" to the CODE column and &F. to assign the value "1".

Figure 4.3-20: SYMGET II: Output in SAS Output (Sample)

```
Name       Sex       Age    Height   Weight  CODE
-------------------------------------------------
Alfred     M          14        69    112.5  0
Alice      F          13      56.5       84  1
Barbara    F          13      65.3       98  1
Carol      F          14      62.8    102.5  1
```

4.3.4 Interfaces II: From PROC SQL to the SAS Macro Facility (INTO)

INTO, already presented in Section 4.1.3, is an important interface to the SAS Macro Facility. This section describes the direction from SQL and the DATA step to the Macro Facility. The previous section, Section 4.3.3, described the opposite direction from the SAS Macro Facility to SQL and DATA step.

Passing to a DATA Step

Macro variables created using SQL can also be passed to the DATA step. This time, the macro variable &MEANSALE is included in a WHERE clause in the DATA step. The DATA step is used to keep all rows whose SALES values range between 0 and the average SALES value. Only one value is written to this macro variable.

Program 4.3-34: Passing to a DATA Step

```
proc sql noprint ;
   select mean(SALES) into :meansale
   from SASHELP.SHOES ;
title "All shops with below average sales
                                    ($ &meansale.).";
data MYDATA ;
   set SASHELP.SHOES ;
      keep Region Subsidiary Product Sales ;
   where SALES ge 0 and SALES le &meansale. ;
run ;
```

Figure 4.3-21: Passing to a DATA Step Output in SAS Output (Abbreviated)

```
All shops with below average sales ($ 85700.17).
```

Region	Product	Subsidiary	Sales
Africa	Boot	Addis Ababa	$29,761
Africa	Men's Casual	Addis Ababa	$67,242
Africa	Men's Dress	Addis Ababa	$76,793
Africa	Sandal	Addis Ababa	$62,819
Africa	Slipper	Addis Ababa	$68,641
Africa	Sport Shoe	Addis Ababa	$1,690

Passing Several Values (%PUT)

In macro variables, not only one, but several numeric values can be passed. In the first example, the first ten values of a column are stored; in the second example, all of them are stored. Principally, this process converts the entries from a data column into a data row. The first variant is restricted by INOBS=; the second variant is not.

Program 4.3-35: Variant1: Storing the First Ten Values	Program 4.3-36: Variant2: Storing all Values
``proc sql noprint inobs=10 ;
select SALES into :sales
   separated by ","
   from SASHELP.SHOES ;
quit;

%put &sales. ;``` | ```proc sql ;
select SALES into :sales
   separated by " "
   from SASHELP.SHOES ;
quit;

%put &sales. ;``` |

Figure 4.3-22: Passing Several Values Output in SAS Log

Example 1:

```
$29.761,$67.242,$76.793,$62.819,$68.641,$1.690,$51.541,$108.942,$21.297,$63.206
```

Example 2 (abbreviated):

```
$29.761 $67.242 $76.793 $62.819 $68.641 $1.690 $51.541 $108.942 $21.297 $63.206 $123.743
$29.198 $64.891 $2.617 $90.648 $4.846 $360.209 $4.051 $10.532 $13.732 $2.259 $328.474 $14.095
$8.365 $17.337 $39.452 $5.172 $42.682 $19.282 $9.244 $18.053 $26.427 $43.452 $2.521 ...
```

Note: Variant 1 stores the first ten values of the SALES variable in the macro variable :SALES. The values are separated by a comma. (See SEPARATED BY.) INOBS=10 limits the number of input rows the first ten. Variant 2 stores all values of the SALES variable in :SALES. The values are separated by a blank. (See SEPARATED BY.) If SEPARATED BY is omitted, only the first value from SALES is stored in the macro variable. %PUT writes all values of the macro variable &SALES to the SAS log.

Outputting all Unique Levels of a Variable (UNIQUE)

The UNIQUE function queries all unique values of the REGION column from SASHELP.SHOES and writes them to the macro variable UNIQ_REG. ORDER BY is used to sort the values alphabetically. Using %PUT, these values are written to the SAS log.

Program 4.3-37: Outputting all Unique Levels of a Variable (UNIQUE)

```
proc sql noprint ;
   select unique(REGION)
      into :UNIQ_REG separated by " , "
   from SASHELP.SHOES
order by REGION ;
quit ;

%put &UNIQ_REG. ;
```

Figure 4.3-23: Outputting all Unique Levels Output in SAS Log

Africa , Asia , Canada , Central America/Caribbean , Eastern Europe , Middle East , Pacific , South America , United States , Western Europe

Writing the First Values of Two Variables

The previous examples stored one or more values of only one variable into a macro variable. PROC SQL can also store values in several macro variables at the same time. The values from the first row of the REGION and SALES columns each are assigned to :REGION and :SALES. If the sequence of variables was reversed after INTO, the first REGION value would be written to :SALES and the first SALES value would be written to :REGION. NOPRINT is used to suppress the output of query results to the SAS output.

Program 4.3-38: Writing the First Values of Two Variables

```
proc sql noprint ;
select REGION, SALES into :REGION, :SALES
from SASHELP.SHOES ;
quit ;
```

Query I: „Only" writing the values to the log:

```
%put &REGION &SALES ;
```

Query II: „Formatted" writing the values to the log:

```
%let REGION = &REGION;
%let SALES = &SALES;
%put In region &REGION sales volumes of &SALES were achieved.;
```

Figure 4.3-24: Unformatted Output in SAS Log (Query I)

```
Africa                          $29,761
```

Note: %PUT only writes the contents of the first two macro variables to the log. The entries "Africa" and "$29,761" are followed by a series of trailing blanks.

Figure 4.3-25: Formatted Output in SAS Log (Query II)

```
In region Africa sales volumes of $29,761 were achieved.
```

Note: The two %LETs result in the trailing blanks being cut off. %PUT then writes the values in the "frame" of the specified sentence to the log. You could also skip the %LET step. However, the price for this would be that any trailing blanks would break the "frame" of the sentence.

Specific Request of Certain Macro Values (Non-aggregated)

With PROC SQL, it is also possible to store individual values from consecutive rows in separate macro variables. For example, each macro variable REGION*n* contains the *n*th value of REGION, and each macro variable SALES*n* contains every *n*th value of SALES. The following example assigns the first four or three values from REGION and SALES to the macro variables REGION1 to REGION4 and SALES1 to SALES3 in the sequence as they are arranged in the query at the time of the query.

Program 4.3-39: Specific Request of Certain Macro Values (Non-aggregated)

```
proc sql noprint ;
   select REGION, SALES
      into :REGION1 - :REGION3, :SALES1 - :SALES4
from SASHELP.SHOES ;
quit ;
```

Program 4.3-40: Requesting Output I

```
%put &REGION1 &SALES1 ;
%put &REGION3 &SALES3 ;
%put          &SALES4 ;
```

Program 4.3-41: Requesting Output II

```
%put &REGION1 &REGION3 ;
%put &SALES1  &SALES3 ;
%put &SALES4 ;
```

Figure 4.3-26: Output I in SAS Log

```
%put &REGION1 &SALES1 ;
Africa $29,761
%put &REGION3 &SALES3 ;
Africa $76,793
%put          &SALES4 ;
       $62,819
```

Figure 4.3-27: Output II in SAS Log

```
%put &REGION1 &REGION3;
Africa Africa
%put &SALES1  &SALES3 ;
$29,761  $76,793
%put &SALES4 ;
$62,819
```

Note: By means of %PUT a selection of the stored values is requested. SAS writes the results to the log, also fewer macro variables are called than were actually created. In examples I and II, the first and third values from REGION and SALES are requested, as well as the fourth SALES value. The only difference is the order of the requested values.

Specific Request of Certain Macro Values (Aggregated and Grouped)

In contrast to the previous example, an aggregation function (SUM) is now applied to the data. The original 395 SALES values are aggregated to 10 SALES values and grouped by REGION. The result of a query would consist of two columns and ten rows and values.

Program 4.3-42: Specific Request of Certain Macro Values (Aggregated and Grouped)

```
proc sql noprint ;
   select REGION, sum(SALES) into
    :REGION1 - :REGION11,
    :SALES1 - :SALES11
   from SASHELP.SHOES
group by REGION ;
quit ;
%put &REGION9  &SALES9 ;
%put &REGION10 &SALES10 ;
%put &REGION11 &SALES11 ;
```

Figure 4.3-28: Query Output

```
Region
---------------------
Africa        2342588
Asia           460231
Canada        4255712
     (output abbreviated)
```

Note: Using INTO, the 10 value pairs are distributed over 11 lines. In this example, more macro variables are created than there are value rows in the query. If the macro variables are queried using %PUT, it is interesting to see how SAS proceeds with the call of the two macro variables REGION11 and SALES 11 that were created last.

Figure 4.3-29: %PUT Output in SAS Log

```
%put &REGION9  &SALES9 ;
United States 5503986
%put &REGION10 &SALES10 ;
Western Europe 4873000
%put &REGION11 &SALES11 ;
WARNUNG: Apparent symbolic reference REGION11 not resolved.
WARNUNG: Apparent symbolic reference SALES11 not resolved.
&REGION11 &SALES11
```

Note: When querying aggregated data, for each cell that contains the aggregated data, the cell of the corresponding aggregating group has to be specified. Any arbitrary arrangement of the %PUT query, such as in example II (see above) would produce a misleading SAS output. The grouping variable (REGION*n*) must always be placed next to the aggregated data variable (SALES*n*). This also applies if the data is grouped according to several aggregation variables, or vice versa, if various data columns are grouped according to the same aggregation variable. Creating more macro variables than there are rows or values in the query does not trigger a warning from SAS at first. Only when %PUT is used to query the values or rows that do not exist in the query, does SAS output a warning in the log. (See above.) In case "open code recursion" problems occur, this is usually because the macro processor cannot read macro statements passed to SAS as intended, which is often as a result of typos.

Requesting Numerous Macro Variables per Macro

If requesting a large number of macro variables is too tedious or prone to typing errors, SAS offers the option of requesting a list of macro variables by macro.

In the following example, two variables are now aggregated and grouped by REGION. Using SUM and GROUP by REGION, the sum of SALES and STORES per REGION is obtained and stored in &SALES*n* and &STORES*n*. The optional NOTRIM prevents blanks from being cut off and thus preserves the columnwise alignment of the values in the SAS log. The result of a query would consist of three columns and nine rows and values. The main difference from the previous examples is the request using the %PUT.

Program 4.3-43: Requesting Numerous Macro Variables per Macro

```
proc sql noprint ;
    select REGION, sum(SALES), sum(STORES) into
      :REGION1 - :REGION9 notrim,
      :SALES1 - :SALES9 notrim,
      :STORES1 - :STORES9 notrim
    from SASHELP.SHOES
group by REGION ;
quit ;
```

Program 4.3-44: Calling by Macro %LIST_IT	**Program 4.3-45: Calling by Manual %PUT Lists**
```%macro LIST_IT ; %do i=1 %to 5; %put REGION&i: &&REGION&i SALES&i: &&SALES&i ; %end; %mend LIST_IT ; %LIST_IT ;```	```%put &REGION1 &SALES1 ; %put &REGION2 &SALES2 ; %put &REGION3 &SALES3 ; %put &REGION4 &SALES4 ; %put &REGION5 &SALES5 ;```

In the %PUT query "by hand," a separate macro variable would have to be entered for each individual macro value to be queried, which would be time-consuming and error-prone. An advantage (especially with smaller lists) might be that macro values can be requested specifically. In the %PUT query of the PUT_IT macro, the macro values are queried rowwise using a loop. After I= and %TO, respectively, only the beginning and the end of the list have to be correctly specified. For example, parameters 2 and 4 instead of 1 and 5 would only result in requesting the values from "Asia" to "Central America/Caribbean". For large lists, the macro version should be efficient and less prone to errors. A disadvantage might be that this listwise request generally enables you to only query all values and not individual values.

Figure 4.3-30: Output of Macro Version in SAS Log

```
REGION1: Africa                     SALES1:   2342588
REGION2: Asia                       SALES2:    460231
REGION3: Canada                     SALES3:   4255712
REGION4: Central America/Caribbean  SALES4:   3657753
REGION5: Eastern Europe             SALES5:   2394940
```

4.4 Application 1: Rowwise Data Update Including Security Check

The INSERT statement enables you to update an empty table rowwise with new variables. However, since there is always a risk that the new rows are unintentionally appended several times to the already created table due to an unintentional repeated submitting of INSERT, the INSERT example is supplemented by a security check in the macro presented. The security check issues a warning as to whether the last row passed already existed in the table.

Program 4.4-1: Create Table EMPTY

```
proc sql ;
create table EMPTY
     (ID num,
       STRING char (30) label="String variable",
       VALUE1  num label="Numerical Variable",
       DATE_VAR  date format=ddmmyy8. label="Date variable");
quit ;
```

Note: Using CREATE TABLE without AS, the empty table EMPTY with the variables ID, STRING, VALUE1, and DATE_ VAR is created first. To this table, INSERT adds the case with the ID 1, the entry "Romeo" in STRING, the value 123 in VALUE1, and the date 15.01.1994 in DATE_VAR . The table to which a case is added "by hand" does not necessarily have to be empty. The INSERT statement is now embedded in a macro.

Program 4.4-2: Macro %INCHECK

```
%macro INCHECK(ID, STRING, VALUE1, DATE_VAR );

 proc sql noprint;
  select count(*)
    into :count
    from EMPTY
   where STRING = &STRING.;
     %if &count >= 1
       %then %do;
           %put Notes: ID &ID. not added, string &STRING. already existed. ;
               select * from EMPTY where STRING = &STRING.;
       %end;
       %else %do;
           insert into EMPTY(ID, STRING, VALUE1, DATE_VAR )
           values(&ID.,&STRING., &VALUE1., &DATE_VAR.);
           %put Notes: ID &ID. has been added for STRING - &STRING..;
       %end;
      quit;

  %mend INCHECK;
```

Note: The macro INCHECK is defined between %MACRO and %MEND. The first SELECT step in PROC SQL queries the number of levels of the variable STRING and stores the value in the macro variable :COUNT. If the value of COUNT is greater than 1, the row in question is not included in the EMPTY table and feedback is output to the SAS log. If the value of COUNT equals 1, the rows are included. The data is passed to SAS rowwise in a similar way to INSERT, but in this variant embedded in the INCHECK macro. In order to process the STRING values, as well as the macro variables, flexibly and easily, they were placed in *single* quotation marks in the macro call (thus slightly different from the INSERT statement example). %INCHECK is the actual call of the macro. In the parentheses, you only need to specify the values to be passed, separated by commas.

Program 4.4-3: Calling %INCHECK

```
%INCHECK(1,'Romeo',123,"15JAN94"d) ;
%INCHECK(2,'Julia',321,"15OCT94"d) ;
%INCHECK(3,'Prospero',124,"02JAN70"d) ;
%INCHECK(4,'Hamlet',111,"02SEP87"d) ;
%INCHECK(5,'Jago',124,"15JUL57"d) ;
%INCHECK(2,'Julia',321,"15OCT94"d) ;
proc print data=EMPTY noobs ;
      run ;
```

After each call of the %INCHECK macro, the note per row is output to the SAS log whether the row in question was included.

Figure 4.4-1: Output of Successful Inserts in SAS Log

```
%INCHECK(1,'Romeo',123,"15JAN94"d);
NOTE: 1 row was inserted into WORK.EMPTY.
Notes: ID 1 has been added for STRING - 'Romeo'.
NOTE: PROCEDURE SQL used (Total process time):
      real time               0.07 seconds
      cpu time                0.00 seconds

%INCHECK(2,'Julia',321,"15OCT94"d) ;
NOTE: 1 row was inserted into WORK.EMPTY.
Notes: ID 2 has been added for STRING - 'Julia'.
NOTE: PROCEDURE SQL used (Total process time):
      real time               0.03 seconds
      cpu time                0.00 seconds
```

If a line was not included because the value of COUNT is greater than 1 (the second "Julia" row), the row in question is not included in the EMPTY table and the macro outputs a corresponding message in the SAS log.

Figure 4.4-2: Output of Rejected Inserts in SAS Log

```
%INCHECK(2,'Julia',321,"15OCT94"d) ;
Notes: ID 2 not added, string 'Julia' already existed.
NOTE: PROCEDURE SQL used (Total process time):
      real time               0.01 seconds
      cpu time                0.01 seconds
```

The data set EMPTY now looks like Figure 4.4-3.

Figure 4.4-3: Contents of Finally Created SAS Data Set in SAS Output

```
ID     STRING      VALUE1     DATE_VAR
1      Romeo        123       15/01/94
2      Julia        321       15/10/94
3      Prospero     124       02/01/70
4      Hamlet       111       02/09/87
5      Jago         124       15/07/57
```

Note: Users can adapt the program in many ways:

- Users can adjust the value of the COUNT variable by setting it to a higher value than 1.
- Users can define the COUNT value as a macro variable and flexibly integrate it into the macro call for each line to be transferred.
- Users can add an ABORT statement to the program, which then explicitly terminates the execution of the program and writes an error message to the SAS log, and so on.

4.5 Application 2: Working with Multiple Files (Splitting)

The two macros SPLIT_DS1 and SPLIT_DS2 split a SAS file into uniformly filtered subsets and simultaneously name the created SAS files according to the levels of the split variables used (see SPLIT_DS1) or according to the split variable used together with a specific ID (see SPLIT_DS2).

A split variable is a variable of type "numeric" or "string," which has at least two missing levels and thus makes it possible to split this data set into two levels. The term "split variable" derives from the function of such a variable, namely, to be able to divide a data set.

Of course, the following two programs also work if the split variable has only one level, but this case can hardly be described as "splitting" a file.

4.5.1 Splitting a Data Set into Uniformly Filtered Subsets (Split Variable is of Type "String")

The following approach enables you to split a data set using a split variable of type "string" and store it in separate data sets, filtered accordingly. The split variable is a string variable with at least two nonmissing levels. The source data set is filtered on these levels, the filtered contents are stored in subsets, and they are given the names of the respective levels of the split variable. If the level of a split variable occurs twice or more often, all data rows belonging together are stored in one file. The prerequisites for such a SAS file are:

- The data set contains a so-called split variable of the type "string" with at least two nonmissing values.
- The levels of the split variables correspond to the conventions of names for SAS data sets, including no blanks or special characters such as "-", "/", and so on.
- Ideally, the split variable does not contain any missing values. Alternatively, the split variable can contain missing values if and when this is irrelevant for splitting according to all other specifications.

Example

The data set SASHELP.SHOES is to be filtered and split into separate subsets using the discretely scaled string variable REGION. REGION is therefore the split variable. In addition, the subsets should have the same name as the split variables.

Program 4.5-1: Defining and Calling %SPLIT_DS1

```
%macro split_ds1 (SPLIT_DS=, SPLIT_BY=);
 proc sql ;
    select count(distinct &SPLIT_BY.)
       into :n
        from  &SPLIT_DS. ;
    select distinct &SPLIT_BY.
        into :&SPLIT_BY.1 - :&SPLIT_BY.%left(&n)
        from  &SPLIT_DS. ;
 quit ;
 %do i=1 %to &n;
      data WORK.&&&SPLIT_BY.&i ;
         set &SPLIT_DS. ;
         if &SPLIT_BY.="&&&SPLIT_BY.&i" ;
      run;
   %end;
 %mend split_ds1 ;

%split_ds1  (SPLIT_DS=SASHELP.SHOES, SPLIT_BY=REGION);
```

What Happens?

The macro SPLIT_DS1 is defined between %MACRO and %MEND. The first SELECT step in PROC SQL queries the number of different values of the split variable and stores the value (10) in the macro variable :N. The second SELECT step queries the different levels and stores them in the macro variable :I. A NOPRINT after PROC SQL could be used to deactivate the display of these levels in the log. However, it is helpful to display the values of the split variables in the log to check whether they comply with the naming conventions for SAS files.

Between %DO %TO, a loop runs from the first to the value :N (10); that is, this loop is run ten times in the example. During each run, a data set with the two-level name *Work.LevelOfSplitvariable* is created for each level (:I) and filtered before using the split variable and the existing levels of the split variable. When the maximum value of 10 is reached, the loop ends and no further data set is created.

%SPLIT_DS1 etc. is the actual call of the macro. After SPLIT_DS and SPLIT_BY, you only need to specify the data set to be split (SASHELP.SHOES) and the desired split variable (REGION). After initializing the macro, SAS outputs the following information in the log.

SAS Log 4.5-1: Successful Processing (Sample)

```
NOTE: There were 395 observations read from the data set SASHELP.SHOES.
NOTE: The data set WORK.ASIA has 14 observations and 7 variables.

NOTE: There were 395 observations read from the data set SASHELP.SHOES.
NOTE: The data set WORK.CANADA has 37 observations and 7 variables.
```

The SPLIT_DS1 macro splits the SASHELP.SHOES data set using the REGION levels and stores the contents, filtered accordingly, in separate data sets that are also given the name of the respective level for better identification. Many (but not all) levels of the split variable in the example correspond to the naming conventions for SAS data sets. SAS can therefore use the split variable levels with the appropriate exceptions to name the SAS data sets. The data set names provide information about which data was stored for which level of the split variable REGION.

The content of the SAS file AFRICA looks like Figure 4.5-1.

Figure 4.5-1: Contents of SAS Data Set (Sample)

Region	Product	Subsidiary	Stores	Sales	Inventory	Returns
Africa	Boot	Nairobi	25	$16,282	$66,017	$844
Africa	Men's Dress	Nairobi	1	$8,587	$20,877	$363
Africa	Sandal	Nairobi	19	$16,289	$47,406	$1,175

Depending on the distribution of the data, the number of measurement values in the respective partial data sets can be the same or different, continuously increasing, continuously decreasing, or have no clear pattern. In the example, the number of measured values gradually decreases over all 33 subsets.

If a level of a split variable does not correspond to the conventions of names for SAS data sets, this triggers an error message. (See SAS Log 4.5-2.)

SAS Log 4.5-2: Failed Processing (Sample)

```
NOTE: Line generated by the macro variable "REGION4".
1      WORK.Central America/Caribbean
                          ---------
                             22
                             76
ERROR 22-322: Syntax error, expecting one of the following: BUFFERED, HEXLISTALL, LIST,
MISSOPT,NESTING, NOLIST, NOMISSOPT, NONESTING, NONOTE2ERR, NOPASSTHRU, NOPMML, NOTE2ERR,
PASSTHRU, PGM, PMML, SESSREF, SESSUUID, THREADS, UNBUFFERED, VIEW.
```

```
ERROR 76-322: Syntax error, statement will be ignored.

NOTE: The SAS System stopped processing this step because of errors.
WARNING: The data set WORK.CENTRAL may be incomplete.  When this step was stopped there
were 0 observations and 7 variables.

WARNING: The data set WORK.AMERICA may be incomplete.  When this step was stopped there
were 0 observations and 7 variables.
NOTE: DATA statement used (Total process time):
      real time                0.04 seconds
      cpu time                 0.01 seconds
```

For example, SAS cannot use the string "Central America/Caribbean" for naming the SAS data set because of the "/". SAS cannot create files with the name "Central America/Caribbean" and store the strings there. The log and *missing* data set names in the target directory (WORK) therefore provide information about which data could *not* be stored for which level of the split variable, also "UNITED STATES", "WESTERN EUROPE", and so on.

If a level of the split variable is "blank" or contains special characters (except for "_", "$" or "#"), no separate data set is created for REGION. However, all other partial data sets are created. Special characters and blanks in the split variable result in subsets not being created. If data are to be stored according to such levels, blanks can be converted into strings before, which correspond to the conventions for SAS file names such as "Missing". On the one hand, special characters can be searched for with the SAS functions TRANWRD and TRANSLATE, as well as the Perl Regular Expression PRXCHANGE, and replaced with more suitable characters. On the other hand, since SAS 9.1, the SAS option VALIDVARNAME with the keyword ANY enables you to give columns almost any name. In Chapter 4, application example 6 in Section 4.9 deals with VALIDVARNAME.

4.5.2 Splitting a Data Set into Uniformly Filtered Subsets (Split Variable is of Type "Numeric")

Using a numerical split variable, the following approach enables you to split and store a data set in separate data sets, filtered accordingly. A split variable is a discretely scaled numeric variable that has at least two missing values and therefore divides or splits this data set into two levels. In contrast to string variables, the content of a numerical variable cannot be used directly for a data set name, since a number does not correspond to the convention for naming a SAS file. In the following example, the subsets created are created in the form *Work.NameOfSplitVariable_Level*. If the value of a split variable occurs twice or more, all data rows that belong together are stored in one file. The requirements for such a SAS file are:

- The data set contains a split variable of type "numeric" with at least two nonmissing values.
- The values of the split variables correspond to the naming conventions for SAS data sets and do not contain minus signs, periods, or other formatting or punctuation that violate the conventions for SAS file names.
- Ideally, the split variable should not contain missing values, but it might contain missing values if and when this is irrelevant for splitting according to all other specifications.
- If the split variable is not nonnegative and discretely scaled, the split variable can be subjected to a pre-processing step in which it is rounded or subjected to further transformations.
- The macro can be supplemented by a TRANWRD step, which eliminates punctuation before the numerical values are processed further.

Example

Using the discretely scaled numeric variable STORES, the data set SASHELP.SHOES is to be split into filtered separate subsets. STORES is the split variable. The subsets contain all rows from SASHELP.SHOES with the same number of stores. In addition, the subsets are created with the names *Work.STORES_10* and so on.

Program 4.5-2: Defining and Calling %SPLIT_DS2

```
%macro split_ds2 (SPLIT_DS=, SPLIT_BY=);
  proc sql ;
     select count(distinct &SPLIT_BY.)
        into :n
         from  &SPLIT_DS. ;
     select distinct &SPLIT_BY.
        into :&SPLIT_BY.1 - :&SPLIT_BY.%left(&n)
        from  &SPLIT_DS. ;
  quit ;
  %do i=1 %to &n;
       data WORK.&SPLIT_BY._&i ;
          set &SPLIT_DS. ;
           if &SPLIT_BY.=&&&SPLIT_BY.&i ;
       run;
     %end;
  %mend split_ds2 ;


%split_ds2 (SPLIT_DS=SASHELP.SHOES, SPLIT_BY=STORES);
%split_ds2 (SPLIT_DS=SASHELP.CLASS, SPLIT_BY=AGE);
```

What Happens?

The macro SPLIT_DS2 is defined between %MACRO and %MEND. The first SELECT step in PROC SQL queries the number of different levels of the split variables and stores the value (33) in the macro variable ":n". The second SELECT step queries the different values themselves and stores it in the macro variable ":i". A NOPRINT after PROC SQL could deactivate the display of these values in the log.

Starting at %DO %TO, a loop runs from the first to the value ":n" (33); this loop is run through 33 times in the example. During each run, a data set with the two-level name *Work.NameOfSplitVariable_Level* is created for each level (":i") and filtered before using the split variable and the existing levels of the split variable. The SAS file STORES_2 created contains all 29 rows for the split variable STORES with the level 2 and so on. Upon reaching the maximum value 33, the loop is terminated and no further data set is created.

%SPLIT_DS2 is the actual call of the macro. After SPLIT_DS and SPLIT_BY, only the data set to be split (SASHELP. SHOES) and the desired split variable (STORES) need be specified. After initializing the macro, SAS outputs the following information in the log.

SAS Log 4.5-3: Successful Processing (Sample)

```
NOTE: There were 395 observations read from the data set SASHELP.SHOES.
NOTE: The data set WORK.STORES_1 has 37 observations and 7 variables.

NOTE: There were 395 observations read from the data set SASHELP.SHOES.
NOTE: The data set WORK.STORES_2 has 29 observations and 7 variables.

NOTE: There were 395 observations read from the data set SASHELP.SHOES.
NOTE: The data set WORK.STORES_3 has 33 observations and 7 variables.
    Abbreviated.
NOTE: There were 19 observations read from the data set SASHELP.CLASS.
NOTE: The data set WORK.AGE_1 has 2 observations and 5 variables.

NOTE: There were 19 observations read from the data set SASHELP.CLASS.
NOTE: The data set WORK.AGE_2 has 5 observations and 5 variables.

NOTE: There were 19 observations read from the data set SASHELP.CLASS.
NOTE: The data set WORK.AGE_3 has 3 observations and 5 variables.
    Abbreviated.
```

Depending on the distribution of the data, the number of measured values in the respective partial data sets can be the same or different, continuously increasing, continuously decreasing, or without any clear pattern. In the example, the number of measured values gradually decreases over all 33 subsets. If the split variable contains missings, no STORES subset is created. If a split variable occurs twice, all data rows belonging together are stored in a file. If a level of the split variable is of the category missing, no separate data set is created for missing values in STORES; however, all other partial data sets are created. Missing values in the split variable result in subsets not being created. This approach assumes that the split variable is numeric, has at least two levels with values in it, and does not contain any missing values. If there are missing values, these must first be converted into a numeric code before this program can be used, for example, 99999 and so on. If the split variable is a string, you can either convert the string into a numeric variable or adapt the macro to the special features of the split string.

4.6 Application 3: Transposing a SAS Table (Stack and Unstack)

Data in tables is usually arranged in rows and columns. In principle, there is no rule regarding specific arrangement of the data structure. On the contrary, not only can the same file content be stored in tables with different structures, it can also be flexibly converted from one structure to another using data management.

Restructuring, rotating, or "looping" a file is a technique for converting tables from one structure to another structure. For example, when transposing, rows and columns of a file are rotated. Rows become columns and columns become rows. Looping works similarly but changes the structure of the table. The contents of the tables are not changed. The arrangement, structure, and data in tables or diagrams are differentiated in stack (long) and unstack (wide). With stack, related values of a case are arranged one below the other in a column. With unstack, related values of a case are arranged next to each other in a row. (See Figures 4.6-1 and 4.6-2.)

4.6.1 Stack

Values belonging to one case are stacked or arranged in a column one below the other. The affiliation to a case is indicated by a common, repeatedly specified key value. (See ID in Figure 4.6-1.) In order to distinguish the individual values row by row, they are usually differentiated by an incrementing integer variable. The three DOSE values 45, 78, and 76 of case ID 1 are kept apart by means of the TIME codes 1, 2, and 3.

A key variable ensures that values of a case that are stored in different rows can be identified as belonging to a case. A counting variable such as TIME ensures that values of a point in time that are stored in different row can be identified as belonging to a common point in time.

4.6.2 Unstack

Values belonging to one case are arranged next to each other in one row. Belonging to one case is indicated by an ID value specified once. (See ID in Figure 4.6-2.) In order to be able to distinguish between the individual values column by column, they are differentiated by identical column names, which differ by an incrementing integer variable as a suffix. The three DOSE values 45, 78, and 76 of case ID 1 are distinguished by the columns DOSE1, DOSE2, and DOSE3.

Figure 4.6-1: Table in STACK Structure (N=2 Cases)				**Figure 4.6-2: Table in UNSTACK Structure (N=2 Cases)**			
ID	TIME	DOSE		ID	DOSE1	DOSE2	DOSE3
01	1	45		01	45	78	76
01	2	78		02	82	83	. . .
01	3	76					
02	1	82					
02	2	83					
02	3	. . .					

Before a conversion from stack to unstack, there are *three* rows with three columns for one case in Figure 4.6-1. *After* that, however, there are four variables, but only one row. The rows and columns are not swapped, but the number of rows is reduced, while the table widens proportionally by the corresponding number of columns. By way of illustration, one could also say that the number of rows is reduced by three times, while the number of columns is increased by three times. The contents of the table remain unchanged; it is only structured differently.

Looping enables you to prepare data in such a way that it can be used for graphical and statistical analysis in the first place. PROC GLM expects tables in a STACK structure, whereas PROC MIXED expects tables in an UNSTACK structure. A side effect of looping from unstack to stack is that a table in a STACK structure becomes clearer in width because the variable row is shorter. A side-effect of looping from stack to unstack is analogous to the fact that a table in an UNSTACK structure becomes clearer in length because the number of rows is smaller. These advantages can be particularly relevant for analyses by visual inspection, that is, for small data sets or representative partial data for testing SQL programs.

Looping is one of the more advanced techniques of data management and is therefore presented below in the form of two macros. The macro in Section 4.6.1 rotates SAS tables from a STACK to an UNSTACK structure. The macro in Section 4.6.2 rotates SAS tables from an UNSTACK into a STACK structure. Both macros are named according to the target structure of the tables to be created. UNSTACK_CLEAN creates a table in an UNSTACK structure. STACK_CLEAN creates a table in a STACK structure. The two macros presented here can be supplemented by further check rules or functions, for example:
- additional or fewer check rules, loops, formatting statements, and so on
- using an ABORT or STOP statement with or without creation of an output table instead of continuing automatically
- automatic corrections of identified errors if possible from the factual or data situation

4.6.3 From Stack to Unstack (From 1 to 3)

The following macro transforms SAS tables from a STACK to an UNSTACK structure. Values in one column (GRP=1, 2, 3) become three columns (GRP_1, GRP_2, GRP_3), hence the mnemonic "from 1 to 3". Do not confuse this with a row-oriented approach; here the reverse is true: three rows become one. The example is based on Schendera (2004, 280-283). In addition, it was written in the form of a macro which also contains various integrity constraints including the output of an audit file. The advantage of the macro is that looping works even if all numerical values are of type "character." The following extremely versatile processes are running in macro UNSTACK_CLEAN:

- The macro outputs a SAS table in an UNSTACK structure.
- The macro carries out various checks. For example, the macro checks IDs for duplicates or missing values. IDs with duplicates or missing values are automatically excluded and stored in the automatically created SAS table AUDIT for future review. In addition, the macro checks for missing values in group variables, event variables, and series of measurements.
- The macro also creates two test documents: a SAS table with the default name AUDIT and an RTF document with the default name *AUDIT_of_&data_in.RTF*, which contains the same information. Users definitely need to review these documents before proceeding with further management or analysis of the data.

After the sample data, you will find two detailed overviews of the individual processes of restructuring, checking, and filtering, as well as the various checks carried out.

Note: After this macro has run, the groups in the output table in the UNSTACK structure are by definition exactly the same size and might differ from the original group sizes in the table in STACK structure.

SAS Demo Data

The demo data in Program 4.6-1 includes missing values and duplicates in the ID.

Program 4.6-1: Demo Data including Missing Values and Duplicates

```
data STACKDbuggy ;
input ID CASECONT REPEATED RESULT @@;
datalines;
 1   1    1 23   1   1    2    7
 1   1    3  .    2   1    1   11
 2   1    2 19   2   1    3    7
 3   1    1 18   3   1    2   14
 3   1    3  .    4   1    2   44
 4   1    3 40   5   1    1    6
 5   1    2 50   5   1    3   31
 6   0    1 37   6   0    2   23
 6   0    3 34   7   0    1    5
 7   0    2 41   8   0    1    3
 8   0    2 56   8   0    3   42
 9   0    1 53   9   0    2   54
 9   0    3 13  10   0    1   13
10   0    2 13  10   0    3   13
10   0    3 13   .   0    3   13
10   0    3 13
      ;
run ;
```

Overview of Executing Processes of Restructuring, Checking, and Filtering

Numerous processes for restructuring, checking, and filtering run in the macro UNSTACK_CLEAN. The individual parameters are explained in detail in the section on calling the macro. The numbers indicate the corresponding positions in the macro.

1. The imported file "STACKDbuggy" is sorted by ID_in in the key variable ID; ID_in is first checked for missing values.
2. The sorted file is sorted by grpwident; meas_var and grpwident are checked for missing values.
3. The sorted file is sorted by ID_in and event; event is also checked for missing values.
4. In the PROC SQL step, the completeness of the measurement series is checked; only measurement series with complete data are kept. ID_in is checked for multiple entries.
5. The DATA step excludes data series in which IDs occur more often than the target value in grp_size. The PROC SQL step stores the excluded rows in the file with the default name AUDIT in the original STACK structure.
6. Using ODS, AUDIT is stored in an RTF document at the end of the ROOT path. The SAS version is displayed by PROC PRINT in the SAS output. If no rows are excluded, AUDIT is created empty.
7. The data prepared via steps 1 to 5 are finally transposed at this point.
8. The cleansed data is rearranged. The unstacked data is stored in the output file UNSTACKD_CLEAN.
9. Auxiliary files are deleted using PROC DATASETS.
10. PROC PRINT displays the cleansed and unstacked data in the SAS output.

Overview of Checks Carried Out

The UNSTACK_CLEAN macro makes the following checks. The numbers indicate the corresponding positions in the macro.

1. Checks ID_in for missing values.
2. Checks meas_var and grpwident for missing values.
3. Checks event for missing values.
4. Checks the completeness of the measurement series and keeps only measurement series with complete data. At the same time, checks ID_in for multiple entries.

Macro UNSTACK_CLEAN

Program 4.6-2: Macro UNSTACK_CLEAN

```
%macro UNSTACK_CLEAN
(data_in=, ID_in=, root=, meas_var=, event=, grpwident=, grp_size=, data_out=);

/* (1) Filtering for missings in ID_in  */
proc sort data=&data_in.  out=prep_01 ;
      by &ID_in. ;
            where &ID_in. is not missing ;
      run ;

/* (2) Filtering for missings in &grpwident. and &meas_var. */
proc sort data=prep_01 out=prep_02 (where=(&meas_var. is not missing)) ;
      by &grpwident. ;
            where &grpwident. is not missing ;
      run ;

/* (3) Filtering for missings in &event. */
proc sort data=prep_02 out=prep_03 (where=(&event. is not missing));
            by &ID_in. &event.;
      run;

/* (4) Filtering for complete measurement series */
proc sql;
      create table prep_04
            as select *, count(*) as count
            from prep_03
            group by &ID_in.
            having count(*) = &grp_size.
            order by &ID_in.;
      quit;

/* (5) Excluding !too frequent! data series. Storing the excluded data in AUDIT */
data prep_05 (drop=count) ;
      set prep_04 ;
      where COUNT =  &grp_size.  ;
run ;

proc sql ;
      create table AUDIT as
            select * from &data_in.
            except all
      select * from prep_05 ;
      quit ;

/* (6) Output of audit files (SAS, RTF)  */
ods rtf file="&root.AUDIT_of_&data_in..rtf" ;
title1 "Compilation of excluded data" ;
title2 "- To be reviewed by user -";

proc print data = AUDIT ;
      run;
title ;
ods rtf close;

/* (7)Transposing cleansed data */
data prep_06 ;
      set prep_05 ;
      by &ID_in.  &grpwident. ;
      array LIST[*] RESULT1 - RESULT&grp_size. ;
```

```
            retain RESULT1 - RESULT&grp_size. ;
            if first.&ID_in. then do i=1 to &grp_size. ;
            LIST[i]=. ;
            end ;
            LIST[&grpwident.]=&meas_var. ;
                    if last.&ID_in. then output ;
        keep &ID_in. &event. RESULT1 - RESULT&grp_size. ;
run ;

/* (8) Arranging cleansed data. Sorting unstacked data in &data_out.  */
data &data_out. ;
        retain &ID_in. &event. RESULT1 - RESULT&grp_size. ;
        set prep_06 ;
run ;

proc datasets  library=work memtype=data nolist;
        delete prep_01 prep_02 prep_03 prep_04 prep_05 prep_06 ;
        run ;

/* (9) Output in SAS Output */
title1 "Unstacked dataset (method: 'automatic cleaning'.)" ;
title2 "Reduced to complete measurement series." ;
proc print data = &data_out. ;
        run;
title ;

%mend UNSTACK_CLEAN;
```

Calling Macro UNSTACK_CLEAN

When passing the parameters to the macro call, make sure to use the correct upper- and lowercase. The comments contain short explanations of the parameters; more detailed explanations are given after the macro call.

Program 4.6-3: Calling Macro UNSTACK_CLEAN

```
%let root = %str(C:\);      /* root: Specify the storage path for the   */
                            /*       RTF output after %STR(), e.g.'C:\'  */

%UNSTACK_CLEAN
 (data_in=STACKDbuggy,      /* data_in: Input dataset, e.g. STACKDbuggy  */
  root=&root.,              /* root: Passing the storage path            */
  ID_in=ID ,               /* ID_in: Key variable in input dataset      */
  meas_var=RESULT,          /* meas_var: Measurement variable (response) */
  event=CASECONT,           /* event: Event, e.g. diagnosis              */
  grpwident=REPEATED,       /* grpwident: Within ID, e.g. REPEATED       */
  grp_size=3,               /* grp_size: Standard size of groups         */
  data_out=UNSTACKD_CLEAN)/* data_out: Output dataset                    */
      ;
```

The following list is an explanation of the parameters in the macro call (to be set by the user):

- **data_in=** Input data set (in the example: STACKDbuggy).
  The input data set contains at least one key variable (numeric), one event variable (numeric), one group variable (numeric), and one variable for the measurement values (numeric). A two-level name in the form LIBNAME.DSNAME can be specified.
- **root=** Storage path for RTF output.
  A user usually does not change this parameter. To set a different output location, a user can pass the desired path to SAS in the *%let* statement in parenthesis after STR.
- **ID_in=** Key variable from the input data set (in the example: ID).

ID_in is numeric. The specification of ID_in is essential for data transformation and quality (5,7). An ID groups either simultaneous measurements on the same batch (assays) or successive measurements on the same case (repeated measurements).

The values in ID_in do not always occur equally frequently. ID_in is checked for missing values and for cases that occur too frequently (1, 5).

- **meas_var=** Variable for the measurement values (in the example: RESULT).

  meas_var is numeric or string. The scale of meas_var and whether it is a dependent or independent variable is irrelevant for the functioning of the macro. Only one variable can be specified for measurement values. meas_var is checked for missing values (2).

- **event=** Variable for the event (in the example: CASECONT).

  event is numeric. Depending on the study design, event can be a coding for a diagnosis, a treatment, or also for a case control group assignment. event is checked for missing values (3).

- **grpwident=** Identification variable within the rows that are grouped by ID (in the example: REPEATED). "grpwident" stands for group-within identifier. grpwident is numeric, discretely scaled, and has at least two values. In order to be able to distinguish values in the order in which they are measured or for an invariant position in a batch, they are generally differentiated for each ID by an incrementing integer. The macro assumes that the measurements have the same frequency, that is, a common maximum (standard size as reference value). The maximum number of levels of grpwident must be known and can also be explicitly specified in grp_size (see below). grpwident is checked for missing values (2).

- **grp_size=** Default size of the output groups within ID.

  For the standardization of the data volume per ID, grp_size defines the standard size of the groups according to the number of measurement values or rows within an ID. grp_size allows to explicitly specify to the system which maximum value the variable grpwident must have constantly or how consistently often an ID value might occur. The specification of grp_size is essential for data transformation and quality (5,7). grp_size is not a maximum value but a standard size. The macro excludes IDs with more or less data or rows than specified by grp_size from the table and stores them in the AUDIT table for closer examination.

- **data_out=** Output data set.

  The output data set contains the input data set (originally STACK structure) now in UNSTACK structure. You can specify a two-level name in the form LIBNAME.DSNAME. The output data set contains a newly created key variable, the event variable, as well as a variable for measurement values for each level of groups (=grp_size).

Output 1 - Data Set UNSTACKD_clean in UNSTACK Structure (Cleansed for Potentially Incorrect Data)

Figure 4.6-3: Data Set UNSTACKD_clean in UNSTACK Structure

```
          Unstacked dataset (method: 'automatic cleaning').
             Reduced to complete measurement series.


 Obs    SAMPLEID    CASECONT    RESULT1    RESULT2    RESULT3

  1         2           1          11         19         7
  2         5           1           6         50        31
  3         6           0          37         23        34
  4         8           0           3         56        42
  5         9           0          53         54        13
```

Note: The data in the UNSTACK structure does not contain any missing values in the ID and measurement series, no duplicate IDs, and groups of exactly the same size.

Output 2 - Data Set AUDIT in STACK Structure

Figure 4.6-4: Data Set AUDIT in STACK Structure

```
            Compilation of excluded data.
              - To be reviewed by user -
```

```
Obs    SAMPLEID    CASECONT    REPEATED    RESULT

 1         .           0           3          13
 2         1           1           1          23
 3         1           1           2           7
 4         1           1           3           .
 5         3           1           1          18
 6         3           1           2          14
 7         3           1           3           .
 8         4           1           2          44
 9         4           1           3          40
10         7           0           1           5
11         7           0           2          41
12        10           0           1          13
13        10           0           2          13
14        10           0           3          13
15        10           0           3          13
16        10           0           3          13
```

Note: The AUDIT table contains all rows that do not comply with the integrity constraints. For example, all rows with missing values in ID and measurement series, IDs that occur too frequently, and so on.

Output 3 - RTF Output

Figure 4.6-5: RTF Output (in the Example under Drive C:\ )

Compilation of excluded data.

- To be reviewed by user -

Obs	SAMPLEID	CASECONT	REPEATED	RESULT
1	.	0	3	13
2	1	1	1	23
3	1	1	2	7
4	1	1	3	.
5	3	1	1	18
6	3	1	2	14
7	3	1	3	.
8	4	1	2	44
9	4	1	3	40
10	7	0	1	5
11	7	0	2	41
12	10	0	1	13
13	10	0	2	13
14	10	0	3	13
15	10	0	3	13
16	10	0	3	13

Note: The RTF output contains the same content as AUDIT, only in RTF instead of SAS format. Of course, you can also transform a table in the opposite direction, from UNSTACK to STACK.

Chapter 4: Macro Programming with PROC SQL **151**

4.6.4 From Unstack to Stack (From 3 to 1)

Values in three columns (GRP_1, GRP_2, GRP_3) become one column (GRP=1, 2, 3), hence the mnemonic "make 1 from 3", which is not to be confused with a row-oriented approach. The example is based on Schendera (2004, 280-283). In addition, it was written in the form of a macro which also contains various integrity constraints including the output of an audit file. The advantage of the macro is also that looping works even if all numerical values are of type "character."

Numerous processes run in the STACK_CLEAN macro:

- The macro outputs a SAS table in a STACK structure.
- The macro carries out various checks. For example, the macro checks IDs for duplicates or missing values. IDs with duplicates or missing values are automatically excluded and stored in the automatically created SAS table AUDIT for review. The macro also checks for missing values in the event variable, as well as the measurement series as a whole.
- The macro creates two test documents: a SAS table with the default name AUDIT and an RTF document with the default name *AUDIT_of_&data_in.RTF*, which contains the same information. Users need to review these documents before proceeding with further management or analysis of the data.

After the sample data, you will find two detailed overviews of the individual processes of restructuring and filtering, as well as of the various checks carried out.

Note: After submitting this macro, the groups in the output table in STACK structure are exactly the same size because the number of group levels is determined by the number of columns. The number of groups in the STACK table cannot differ from the original group sizes in the table in UNSTACK structure.

SAS Demo Data

The demo data contains missing values and duplicates in the ID, as well as incomplete measurement series.

Program 4.6-4: Demo Data including Missing Values, Duplicates, and Incomplete Data Series

```
data USTACKDbuggy ;
input ID DIAGNOSIS  RESULT1   RESULT2    RESULT3 ;
datalines;
  .         1        23          7         24
  1         1        23          .         24
  2         0         .         19          7
  3         0        18         14         46
  4         1        21         44         40
  5         1         6         50         89
  5         1         6         50         51
  ;
run ;
```

Overview of Executing Processes of Restructuring and Filtering

The macro STACK_CLEAN runs numerous processes for restructuring, checking, and filtering. The individual parameters are explained in detail in the section on calling the macro. The numbers indicate the corresponding positions in the macro.

1. In the input file USTACKDbuggy, the measurement series defined by list_b and list_e (RESULT1 to RESULT3) are checked for missing values; only measurement series with complete data are kept.
2. Event (DIAGNOSIS) is checked for missing values. Rows with missing values in DIAGNOSIS are excluded.
3. The auxiliary file is sorted by ID; rows with missing values and duplicates in ID are excluded.
4. The PROC SQL step stores excluded rows in the file with the standard name AUDIT in the original UNSTACK structure. Deleting of first auxiliary files.

5. Outputting of the audit files (SAS, RTF). Before this, the SAS file AUDIT is stored in an RTF document under the specified path using ODS. PROC PRINT displays AUDIT itself in the SAS output. If no rows are excluded, AUDIT is created empty.
6. The table prepared by steps (1) to (5) is restructured from unstack to stack. In contrast to the UNSTACK_CLEAN macro, the variable GRPWIDENT is created automatically. (See also Section 4.6.1.) GRPWIDENT stands for group-within identifier.
7. The stacked data is stored in the output file STACKD_CLEAN.
8. The auxiliary files are deleted using PROC DATASETS.
9. PROC PRINT displays the cleaned and stacked data in the SAS output.

Overview of Checks Carried Out

The checks below are made in the following macro. The numbers indicate the corresponding positions in the macro.

1. The measurement series specified by list_b and list_e are checked for missings; only measurement series with complete data are kept.
2. Event (DIAGNOSIS) is checked for missing values. Rows with missing values in DIAGNOSIS are excluded.
3. ID is checked for missing values and duplicates. Rows with missing values and duplicates are excluded. A PROC SORT with NODUPKEY, instead of the SQL filter used, would not exclude all rows with multiple IDs, but all rows except the first one found.

Macro STACK_CLEAN

Program 4.6-5: Macro STACK_CLEAN

```
%macro STACK_CLEAN
(data_in=, root=, event=, id_in=, list_b=, list_e=, nvar=, outvar=, data_out=);

/* (1) Filtering for incomplete measurement series */
/* (2) Filtering for missing values in event            */
data prep_01 (where=(&event. is not missing));
      set &data_in. ;
            array MISSKILL{&nvar.} &list_b. - &list_e. ;
            do i=1 to dim(MISSKILL);
                  if MISSKILL(i) = . then delete;
                  end;
      run ;

/* (3) Filtering for missing values and duplicates in ID */
proc sql ;
      create table prep_02 as
          select *, count(&id_in.) as FILTER
            from prep_01
          group by &id_in.
          having FILTER = 1 ;
          quit ;

/* (4) Storing the excluded data in AUDIT */
data prep_03 ;
      drop i FILTER ;
      set prep_02 ;
run ;
proc sql ;
      create table AUDIT as
            select * from &data_in.
            except all
      select * from prep_03 ;
quit ;
```

```
/* (5) Output of audit files (SAS, RTF) */
ods rtf file="&root.AUDIT_of_&data_in..rtf" ;
title1 "Compilation of excluded data" ;
title2 "- To be reviewed by user -";
proc print data = AUDIT ;
run;
title ;
ods rtf close;

/* (6) Transposing cleansed data */
data prep_04 ;
set prep_03 ;
      array LIST[&nvar.] &list_b. - &list_e. ;
            do GRPWIDENT=1 to &nvar. ;
            RESULTS=LIST[GRPWIDENT] ;
            output ;
            end ;
      drop &list_b. - &list_e. ;
run ;
/* (7) Storing stacked data in &data_out. */
data &data_out. ;
set prep_04 ;
run;

/* (8) Deleting auxiliary files using PROC DATASETS.*/
proc datasets  library=work memtype=data nolist;
delete prep_01 prep_02 prep_03 prep_04 ;
run ;

/* (9) Output in SAS Output */
title1 "Stacked dataset (method: 'automatic cleaning')" ;
title2 "Reduced to complete measurement series." ;

proc print data = &data_out.;
run ;
title ;


%mend STACK_CLEAN ;
```

Calling Macro STACK_CLEAN

When passing the parameters to the macro call, make sure to use the correct upper- and lowercase. The comments contain short explanations of the parameters; more detailed explanations are given after the macro call.

Program 4.6-6: Calling Macro STACK_CLEAN

```
%let root = %str(C:\);   /* root: Specify the storage path for the   */
                         /*       RTF output after %STR(), e.g. 'C:\'*/
%STACK_CLEAN
 (data_in=USTACKDbuggy,   /* data_in: Input Dataset, e.g. USTACKDbuggy */
  root=&root.,            /* root: Passing the storage path           */
  event=DIAGNOSIS,        /* event: Event, e.g. diagnosis             */
  ID_in=ID ,              /* ID_in: Key variable in input dataset     */
  list_b=RESULT1,         /* Column at list start, e.g. RESULT1       */
  list_e=RESULT3,         /* Column at list start, e.g. RESULT3       */
  nvar=3,                 /* Number of columns (vars) in list, e.g. 3 */
  outvar=RESULTS,         /* Name of output variable (will contain    */
                          /* the stacked values), e.g. RESULTS        */
   data_out=STACKD_CLEAN) /* data_out: Output dataset                 */
       ;
```

The following list is an explanation of the parameters in the macro call set by the user:

- **data_in=** Input data set (in the example: USTACKDbuggy).
  The input data set contains at least one key variable (numeric), one event variable (numeric), one group variable (numeric), and at least two measurement values (numeric; see list_b, list_e: RESULT1, RESULT2). A two-level name in the form LIBNAME.DSNAME can be specified.
- **root=** Storage path for RTF output.
  A user usually does not change this parameter. To set a different output location, a user can pass the desired path to SAS in the *%let* statement in parenthesis after STR.
- **ID_in=** Key variable from the input data set (in the example: ID).
  ID_in is numeric. The specification of ID_in is essential for data transformation and quality (6). An ID groups either simultaneous measurements on the same batch (assays) or successive measurements on the same case (repeated measurements). The values in ID_in do not always occur equally frequently. ID_in is checked for missing values (3).
- **event=** Variable for the event (in the example: CASECONT).
  event is numeric. Depending on the study design, event can be a coding for a diagnosis, a treatment, or also for a case control group assignment. event is checked for missing values (2).
- **list_b=** Variable at the beginning of a list of numeric measurement variables (2,6).
  The variable has a numeric suffix. Ideally, the measurement variables are differentiated by an incrementing integer in the suffix. Between list_b and list_e there are either more or no measurement variables. If there are measurement variables between list_b and list_e, they have a numerical suffix that is not less than list_b and not greater than list_e.
- **list_e=** Variable at the end of a list of numeric measurement variables (2,6).
  The variable has a numerical suffix. Between list_b and list_e there are either more or no measured value variables. Variables, which are not to be stacked, must *not* occur between list_b and list_e in the table. The measurement series formed by list_b and list_e are checked for completeness. Rows with missing values are excluded.
- **nvar=** Explicit specification to the system of how many variables form the variable list.
  This specification is essential for data transformation (2,6).
- **outvar=** Output variable in which the values of the stacked measurement variables are stored (6). It is often helpful if the name of outvar is similar to the names of the variables in list_b and list_e.
- **data_out=** Output data set.
  The output data set contains the input data set (originally in UNSTACK structure) now in STACK structure. A two-level name in the form LIBNAME.DSNAME can be specified. The output data set contains a key variable, the event variable, GRPWIDENT as group-within identifier, and a newly created variable outvar for the measured values.

Output 1 - Data Set STACKD_clean in STACK Structure

Figure 4.6-6: Data Set STACKD_clean in STACK Structure (Cleansed for Potentially Incorrect Data)

```
           Stacked dataset (method: 'automatic cleaning')
              Reduced to complete measurement series.


     Obs      ID     DIAGNOSIS     GRPWIDENT      RESULTS

      1        3         0             1             18
      2        3         0             2             14
      3        3         0             3             46
      4        4         1             1             21
      5        4         1             2             44
      6        4         1             3             40
```

Note: The data in the STACK structure contain complete measurement series and neither missing values nor duplicates in the ID column.

Output 2 - Data Set AUDIT in Original UNSTACK Structure

Figure 4.6-7: Data Set AUDIT in Original UNSTACK Structure

```
                   Compilation of excluded data.
                    - To be reviewed by user -

      Obs    ID    DIAGNOSIS    RESULT1    RESULT2    RESULT3

       1     .        1           23          7         24
       2     1        1           23          .         24
       3     2        0            .         19          7
       4     5        1            6         50         51
       5     5        1            6         50         89
```

Note: The AUDIT table contains all rows that do not comply with the integrity constraints, for example, all rows with missing values or duplicates in column ID, measurement series with missing values, and so on.

Output 3 - RTF Output

Figure 4.6-8: RTF Output (in the Example Under Drive C:\ )

Compilation of excluded data.

- To be reviewed by user -

Obs	ID	DIAGNOSIS	RESULT1	RESULT2	RESULT3
1	.	1	23	7	24
2	1	1	23	.	24
3	2	0	.	19	7
4	5	1	6	50	51
5	5	1	6	50	89

Note: The RTF output contains the same content as AUDIT, only in RTF instead of SAS format. Of course, you can also transform a table in the opposite direction, from STACK to UNSTACK.

4.7 Application 4: Macros to Retrieve System Information

This section presents various macros for retrieving information about SAS tables or from SAS Dictionaries. SAS Dictionaries contain extensive information about directories, SAS tables and views, and SAS catalogs, and are presented in Section 8.7.

- Macros (VARLIST, VARLIST2) to query the contents of SAS tables (DATA step approach).
- Macros (DO_VAR_EX, DO_VAR_EX2) to search a SAS table for a specific column (PROC SQL approach, DATA step approach).
- Macro IS_TERM and PROC OPTION to find something in the Dictionary Options (PROC SQL approach).

4.7.1 Query the Contents of SAS tables (VARLIST, VARLIST2)

The VARLIST macro lists the complete contents of SAS tables. VARLIST is based on the DATA step and the use of the SAS function SYSFUNC. SYSFUNC was already introduced in Section 4.3.1. When calling the VARLIST macro, you only need to specify the name of the SAS table whose variable names are to be output in the form of a list.

In the call, the table SASHELP.CLASS is specified; the variables of this table are to be output as a list.

Program 4.7-1: Query the Contents of SAS Tables (%VARLIST)

```
%macro VARLIST(data_in);
  %let FIELDS=;
  %let DINFO=%sysfunc(open(&data_in.));
  %let N=%sysfunc(attrn(&DINFO,nvars));
  %do i = 1 %to &N;
      %let FIELDS=&FIELDS %sysfunc(varname(&DINFO,&i));
  %end;
  %let rc=%sysfunc(close(&DINFO));
  %put Variables in &data_in. are: &FIELDS ;
%mend VARLIST ;
%VARLIST(sashelp.class) ;
```

Figure 4.7-1: %VARLIST Output in SAS Log

```
Variables in sashelp.class are: Name Sex Age Height Weight
```

The SAS log displays the names of five variables from the table SASHELP.CLASS.

The VARLIST2 macro also lists the complete contents of SAS tables. VARLIST2 is also based on the DATA step and the use of SYSFUNC. In addition to the variable names in list form, macro VARLIST2 also outputs the number of rows and columns.

In the call, the table SASHELP.CLASS is specified; the parameters of this table are to be output.

Program 4.7-2: Query the Contents of SAS Tables (%VARLIST2)

```
%macro VARLIST2 (data_in);
  %let FIELDS= ; %let N_ROWS= ; %let N_FIELDS= ;
  %let DINFO=%sysfunc(open(&data_in.));
  %let N_FIELDS=%sysfunc(attrn(&DINFO,nvars));
  %do i = 1 %to &N_FIELDS;
      %let FIELDS=&FIELDS %sysfunc(varname(&DINFO,&i));
      %let N_ROWS=%sysfunc(attrn(&DINFO,nobs));
  %end;
  %let rc=%sysfunc(close(&DINFO));
      %put Variables in &data_in. are: &FIELDS. ;
      %put Number of rows in &data_in. is: &N_ROWS. ;
      %put Number of columns in &data_in. is: &N_FIELDS. ;
%mend VARLIST2 ;
%VARLIST2(sashelp.class) ;
```

Figure 4.7-2: %VARLIST2 Output in SAS Log

```
Variables in sashelp.class are: Name Sex Age Height Weight
Number of rows in sashelp.class is: 19
Number of columns in sashelp.class is: 5
```

In the SAS log, the number of rows and columns is also displayed after the names of the variables from the table SASHELP.CLASS.

To search a SAS table for a *particular* variable, the following approach might be more appropriate for very large tables. These two DATA step approaches appear to be somewhat more flexible than the PROC SQL approach presented below.

4.7.2 Searching a SAS Table for a Specific Column (DO_VAR_EX and DO_VAR_EX2)

The two macros DO_VAR_EX and DO_VAR_EX2 search a selected SAS table to see whether it contains a specific variable. The macro DO_VAR_EX operates with an access to SAS dictionaries, whereas the macro DO_VAR_EX2 operates with DATA step functionalities.

DO_VAR_EX Macro

As mentioned above, the DO_VAR_EX macro uses dictionaries to search a selected SAS table to see whether it contains a specific variable. In the call of the DO_VAR_EX macro, you specify the directory in which the selected SAS table is supposed to be located, the name of the SAS table, and the variable to search for.

The exemplary call of DO_VAR_EX specifies the directory (SASHELP), the name of the SAS table to be searched (SHOES), and the variable to be searched (Sales).

Program 4.7-3: Searching a SAS Table for a Specific Column (%DO_VAR_EX)

```
%macro DO_VAR_EX(LIBRARY, MEMNAME, NAME);
proc sql;
       select LIBNAME, MEMNAME, NAME, MEMTYPE
       from DICTIONARY.COLUMNS
       where LIBNAME="&LIBRARY." and
             MEMNAME="&MEMNAME." and
             NAME="&NAME." and MEMTYPE="DATA";
quit;
%MEND DO_VAR_EX ;
%DO_VAR_EX(SASHELP,SHOES,Sales);
```

The result appearing in the SAS output window confirms that the searched variable could be found in the specified table in the specified directory.

Figure 4.7-3: Output: Searching a SAS Table for a Specific Column

```
Library Name            Member Name           Column Name          Member Type
-------------------------------------------------------------------------------
SASHELP                 SHOES                 Sales                DATA
```

If the search was unsuccessful, the following message appears in the SAS log:

SAS Log 4.7-1: Output in SAS Log

```
NOTE: No rows were selected.
```

This note should be understood to mean that the specified variable either does not exist at all (or at least not in the specified directory or table), or that one or more of these specifications have been passed to SAS incorrectly written.

DO_VAR_EX2 Macro

The DO_VAR_EX2 macro uses DATA step functionalities to search a selected SAS table to see whether it contains a specific variable. In the call of the DO_VAR_EX2 macro, you specify the SAS table that is supposed to contain the specified column and the variable to search for. A limitation of this program, apart from its somewhat more complex programming compared to the PROC SQL and dictionaries approach, is that this variant only works for the correct capitalization of column names ("NAME" versus "Name").

In the call of DO_VAR_EX2, the name of the SAS table to be searched (SASHELP.CLASS) and the variable to be searched ("name") are specified.

Program 4.7-4: Searching a SAS Table for a Specific Column (%DO_VAR_EX2)

```
%macro DO_VAR_EX2(data_in, var);
    %let CHECK= ;
    %let DINFO=%sysfunc(open(&data_in));
    %let N_FIELDS=%sysfunc(attrn(&DINFO,nvars)) ;
%do i=1 %to &N_FIELDS ;
```

```
      %let FIELDS=%sysfunc(varname(&DINFO,&i)) ;
       %if &FIELDS=&var %then %do ;
         %let CHECK=Variable found ;
        %let i=&N_FIELDS ;
       %end;
       %else %let CHECK=Variable not found ;
    %end ;
  %put CHECK:  &CHECK;
  %let rc=%sysfunc(close(&DINFO));
  %mend ;
  %DO_VAR_EX2(SASHELP.CLASS,Name) ;
  %DO_VAR_EX2(SASHELP.CLASS,NAME) ;
```

Figure 4.7-4: %DO_VAR_EX2 Output in SAS Log

For each call, the SAS log reports whether the searched variable could be found in the specified table or not.

```
nnn      %DO_VAR_EX2(SASHELP.CLASS,Name) ;
CHECK:  Variable found
nnn      %DO_VAR_EX2(SASHELP.CLASS,NAME) ;
CHECK:  Variable not found
```

If a variable was not found, this means that the specified variable either does not exist at all in the specified table, or that one or more of these specifications were passed to SAS in a misspelled form. The SQL approach seems to be a bit more elegant compared to the DATA step approaches presented previously.

4.7.3 Search in the Dictionary.Options (IS_TERM)

The fact that SAS can provide a lot of detailed information is an advantage that can turn into a disadvantage if you don't know exactly what you want to find or don't know the exact term for a SAS option. The following approach shows how to find the required information in three steps, from the first keywords (in whole or in part) to the exact information about options, settings, and descriptions. In the first step, the name of an option, setting, or description is specified in the IS_TERM macro. In later steps, the information can be narrowed down further.

Step I: Search for the Name of an Option, Setting, or Description

In the macro IS_TERM, the name of an option, setting, or description is specified. The specified search term can also be part of the name, for example, "Macr" instead of "Macro". The macro IS_TERM is case sensitive.

Program 4.7-5: Step I: Searching DICTIONARY.OPTIONS (%IS_TERM)

```
%macro IS_TERM(FINDME) ;
proc sql ;
        select group, OPTNAME ,SETTING, OPTDESC
          from   DICTIONARY.OPTIONS
          where  index(upcase(OptName),upcase("&FINDME."))
            or   index(upcase(Setting),upcase("&FINDME."))
            or   index(upcase(OptDesc),upcase("&FINDME."))  ;
         quit ;
%mend IS_TERM ;

%IS_TERM(macr) ;
```

The call of IS_TERM in Program 4.7-5 asks the macro to search for "Macr" in options, settings, or descriptions, regardless of upper- and lowercase. The hits from IS_TERM are output in the SAS output.

Figure 4.7-5: Output: Step I: Searching DICTIONARY.OPTIONS

```
Option Group    Option Name    Option Setting   Option Description
-----------------------------------------------------------------------------------
MACRO           MACRO          MACRO            Enables the macro facility.
MACRO           MAUTOSOURCE    MAUTOSOURCE      Enables the macro autocall feature.
MACRO           MCOMPILE       MCOMPILE         Allows new macro definitions.
MACRO           MERROR         MERROR           Issues a warning message for an unresolved
                                                macro reference.
MACRO           MEXECNOTE      NOMEXECNOTE      Displays the macro execution information
                                                in the SAS log when the macro is invoked.
Output massively abbreviated. MEXECNOTE highlighted by author.
```

The column on the left of Figure 4.7-5 shows the higher-level option groups (the SAS Option Group MACRO). The next column shows the name of a specific option from this group (for example, MEXECNOTE). Other higher-level option groups are DATAQUALITY, GRAPHICS, or PERFORMANCE; they are not shown because they do not meet the search criterion. The names for SAS Option Groups can be the same as specific options. For example, the option MACRO belongs to the higher-level group MACRO, but other SAS options also belong to this group, such as MEXECNOTE. The individual options are described on the right. The next column shows the current setting of the respective option in SAS (for example, NOMEXECNOTE). The next column displays descriptions of the options. For space reasons, the original output is only shown in excerpts.

The next two steps now enable us to take a closer look at the current settings in the options within a SAS Option Group.

Step II: Look into a Group of Options (SAS Option Group Name)

In a second step, you can pass the name of a higher-level option group (for example, MACRO) to PROC OPTIONS (after GROUP=). PROC OPTIONS lists the current system option settings to the SAS log and provides you with an easy overview of all current settings within the specified SAS Option Group.

The specification of a concrete SAS Option Name such as MEXECNOTE does not yet work with the SAS sample code for Step II and is instead illustrated in Step III.

Program 4.7-6: Step II: Look into a Group of Options (SAS Option Group)

```
proc options group = MACRO ;
run ;
```

The information from PROC OPTIONS GROUP= is output in the SAS log.

Figure 4.7-6: SAS Option Group Output in SAS Log (SAS 9.4)

```
        SAS (r) Proprietary Software Release 9.4   TS1M7

Group=MACRO
NOCMDMAC           Does not check window environment commands for command-style macros.
NOIMPLMAC          Does not check for statement-style macros.
MACRO              Enables the macro facility.
MCOMPILE           Allows new macro definitions.
MAUTOSOURCE        Enables the macro autocall feature.
NOMFILE            Does not write MPRINT output to an external file.
NOMAUTOCOMPLOC     Does not display the autocall macro source location in the SAS log
                   when the autocall macro is compiled.
NOMAUTOLOCDISPLAY  Disables the macro facility from displaying the autocall macro
                   source location in the log.
Output massively abbreviated.
```

On the left, the names of the current system option settings of the SAS Option Group MACRO are shown, on the right their description.

Step III: Looking into a Specific SAS Option Name

After you have obtained an overview of all groups and names of SAS options in the previous steps, in Step III you can now take a closer look at an option found. You simply specify a SAS Option Name (MEXECNOTE) after OPTION=.

Program 4.7-7: Step III: Look into a Specific SAS Option Name

```
proc options define value option = MEXECNOTE ;
run ;
```

The keyword DEFINE requests a short description of the option, the SAS Option Group, and the option type. It also indicates whether the option can be set or restricted. VALUE shows the value and scope of the specified option and also how the value was pre-set. The information about the PROC OPTIONS OPTION = side is also output in the SAS log. The output of the example shows that the SAS Option Name MEXECNOTE (of the Option Group MACRO) is currently set to NOMEXECNOTE.

Figure 4.7-7: SAS Option Name Output in SAS Log (SAS 9.4)

```
Option Value Information For SAS Option MEXECNOTE
     Value: NOMEXECNOTE
     Scope: Default
     How option value set: Shipped Default

Option Definition Information for SAS Option MEXECNOTE
     Group= MACRO
     Group Description: SAS macro language settings
     Description: Does not display the macro execution information in the SAS log
               when the macro is invoked.
     Type: The option value is of type BOOLEAN
     When Can Set: Startup or anytime during the SAS Session
     Restricted: Your Site Administrator can restrict modification of this option
     Optsave: PROC Optsave or command Dmoptsave will save this option
```

4.8 Application 5: Creating Folders for Data Storage

As indicated by the macro for looping SAS files, it can be very handy to specify a storage path using a macro variable. The approach presented there requires that the desired storage paths are known and, above all, exist. This centrally controlled definition of the storage directory saves a lot of entering work, especially with very long paths, due to its uniformity. The central definition prevents the data from being inadvertently stored somewhere else. The central definition of desired but existing paths can be topped by a SAS macro that also allows creating the desired storage paths at the same time.

The following SAS macro combines both requirements: specifying and creating storage paths and ensuring that the data is stored where it should be stored. This approach also assigns a time stamp. This time stamp causes different paths to be created if the program is executed on different days. As a result, different result versions are not easily overwritten if several users work with the same data or programs.

This approach might have certain limitations. First, there could be physical limitations depending on whether certain storage locations are physically available at all. For example, you cannot simply define a path to the drive "D:\" if this drive does not exist. Depending on the operating-system environment, there might also be restrictions on the length of the path. The following program was successfully tested on several Windows operating systems.

4.8.1 Macro MYSTORAGE and Call

Program 4.8-1: Application: Creating Folders for Data Storage

```
/* I. Definition of the storage directory to be created  */
%let MYPATH=%str(C:\Tables\RUN_&sysdate9.\);

/* II. Command to create the directory  */
%sysexec md "&MYPATH.";

/* III. Command to store data in the above directory */
%macro MYSTORAGE(MYPATH=);
ods listing close;
ods TAGSETS.CSVNOQ file="&MYPATH2.TABLE_&sysdate9..txt" ;
   proc print data=SASHELP.CLASS noobs ;
   run;
ods _ALL_ close ;
ods listing ;

%mend MYSTORAGE ;

%MYSTORAGE(MYPATH2=&MYPATH.) ;
```

Explanation

- **Step I** defines the desired path. If all paths are defined early and uniformly, even this seemingly inconspicuous measure can increase the performance of SAS a little.
- **Step II** creates the directory for the desired path. The approach using %SYSEXEC is only one of several variants offered by SAS. A combination of %SYSFUNC and the SAS function DCREATE to create an external directory is also possible. Both path specifications include the automatic SAS macro variable SYSDATE9. as timestamp. The resulting time stamp in the path or directory has the form DDMMMYYYY, for example, 12NOV2010. This specification has the effect that different paths are created if the program is executed on different days, for example, in the form: "C:\Tables\RUN_10NOV2010\" or "C:\Tables\RUN_12NOV2010\" and so on. The consequence is that different result versions are not easily overwritten if several users work with the same data or programs, but that the data is stored in different directories created on a daily basis. Further possibilities for the definition of paths or directories are the explicit inclusion of an abbreviation for the concrete user in the definition of directory, path, or table or, if necessary, the implicit logging of the respective SAS user via the SAS macro variable SYSUSERID. Depending on the operating system, a DOS window might open. This window can easily be closed again with the command "EXIT".
- **Step III** is a command to store data via &MYPATH2. includes the path specification. The time stamp &SYSDATE9. is also used here to name the SAS tables to be stored, for example, "TABLE_12NOV2010.txt". The TAGSETS approach is explained in Section 9.3.

4.9 Application 6: Consecutive "Exotic" Names for SAS Columns ("2010", "2011", ...)

Data often has to be stored in columns with names especially for export that do not correspond to SAS conventions. Although SAS requires various conventions for defining column names such as a maximum length of 32 characters, the first character is a letter ("A", "B", "C", ..., "Z") or underscore ("_"), without special characters (except for "_", "$", "#" or "&"), and without blanks, SAS is flexible enough to create columns and data in a format that can be called almost arbitrary.

Since version 9.1, SAS provides the SAS option VALIDVARNAME. The VALIDVARNAME option with the keyword ANY enables you to give columns almost arbitrary names. For more information about VALIDVARNAME, see Section 9.3. The main functions of the EXOTICS macro are:

- Creating a single "exotic" variable (string) with the name "!?".
- Filling of EXOTICVAR (see above) with a constant string entry ("*/|").
- Creating a list of consecutive, "exotic" variables (numerical) with the names "2010", "2011", and so on.
- Filling the consecutive "exotic" variables with a constant ("0").
- Linking the created variables "left" and "right" of SASHELP.CLASS.
- Completing several checks on the correctness of the inputs in the macro call.

Further notes can be found at the end of the program.

4.9.1 Sample Output

Figure 4.9-1: Sample Output: Consecutive "Exotic" Names for SAS Columns

```
!? NAME SEX AGE HEIGHT WEIGHT 2010 2011 2012 2013 2014 2015 2016 2017 2018 2019 2020

*/| Alfred  M 14  69.0  112.5     0    0    0    0    0    0    0    0    0    0    0
*/| Alice   F 13  56.5   84.0     0    0    0    0    0    0    0    0    0    0    0
*/| Barbara F 13  65.3   98.0     0    0    0    0    0    0    0    0    0    0    0
*/| Carol   F 14  62.8  102.5     0    0    0    0    0    0    0    0    0    0    0
*/| Henry   M 14  63.5  102.5     0    0    0    0    0    0    0    0    0    0    0
```

4.9.2 Macro EXOTICS

Program 4.9-1: Application: Consecutive "Exotic" Names for SAS Columns

```
/* I. Definition of storage location */
%let ROOT=%str(C:\My data\Tables\);

%macro EXOTICS (INDATA=, START=, STOPP=, BASEVAL=, OUTDATA=, EXOTICVAR=, ROOT2= );

/* II. Option for creating "exotic" column names */
options validvarname=any ;

/* III. Checks on correctness of inputs in macro call */
%let check1=%sysevalf(&STOPP. - &START.) ;
    %if &check1. <= 0 %then %put
"--------- ERROR: Start/Target year do not match ---------" ;
%let check2=%sysevalf(&STOPP. - 2000) ;
    %if &check2. <= 0 %then %put
"--------- ERROR: Target year too low ---------" ;

%let MAXVAL=%sysevalf(&STOPP. - 2000) ;

%if &START. >= 2010 %then %do ;
      data STEP_1 ;
        retain "&EXOTICVAR."n NAME SEX AGE HEIGHT WEIGHT ;
              drop  i ;

/* IV. Creating the consecutive list, still with "_" prefix,
      as 2-digit year values _2010 to _20maxVAL */
                array _20[&maxVAL.];
                  do i = 10 to dim(_20);
                    _20[i] = &BASEVAL. ;
                  end ;
                drop _201 -- _209 ;
                set &INDATA. ;
```

```
/* V. Creating the individual exotic variables */
            "&EXOTICVAR."n= '*/|';
            run;
            %end;

/* VI. Renaming the columns in the list in "exotic" names */
        proc datasets library=work;
          delete &OUTDATA. ;
        quit;
         proc sql noprint ;
            select NAME||"="||"'"||trim(substr(NAME,2,4))||"'"n"
            into :RENAMEV separated by " "
            from SASHELP.VCOLUMN
            where Libname="WORK" and memname="STEP_1" and
                prxmatch("/_20/", NAME) ;
        quit;
            proc datasets library=work memtype=data nolist;
              change STEP_1=&OUTDATA. ;
              modify &OUTDATA. ;
              rename &RENAMEV ;
            quit;

            proc print data=&OUTDATA. (obs=5) noobs ;
            run;

%mend EXOTICS ;
```

4.9.3 Explanation

- **Step I** defines the desired path.
- **Step II** initializes the SAS option for creating "exotic" column names.
- **Step III** contains various checks for the correctness of the inputs in the macro call, for example, whether the START year is smaller than the STOPP year. The MAXVAL parameter is obtained to determine the end (=maximum) of the consecutive list of "year" variables created.
- **Step IV** creates the consecutive list of "year" variables using the "_" prefix. Two-digit "year" numbers are created in the form "_2010" to "_20maxVAL" (the program is written for years between 2010 and 2099). Additional arrays can be created for additional columns.
- **Step V** creates the single "exotic" variable at the position of the macro variable &EXOTICVAR. Alternatively, users could control the filling of &EXOTICVAR. by means of "*/|" by another macro variable in the call.
- **Step VI** renames the columns from the list (still with "_" prefix) to "exotic" names without a prefix. The result is column names in the form "2010" to "_20maxVAL". The specific functionality of this renaming step using a SAS view is explained in Section 9.2 on working with SAS dictionaries.

4.9.4 Sample Call

Program 4.9-2: Sample Call

```
%EXOTICS (START=2010,
        STOPP=2020,
        BASEVAL=0.0,
        INDATA=SASHELP.CLASS,
        OUTDATA=EXOTICDATA,
        EXOTICVAR=!?,
        ROOT2=&ROOT.) ;
```

The entries in START and STOPP request a total of eleven columns with names from "2010" to "2020". All variables in this list are filled with 0.0. SASHELP.CLASS is specified as the SAS file to be subjected to all these transformations.

The temporary SAS file EXOTICDATA is created as the output file, into which all these columns with "exotic" column names are to be saved. The single "exotic" column is named "!?". The desired path from I. is included under ROOT.

4.10 Application 7: Converting Entire Lists of String Variables to Numeric Variables

After importing external data into SAS, you might find out that the columns for numerical entries had already been defined incorrectly as being of type String before the import. Redefining the type of hundreds or thousands of columns "by hand" is, of course, out of the question. The following PROC SQL application enables you to convert complete SAS files of type String into a numeric format. The values of the former string variables are kept.

The program assumes two requirements:

1. One numerical format is sufficient as a **uniform** format; in other words, the same numerical format is assigned to all former string variables. If different numerical formats are to be assigned, it is necessary to subdivide the string variables into a subset for each required numerical format and to subject each to a separate conversion run.
2. All variables in the SAS file are converted to a **numeric** format; in other words, SAS also tries to assign a numeric format to "real" string variables with text entries. For the original text entries, this results in missing values without exception. If this is not desired, it is necessary to store the "real" string variables in a separate SAS file to ensure that they do not undergo any conversion and to merge them to the file with the converted data later, if necessary. In order to be able to merge the separate SAS files easily, it is helpful to create a key variable such as ID, if not already available.

4.10.1 Explanation

In principle, the following program passes through six steps:

1. Creating a SAS file for the data to be converted. This step is not necessary if all variables in the SAS file are to be converted.
2. Using PROC CONTENTS, a SAS file VAR_LIST is created, which contains a list of the names and types of the columns from step (1).
3. Using the following DATA step, only the names of the string variables are kept; then a copy is created for each column name with the additional suffix "_n" at the end of the column name. The format of the variable names with the suffix "_n" is also of type String.
4. PROC SQL then creates three macro variables: CHAR_LIST as a list of the string variables to be converted, NUM_LIST as a list of the newly created variable names with the suffix "_n", and RNAME_LIST as a statement for the pairwise renaming in the RENAME statement of the subsequent DATA Step.
5. Using the entries in the macro variables from (4), the actual "conversion" of string is performed in the DATA step in two steps: First, all entries from the original string variables are converted into the specified numerical format and stored in the corresponding columns with the suffix "_n". Then the now numerical columns with the "_n" suffix are renamed to the names of the original string variables. This completes the "actual" conversion.
6. Finally, the converted data are remerged with the data not to be converted. This step is not necessary if all variables in the SAS file have been converted. Further instructions can be found at the end of the program.

4.10.2 Application

Program 4.10-1: Application: Converting Entire Lists of String Variables to Numeric Variables

```
/* (1) Separate SAS files for data (not) to be converted */
```

```
data ALL_BEFORE ;
input id var_a $ var_b $ var_c $
         var_d var_x $ var_y $ var_z $;
datalines;
1 50 11 1 222 22 1 8
2 35 12 2 250 25 2 8
3 75 13 3 990 99 3 8
;
run ;


/* Data not to be converted */
data DATAkeep
        (drop= VAR_A VAR_B VAR_C
               VAR_X VAR_Y VAR_Z );
set ALL_BEFORE ;
run ;
/* Data to be converted */
data DATAcvrt
      (keep=ID VAR_A VAR_B VAR_C
            VAR_X VAR_Y VAR_Z );
set ALL_BEFORE (rename=(ID=IDnum));
        ID=put(IDnum,4.) ;
run ;


/* (2) Create SAS file VAR_LIST with name and type of columns from step (1) */

proc contents data=DATAcvrt
                out=VAR_LIST(keep=NAME TYPE)
noprint ;
run ;


/* (3) Keep the string variables; create a copy of column names with additional suffix "_n"
*/

data VAR_LIST ;
   set VAR_LIST ;
   if TYPE=2 and NAME ne 'id';
   NEUNAME=trim(left(NAME))||"_n";
run ;


/* (4) Create three macro variables CHAR_LIST, NUM_LIST, and RNAME_LIST */

options symbolgen;
proc sql noprint;
select trim(left(NAME)), trim(left(NEUNAME)),
      trim(left(NEUNAME))||'='||trim(left(NAME))
                 into :CHAR_LIST separated by ' ',
                      :NUM_LIST separated by ' ',
                      :RNAME_LIST separated by ' '
      from VAR_LIST ;
quit ;


/* (5) Store in a numeric format and rename to the original column names */

data DATAcvrt_2 ;
    set DATAcvrt  ;
   array chrvar(*) $ &CHAR_LIST ;
   array numvar(*) &NUM_LIST ;
          do i = 1 to dim(chrvar) ;
                     numvar(i)=input(chrvar(i),8.) ;
           end ;
   drop i &CHAR_LIST ;
```

```
    rename &RNAME_LIST ;
run ;

/* (6) Merge converted and not converted data   */

data ALL_AFTER ;
    retain ID VAR_A VAR_B VAR_C VAR_D VAR_X VAR_Y VAR_Z ;
    merge DATAkeep DATAcvrt_2 ;
    by ID ;
run ;
```

4.10.3 Requesting Outputs Before and After the Conversion (PROC CONTENTS)

Program 4.10-2: Requesting Outputs Before and After the Conversion (PROC CONTENTS)

```
proc contents data=ALL_BEFORE ;
run ;

proc contents data=ALL_AFTER ;
run ;
```

Figure 4.10-1: Output: WORK.ALL_BEFORE (Extract)

```
The CONTENTS Procedure
Data Set Name           WORK.ALL_BEFORE
Alphabetic List of Variables and Attributes

        #    Variable    Type    Len

        1    id          Num       8
        2    var_a       Char      8
        3    var_b       Char      8
        4    var_c       Char      8
        5    var_d       Num       8
        6    var_x       Char      8
        7    var_y       Char      8
        8    var_z       Char      8
```

Figure 4.10-2: Output: WORK.ALL_AFTER (Extract)

```
The CONTENTS Procedure
Data Set Name           WORK.ALL_AFTER
Alphabetic List of Variables and Attributes

        #    Variable    Type    Len

        1    id          Num       8
        2    VAR_A       Num       8
        3    VAR_B       Num       8
        4    VAR_C       Num       8
        5    VAR_D       Num       8
        6    VAR_X       Num       8
        7    VAR_Y       Num       8
        8    VAR_Z       Num       8
```

4.10.4 Further Notes on the Six Steps Presented

1. The syntax for splitting the ALL_BEFORE file into data to be converted and data not to be converted differs in two points: KEEP for the data to be converted and DROP for the data not to be converted. However,

depending on the data volume, reverse use of DROP and KEEP might be useful. Depending on the type of the key variable, it is either kept or excluded. If no key variable exists, it is recommended to create one, either via PROC SQL (MONOTONIC) or the DATA step (RETAIN). (See Section 8.3.)

2. The created auxiliary file VAR_LIST does not itself contain any data, but it contains metadata such as the list of names (NAME) and type (TYPE) of the data to be converted from step (1). You must ensure that this file actually contains only those string variables that are to be converted. Otherwise, text entries are converted into missing values without exception, or completely filled text columns are converted into numeric columns filled with missing values without exception.

3. In the DATA step, a copy is created for each name of a string variable with the additional suffix "_n" at the end of the column name.

4. The previously set macro system option SYMBOLGEN enables you to view the contents of the macro variables CHAR_LIST, NUM_LIST, as well as RNAME_LIST in the SAS log.

5. This double step is based on two arrays for creating new variables, as well as a DO/END loop, which converts the string entries of the variables in the first array of string variables (original, without "_n") into numeric entries of the variables in the second array (with "_n"). Finally, the numeric variables are renamed to the names of the variables in the first array (without "_n"). This double step could be simplified if renaming back to the names of the original string variables is not necessary.

Chapter 5: SQL for Geodata

Business intelligence and data analysis commonly include the calculation of distances and related parameters such as time or costs. In addition to these basic calculations, you might need to calculate the shortest, fastest, or even cheapest distance. Logistics companies calculate delivery routes where a driver does not have to change lanes while driving, always can exit the vehicle on the sidewalk, and deliver as much general cargo as possible with as little fuel as possible. More sophisticated analyses using GIS systems are easily able to take into account numerous other parameters such as the time of day and the usual traffic volume, maximum permitted speeds, slopes and differences in altitude, as well as weather and road conditions.

Section 5.1 introduces basic approaches to the analysis of geodata and the calculation of geographical distances. This type of analysis is recommended if calculating approximate data is sufficient, and details of routes, tracks, and distances can be neglected at first. Section 5.2 introduces several SQL queries for coordinate data using subqueries, views, inner joins, and the like. Section 5.3 introduces visualizing data by means of maps and the interaction between the new PROC SGMAP and SQL, handling response and map data. Visualizations are recommended when you need to prioritize the efficient communication of complex information over the mathematical precision of individual key figures.

5.1 Geographical Data and Distances

These SQL and SAS examples can be extended by further parameters, more complex formulas, or completely switched to special GIS systems (provided that the required information content in the form of the data is given). SAS offers users numerous additional analysis options such as the new Geo-SQL used in combination with Teradata.

Section 5.1.1 introduces the calculation of distances in flat, two-dimensional space. Section 5.1.2, on the other hand, presents the determination of distances in spherical space.

The pedantic reader will notice at this point that there are smooth transitions when choosing between approaches for two- and three-dimensional space:

- A distance should be calculated using formulas for **spherical space**, the greater the distance and the greater the curvature is between two or more points.
- A distance should be calculated using formulas for **two-dimensional** space if, regardless of the extent of the distance, there is no curvature between two or more points.
- However, the **shorter a distance** between two points in curved (spherical) space, the more likely it is to be obtained by a direct line, that is, a formula for two-dimensional space. The reasoning behind this is that the effect of the curvature decreases with increasing relative shortness of the distance between two points.

In comparison to the "usual" analysis of metric data, no distinction is made between N=1 and N=n approaches in the analysis of geodata. In the case of geodata, the aim is to optimally calculate each route, regardless of whether it is many routes or a single route (which could possibly consist of many partial routes). A simple example for a single case analysis (N=1) would be to determine the optimal route for the journey to work. Examples of mass data analyses (N=n) are, for example, the average travel distance of all customers of a shopping center or the distances of patients to a care center. This section introduces two basic approaches to calculating distances.

5.1.1 Distances in Two-Dimensional Space (Basis: Metric Coordinates)

Imagine that there is a checkered piece of paper lying flat in front of you. On this sheet of paper there are two crosses. For the sake of simplicity, each cross is located at one of the intersections of the horizontal and vertical lines of the checkered sheet. The two points can represent any two objects. The distance on the sheet of paper is the distance to be calculated between the two crosses. One of the points is a starting point and the other point is a destination.

Possible examples from everyday life might be:

- **Example 1:** Point 1: Departure point; Point 2: Arrival point; Distance: Approach route.
- **Example 2:** Point 1: Location of dispatch; Point 2: Location of receipt; Distance: Dispatch route.
- **Example 3:** Point 1: Getting into the taxi; Point 2: Getting out of the taxi; Distance: Route.
- **Example 4:** Point 1: Start of mobile media consumption by switching on a corresponding app; Point 2: Update of mobile media consumption; Distance: media consumption distance.

The calculated distances become very significant in combination with one more variable:

- When traveling to work, for example, the **mileage allowance** per kilometer of the calculated distance often plays a role.
- The expected **duration** plays a role in planning the shipping route (especially of perishable products).
- In the case of taxis, the expected distance plays a role in the preliminary **costing**, for example, whether you want to cover the distance on foot or whether you would rather afford a taxi.

Experience shows that applications of distances are much more complex than these examples. When driving to work, the departure time is added to the calculation, as well as whether one has a sufficient amount of fuel.

In the case of mobile media consumption, in addition to the route of media consumption, the mode (mobile phone, radio, TV, laptop, and so on) and content of the media consumed (programs, genres, advertising, stations, and so on) can be recorded and analyzed. For broadcasters, the range of their programs is often also of interest, for example, how far local TV or radio stations are listened to or seen.

For two or more points that are on a plane, there are numerous ways to calculate the distance between them. (Points in curved space are discussed in Section 5.1.2.) In the following section, three approaches are presented: straight line, City-Block, and Minkowski.

Straight Line (Euclidean Distance)

A straight line (Euclidean distance) measures the distance between two points on a plane.

Assumptions: The Euclidean distance can be calculated if (a) the distance actually corresponds to a real, direct, and straight line in empirical reality or (b) no object in empirical reality would block the drawing of an imaginary line.

The calculation of the Euclidean distance according to Pythagoras is based on the hypotenuse of a right-angled triangle.

City-Block Measure (Block Metric or Manhattan Distance)

The City-Block metric measures an approximately straight line between two points on a plane.

Assumptions: The City-Block metric can be calculated if the distance is approximately a straight line. The distance does not need to correspond to a real or imaginary direct straight line in empirical reality, in case an object such as a building blocks a real or imaginary line. The distance from start to destination can be covered by walking or driving "around the block," hence the term "City-Block" metric.

In contrast to the Euclidean distance, the calculation of the City-Block metric is based on the two sides of the right-angled triangle.

Minkowski Measure (L(p))

The Minkowski metric for two-dimensional spaces uses the property that the Euclidean distance and the City-Block metric are mathematically closely related. The Euclidean distance is based on the direct path to the target (so to speak "through walls"). The City-Block metric is based on the direct path to the target "around walls." If a taxi on the way to a destination regularly turns left and right and the distances covered are of equal length (and add up to equal parts), the City-Block measure approaches the ideal of the direct straight line of the Euclidean distance with infinitely frequent and fine regular turns. The formulas of Euclidean distance and City-Block measure can be transformed into a common formula based on the Minkowski metric in which only the exponent k is varied.

Assumptions: In contrast to the Euclidean Distance and City-Block measures, the Minkowski metric is a mathematically calculated estimate, the result of which cannot necessarily be observed in the two-dimensional space of empirical reality (especially with k>2).

For computing distances, SAS offers several options:

- User-defined approaches using SAS 9.4 (see Section 5.1.1.1)
- PROC DISTANCE (see Section 5.1.1.2)
- SAS functions like EUCLID, GEODIST, or ZIPCITYDISTANCE. (See Section 5.1.1.3.)

The following discussion starts with the user-defined approach because it has the advantage to see how distances are calculated.

5.1.1.1 Rowwise Calculation of Distance Measures (SAS 9.4)

Frank Ivis (2006) kindly allowed the reproduction of his macros presented at NESUG 26 for the calculation of distances by means of Euclidean distance, City-Block metric, and Minkowski metric. The macro DISTANCE2D contains five parameters: x1/y1 represent the coordinates of the starting point in 2-dimensional space. x2/y2 represent the coordinates of the destination point in 2-dimensional space. The k value (exponent) is passed to SAS as the fifth parameter. k corresponds to the exponent adjustable by the user:

- k=**1**: Block measure.
- k>**1**&<**2**: Minkowski metric1 (result between City-Block metric and Euclidean distance).
- k=**2**: Euclidean distance.
- k>**2**: Minkowski metric2 (result corresponds to the longer one of two sides of the City-Block metric).

A close look reveals that Ivis' macro contains two equations. If the exponent is 1 <= k <=2, the simple Minkowski metric is calculated. With k > 2, an alternative formula for computationally intensive calculations is used.

Program 5.1-1: Defining the DISTANCE2D Macro

```
%macro DISTANCE2D (x1=, y1=, x2=, y2=, k=) ;
    ( max( abs(&x2 - &x1), abs(&y2 - &y1) ) * (&k > 2) )
  + ( ( abs(&x2 - &x1)**&k + abs(&y2 - &y1)**&k )**(1/&k) )
  * ( 1 <=&k <=2 )
%mend DISTANCE2D ;
```

When calculating distances, the values in the x and y coordinates must be in the same scale and unit (Schendera, 2010, Chapter 1.2). Different measures such as miles and kilometers or different units such as 100 m and 0.1 km should be standardized before the calculation.

In the following example, the attributes for the columns ID to DEST_Y are first defined in a DATA step to create the SAS data set POINTS. Then the macro DISTANCE2D is called several times to calculate the desired distances. Apart from the four parameters x1 to y2 for the coordinates of the target or receiving points in two-dimensional space, DISTANCE2D contains also the parameter k. The values for k were varied "manually" in order to be able to calculate distances as Euclidean distance, City-Block metric, and Minkowski metric. As soon as the DATA step is executed, the requested distance values are calculated and output in the SAS output while the data for the coordinates are read in.

Program 5.1-2: Data Set POINTS (x/y System) Calling the DISTANCE2D Macro

```
data POINTS ;
input id $2. +1 strt_x 2. +1 strt_y 2. +1 dest_x 2. dest_y +1 ;
   BLOCK       = %DISTANCE2D (x1=strt_x, y1=strt_y, x2=dest_x, y2=dest_y, k=1) ;
   MINKOWSKI1  = %DISTANCE2D (x1=strt_x, y1=strt_y, x2=dest_x, y2=dest_y, k=1.2) ;
   EUCLID      = %DISTANCE2D (x1=strt_x, y1=strt_y, x2=dest_x, y2=dest_y, k=2) ;
   MINKOWSKI2  = %DISTANCE2D (x1=strt_x, y1=strt_y, x2=dest_x, y2=dest_y, k=3) ;
format BLOCK MINKOWSKI1 EUCLID MINKOWSKI2 8.2 ;
   put 'City-Block=' BLOCK           'Minkowski1=' MINKOWSKI1
       'Euclidian Distance=' EUCLID  'Minkowski2=' MINKOWSKI2 ;
datalines ;
1 11 53 19 45
2 82 74 68 58
3 20 20 15 33
4 13 61 63 17
5 67 31 92 41
;
run ;
```

While reading in the coordinate data, the DATA step calculates the distances by calling the DISTANCE2D macro each time. The column pair strt_x/strt_y contains the coordinates of the starting point. The two values 11 and 53 define the position of the first starting point in two-dimensional space. Similarly, the column pair dest_x/dest_y contains the coordinates of the first destination point (19, 45). k corresponds to the values set by the user.

From the SAS output for the five coordinate pairs passed, City-Block results in significantly different distances for the same coordinate.

Figure 5.1-1: DISTANCE2D Macro Output in SAS Log

```
City-Block=50.00 Minkowski1=46.73 Euclidian Distance=42.76 Minkowski2=42.00
City-Block=60.00 Minkowski1=57.20 Euclidian Distance=54.33 Minkowski2=54.00
City-Block=38.00 Minkowski1=35.83 Euclidian Distance=33.38 Minkowski2=33.00
City-Block=16.00 Minkowski1=16.00 Euclidian Distance=16.00 Minkowski2=16.00
City-Block=45.00 Minkowski1=42.73 Euclidian Distance=40.31 Minkowski2=40.00
NOTE: The data set WORK.POINTS has 5 observations and 9 variables.
```

For the pair of coordinates 11/53 and 19/45, City-Block achieves 50, Minkowski1 47.73, Euclidian Distance 42.76, and Minkowski2 42.00. The City-Block values tend to be larger than Minkowski1, both are larger than the Euclidian Distance, and the latter is at least as large as Minkowski2. When evaluating a distance, the results of the individual measures can differ massively. In the example, City-Block is about 25% larger than the Euclidian Distance or Minkowski2.

The distance measure that should be used can be decided from two points of view:

- From a theoretical point of view, the measures should be used that are **most appropriate** for the measured distance. Since the measures themselves are often a simplification of empirical reality and its measurement, you cannot accurately assess this in practice. As a result, it is not clear which of the metrics best measures empirical reality.
- However, it might be possible that users want to answer this question from the practical side, namely by which values of which measure they will achieve the **best results** in practice, in a simulation, or a prediction.

- The decisive difference is the **interpretation**. The chosen modeling of the geographical reality with measure x achieves the best results in practice and not that metric x measures the geographical reality best. However, as a general rule, the longer the distances, the more similar the estimates obtained should be.

5.1.1.2 Matrix-wise Calculation of Distance Measures (PROC DISTANCE)

PROC DISTANCE is a SAS procedure designed to calculate numerous measures of distance and similarity or dissimilarity between the observations in a SAS data set. The special features of the DISTANCE procedure are:

- A wide range of distance measures, for example, Euclidean distance, City-Block metric, Minkowski (here called L(p)), Chebychev, Squared Euclidean distance, Size distance, Shape distance, Binary Lance and Williams nonmetric (Bray-Curtis coefficient), and several more.
- Options for transformations to handle different levels of measurement such as ratio, interval, and so on.
- Options for standardization. Standardization is required if there is more than one level of measurement, for example, numeric or character variables at the same time.
- Options for output. The calculated distance measures can be stored as a square or triangular matrix.

Using the POINTS data set, the following example computes the Euclidean distance, City-Block metric, and Minkowski measures.

Program 5.1-3: Calling PROC DISTANCE as a Macro

```
%macro PROC_DISTANCE (measure=, out=) ;
title "Distance Measure: '&measure.'";
proc distance data=POINTS
            method=&MEASURE. shape=square  out=&OUT.;
    var ratio(strt_x--dest_y) ;
    id ID ;
  run;
proc print data=&OUT.;
  run ;
%mend PROC_DISTANCE ;
%PROC_DISTANCE (measure=cityblock, out=cityblock) ;
%PROC_DISTANCE (measure=%str(L(1.2)), out=minkowski2) ;
%PROC_DISTANCE (measure=euclid, out=euclid) ;
%PROC_DISTANCE (measure=%str(L(3)), out=minkowski3) ;
```

The PROC DISTANCE macro treats the POINTS data as a ratio scale that is numeric, nonnegative, and on a higher level than interval levels. A standardization is not requested because all values share a common measurement level. SQUARE requests an output table in square shape. (The TRI option requests a lower triangular shape). L(n) requests the Minkowski measure; p is a positive numeric value. According to SAS documentation, typical values of p include 1 and 2. Very small or large values of p might cause floating-point overflow.

The following matrices (square) show the generated distances. Selected distance values are highlighted for comparison with the first example.

Figure 5.1-2: PROC_DISTANCE Macro Output in SAS Output

```
           Distance Measure: 'cityblock'

     Obs     id     _1    _2    _3    _4    _5

      1      1       0    16    20    38    19
      2      2      16     0    34    50    31
      3      3      20    34     0    22    19
      4      4      38    50    22     0    29
      5      5      19    31    19    29     0
```

```
                    Distance Measure: 'L(1.2)'

Obs     id       _1          _2          _3          _4          _5

 1       1     0.0000     14.4804     16.8961     33.5364     15.3435
 2       2    14.4804      0.0000     29.8621     45.5644     25.6942
 3       3    16.8961     29.8621      0.0000     19.3022     15.6814
 4       4    33.5364     45.5644     19.3022      0.0000     26.7404
 5       5    15.3435     25.6942     15.6814     26.7404      0.0000
```

```
                    Distance Measure: 'euclid'

Obs     id       _1          _2          _3          _4          _5

 1       1     0.0000     13.1149     13.0384     28.7750     10.2470
 2       2    13.1149      0.0000     25.5734     41.4246     18.9473
 3       3    13.0384     25.5734      0.0000     16.4317     11.0905
 4       4    28.7750     41.4246     16.4317      0.0000     24.3516
 5       5    10.2470     18.9473     11.0905     24.3516      0.0000
```

```
                    Distance Measure: 'L(3)'

Obs     id       _1          _2          _3          _4          _5

 1       1     0.0000     13.0059     12.2093     28.0983      8.5772
 2       2    13.0059      0.0000     25.0527     41.0303     17.4143
 3       3    12.2093     25.0527      0.0000     16.0467      9.5937
 4       4    28.0983     41.0303     16.0467      0.0000     24.0376
 5       5     8.5772     17.4143      9.5937     24.0376      0.0000
```

For the coordinates 11/53 and 19/45, the City-Block result (50) is matching. The Minkowski measures and Euclidian Distance differ slightly (PROC DISTANCE: Minkowski1: 47.73, Euclidian Distance: 42.76, and Minkowski2: 42.00). Different formulas and parameters, rounding errors, and so on can explain these divergences.

The calculation of geodetic distances, that is, the distance between two or more points based on x/y coordinates, should not be confused with values that represent the concept of similarity between two or more individual values. (See Schendera, 2010.)

5.1.2 Distances in Spherical Space (Basis: Longitudes and Latitudes)

Now imagine that the checkered sheet of paper is no longer lying flat on a table, but that it is curved like a globe. On this paper, there are two crosses. Each cross is located at the intersection of the horizontal and vertical lines of the checkered pattern, the so-called longitudes and latitudes. These points can also represent any two objects. Their distance is to be calculated like the distance between two crosses. What is special now is that the connecting line between two points is not straight but curved. For very long distances in two-dimensional space (for example, flight distances) the consideration of the earth's curvature is indispensable to achieve precise results. A few real-life examples could be:

- **Example 1:** Point 1: Big city 1; Point 2: Big city 2; Distance: Geodetic distance.
- **Example 2:** Point 1: Departure airport; Point 2: Destination airport; Distance: Flight route.
- **Example 3:** Point 1: Marking of an animal; Point 2: Location; Distance: Migration distance.
- **Example 4:** Point 1: Location of an event; Point 2: Location; Distance: Effect range such as environmental damage.

The analysis of distances in spherical space is also very important in combination with one or more variables.

- For **flights**, the expected distance plays an essential role for the calculation of refueling. In addition to the distance and degree of refueling, the expected duration of the flight depends on the cruising speed and wind direction and speed. Experience shows that the application of distances is much more complex than these examples.
- In **biology**, the active marking of migrating animals is common practice for animals like pigeons, storks, whales, or sharks. Typically, an initial locating is carried out at the time of marking. The second locating gives information about direction, distances, and with the help of other parameters, duration and speed of the migration.
- In **environmental protection** there is often passive marking. For example, a first position is the starting point of environmental pollution. A second position then shows the spatial distribution of the environmental damage. Examples include the activity of the Icelandic volcano Eyjafjallajökull, which continued over a period of several weeks, or the spatial extent of the environmental disaster caused by the explosion of the drilling platform "Deepwater Horizon" in the Gulf of Mexico. The migration routes of rubber ducks in the Atlantic Ocean are another well-known example. In 1992, a freighter lost a cargo of rubber ducks during a heavy storm. The beaches where the rubber ducks washed ashore to this day allow conclusions to be drawn about the distribution of waste in the oceans (Ebbesmeyer and Ingraham, 1994).

The calculation of very long distances in two-dimensional space is generally carried out with the aid of longitudes and latitudes, taking into account the curvature of the earth (Slocum et al., 2009). Here are three examples for the specification of locations in geographic coordinates in the sexagesimal system:

London: 51° 30' 33.8" N, 0° 7 5.95" W

Nanjing: 32° 3 0" N, 118° 46' 0" O

Sydney: 33° 52' 50" S 151° 13' 54" O

Besides the classical sexagesimal system, there are numerous other coordinate systems, including Universal Transverse Mercator (UTM), World Geographic Reference System (GEOREF), Global Area Reference System (GARS), and Military Grid Reference System (MGRS).

A special feature of working with longitude and latitude is the conversion of coordinates from the sexagesimal system into the decimal system, the distinction between north and south, as well as east and west, either by N/S or W/E signs. The specification of seconds might introduce a certain inaccuracy into analyses when localizing large-area cities because they only refer to a single official measurement center but not to the deviation of individual districts from it and are therefore excluded from some analyses.

In Berlin, the geodetic center is located near the Berlin television tower in Berlin Alexanderplatz. Places in the districts of Köpenick or Pankow, for example, are already several kilometers away from this "center." In the case of the Chinese city of Chongqing (approx. 28.6 million inhabitants) the scales are even more extreme. The administrative city limits of Chongqing cover an area of approximately 82,000 square kilometers. This makes Chongqing almost as large as the state of Austria (84,871 square kilometers). Given these dimensions, a single geodetic center alone is certainly no longer sufficient to describe satisfactorily precise distances.

The following example runs through four steps:

- In the first step, the data is read into the sexagesimal system (1.). The coordinates are rounded so that they can do without decimal places. (See London.)
- In the second step, the coordinates are converted from the degrees, minutes, and seconds of the sexagesimal system into the decimal system (2.).
- In the third step, the macro SPH_DISTANCE converts the decimal values to radial values.
- In the final, fourth step, the SQL query calculates the distances between the cities based on data in the decimal system.

Program 5.1-4: Demo Data Set COORDINATES (Sexagesimal System)

```
data COORDINATES ;
* 1. Reading the data * ;
input SD $1. +1 CITY $char14. +2 CONTINENT $char2. +1
_LA_Grd 2. +2 _LA_Min 2. +2 _LA_Sec 2. +2 _LA_direction $1. +1 _LO_Grd 3. +2 _LO_Min 2. +2
_LO_Sec 2. +2 _LO_direction $1.;
* 2. Conversion to decimal format * ;
Lat_Dec = ( _LA_Grd + _LA_Min/60 + _LA_Sec/3600 );
 if _LA_direction = 'S'
    then Lat_Dec = Lat_Dec* -1 ;
Long_Dec = ( _LO_Grd + _LO_Min/60 + _LO_Sec/3600) ;
 if _LO_direction = 'W'
    then Long_Dec = Long_Dec* -1 ;
drop _: ;
datalines ;
S London        EU 51° 30' 34" N   0°  7'  6" W
D Stockholm     EU 59° 21' 30" N  18°  4' 15" E
S Frankfurt     EU 50°  6' 43" N   8° 41'  9" E
D Haikou        AS 20°  2' 34" N 110° 20' 30" E
S Nanjing       AS 32°  3'  0" N 118° 46'  0" E
S Berlin        EU 52° 31'  7" N  13° 24' 29" E
D Rome          EU 41° 53' 17" N  12° 38' 21" E
D Miami         NA 25° 47' 16" N  80° 13' 27" W
S Zurich        EU 47° 22' 45" N   8° 32' 28" E
S Sydney        AU 33° 52' 50" S 151° 13' 54" E
D Lucerne       EU 47°  3'  0" N   8° 18'  0" E
S Shanghai      AS 31° 14'  0" N 121° 28'  0" E
S Guangzhou     AS 23°  7' 44" N 113° 15' 32" E
D Johannesburg  AF 26° 11' 12" S  28°  3'  0" E
    ;
run ;
proc print;
run ;
```

In the first step, the data is read into the sexagesimal system. Under SD, the system stores whether it is a starting (S) or destination (D) point. This information will be used in some analysis examples later. Under CITY, the name of the city is read in. Under CONTINENT the continent is read in. "AF" stands for Africa, "AS" for Asia, and so on. The degrees, minutes, and seconds for the geographical latitude are read in in the columns with the prefix _LA (_LA_Grd, _LA_Min, _LA_Sec). The degrees, minutes, and seconds for the geographical longitude are read in in the columns with the prefix _LO (_LO_Grd, _LO_Min, _LO_Sec). By means of _LA_direction and _LO_direction, the necessary distinction between north and south and east and west respectively is passed to SAS.

In the second step, the coordinates for the geographical longitude and latitude are converted from the sexagesimal system (degrees, minutes, or seconds) into the decimal system, and the negative signs are assigned for west and south. During the conversion, minutes are divided by 60 and seconds by 3600. Then, degrees, minutes, and seconds are added up and given the appropriate sign. At the end of this step, all imported fields with an underscore ("_") are removed from the analysis data set.

Program 5.1-5: Macro SPH_DISTANCE: Conversion to Radial Format

```
%macro SPH_DISTANCE
        (latitude1=,longitude1=,latitude2=,longitude2=, unit= );
%local CNSTNT ;
%let CNSTNT = constant('pi')/180 ;
%if %upcase(&unit) = KM %then %let radius = 6371 ;
    %else %if %upcase(&unit) = MI
          %then %let radius = 3959 ;
&radius * ( 2 * arsin(min(1,sqrt( sin( ((&latitude2 - &latitude1)*&CNSTNT)/2 )**2
+ cos(&latitude1*&CNSTNT) * cos(&latitude2*&CNSTNT) * sin(((&longitude2-
&longitude1)*&CNSTNT)/2)**2))) )
%mend SPH_DISTANCE ;
```

In a third step, the SPH_DISTANCE macro converts the data from decimal format to radial values. In this conversion, the decimal values are simply multiplied by π/180. Under UNIT, you specify whether the calculated distances are to be output in the unit miles (MI) or kilometers (KM). By the way, the SPH_DISTANCE macro is based on the Haversine formula. (See Ivis, 2006, 5-6.) The calculation of the distance is based on the simplistic assumption that the earth is a perfect sphere. (In reality, it looks more like a mandarin orange fruit.) The macro is called in the SQL query under IV. The examples in the next section also use this macro.

Program 5.1-6: Calculating Spherical Distances between Cities

```
proc sql number;
select S.City, D.City,
       D.Lat_Dec  as DEST_latitude,
       D.Long_Dec as DEST_longitude,
       S.Lat_Dec  as START_latitude,
       S.Long_Dec as START_longitude,
%SPH_DISTANCE
    (latitude1= START_latitude, longitude1= START_longitude,
     latitude2= DEST_latitude,  longitude2= DEST_longitude,
     unit = KM)
       as Distance format = 8.2
          from COORDINATES (where = (SD eq 'D')) as D,
               COORDINATES (where = (SD eq 'S')) as S
       order by S.City, D.City ;
quit ;
```

The SQL query calculates the distances between the cities. The FROM statement with the WHERE clause specifies that the distance should be calculated for each combination of starting and destination locations. To do this, the names of the columns in which the corresponding values for longitudes and latitudes are stored are specified in the call to the SPH_DISTANCE macro, and the distances are to be output in kilometers (UNIT). To include them in the macro, the variable names are renamed in the SELECT statement before. Finally, the query is sorted first by the starting city, then by the destination city.

Figure 5.1-3: Output SPH_DISTANCE Macro: Spherical Distances between Cities (Sample)

Row	CITY	CITY	DEST_ latitude	DEST_ longitude	START_ latitude	START_ longitude	Distance
1	Berlin	Haikou	20.04278	110.3417	52.51861	13.40806	8705.45
2	Berlin	Johannesburg	-26.1867	28.05	52.51861	13.40806	8866.64
3	Berlin	Lucerne	47.05	8.3	52.51861	13.40806	709.70
4	Berlin	Miami	25.78778	-80.2242	52.51861	13.40806	7996.00
5	Berlin	Rome	41.88806	12.63917	52.51861	13.40806	1183.47
6	Berlin	Stockholm	59.35833	18.07083	52.51861	13.40806	813.61
7	Frankfurt	Haikou	20.04278	110.3417	50.11194	8.685833	9104.59
8	Frankfurt	Johannesburg	-26.1867	28.05	50.11194	8.685833	8696.67
9	Frankfurt	Lucerne	47.05	8.3	50.11194	8.685833	341.65
10	Frankfurt	Miami	25.78778	-80.2242	50.11194	8.685833	7764.86
11	Frankfurt	Rome	41.88806	12.63917	50.11194	8.685833	963.73
12	Frankfurt	Stockholm	59.35833	18.07083	50.11194	8.685833	1189.14
13	Guangzhou	Haikou	20.04278	110.3417	23.12889	113.2589	456.84
14	Guangzhou	Johannesburg	-26.1867	28.05	23.12889	113.2589	10673.99
15	Guangzhou	Lucerne	47.05	8.3	23.12889	113.2589	9204.14
16	Guangzhou	Miami	25.78778	-80.2242	23.12889	113.2589	14385.36
17	Guangzhou	Rome	41.88806	12.63917	23.12889	113.2589	9137.77
18	Guangzhou	... (abbreviated)					

Geographical coordinates can also be processed in two-dimensional space. For conversion into two-dimensional coordinates, projections and the SAS procedure GPROJECT can be used. Frank Ivis (2006, 7-8) subjects the longitudes and latitudes (sexagesimal system) of selected North American cities to an Albers projection and thereby projects them into the two-dimensional coordinate space. Then, Ivis calculates the distances between the cities once using the Euclidean distance and then also using the measure of a spherical distance. The absolute difference between the two measures is less than 5 miles. The following section presents two SAS functions for the calculation of geodetic distances, plus some notes about EUCLID.

5.1.3 SAS Functions for the Calculation of Geodetic Distances Plus Notes about EUCLID

SAS offers two SAS functions for calculating geodetic distances between paired coordinate data:

- GEODIST for distances between longitudes and latitudes.
- ZIPCITYDISTANCE for distances between US ZIP codes in miles.

The EUCLID function returns the non-geodetic Euclidean norm of nonmissing arguments. The EUCLID function will be presented after the GEODIST example below. The ZIPCITYDISTANCE function is presented in Section 5.3.

GEODIST Function

GEODIST calculates the distance between a pair of longitudes and latitudes in miles and kilometers. GEODIST accepts input values expressed in degrees and radians. If the values are expressed in degrees, they must be between 90 and -90. If the values are expressed in radians, they must be between pi/2 and -pi/2. The GEODIST() function should not be confused with the %GEODIST macro, which can also be used to calculate the distance between two ZIP code centroids. (See Hadden et al., 2007.)

The arguments in the GEODIST function are specified in this sequence: Point1Latitude, Point1Longitude, Point2Latitude, and Point2Longitude; an inconsistent or incomplete sequence of variables or values might produce missing values or wrong results.

The following example computes geodetic distances in kilometers and miles between paired coordinate data of all possible city combinations (Cartesian product). The WHERE clause keeps only city pairs whose distance is larger than 10.000 kilometers. The COORDINATES data set was created in Section 5.1.2.

Program 5.1-7: GEODIST Function: Calculating Spherical Distances between Cities

```
proc sql ;
select S.City as Start, D.City as Destination,
geodist(S.Lat_Dec,S.Long_Dec,D.Lat_Dec,D.Long_Dec, 'K') as distance_km format = 8.2,
geodist(S.Lat_Dec,S.Long_Dec,D.Lat_Dec,D.Long_Dec, 'M') as distance_m  format = 8.2
        from COORDINATES (where = (SD eq 'D')) as D,
            COORDINATES (where = (SD eq 'S')) as S
            where calculated distance_km > 10000
        order by S.City, D.City ;
quit ;
```

Figure 5.1-4: GEODIST Function Output: Spherical Distances between Cities

Start	Destination	distance_km	distance_m	
Guangzhou	Johannesburg	**10670.33**	6630.24	
Guangzhou	Miami	14402.03	8949.01	
Nanjing	Johannesburg	11569.80	7189.14	
Nanjing	Miami	13295.65	8261.53	
Shanghai	Johannesburg	11774.48	7316.32	
Shanghai	Miami	13288.51	8257.10	
Sydney	Johannesburg	11062.30	6873.80	
Sydney	Lucerne	16593.24	10310.56	
Sydney	Miami	15025.21	9336.23	
Sydney	Rome	16311.25	10135.34	
Sydney	Stockholm	15587.18	9685.42	(abbreviated)

These GEODIST results are quite close to those of the SPH_DISTANCE macro in Section 5.1.2. The SPH_DISTANCE distance between Guangzhou and Johannesburg was 10673.99 km.

EUCLID Function

The EUCLID function returns the non-geodetic Euclidean norm of nonmissing arguments. The Euclidean norm (2-norm) corresponds to the Euclidean distance of a vector from the origin. This sounds similar to the Euclidian distance in the rowwise calculation of distance measures (see Section 5.1.1), but there are three fundamental differences:

- The EUCLID arguments represent vector points; in PROC DISTANCE (EUCLID method) they represent x/y coordinates.
- The number of arguments thus differs: PROC DISTANCE (EUCLID method) requires a set of four arguments rowwise (see the data set POINTS), EUCLID function any number.
- The calculus differs; the EUCLID method computes geodetic distances, the EUCLID function non-geodetic norms.

The EUCLID function simply squares each value of its arguments, sums them up, and then takes the root of the total. The arguments 1, 2, 2, and 4, result in the squares 1, 4, 4, and 16, which sum up to 25, of which 5 is the square root.

The following SAS program illustrates the EUCLID function approach in the first line of code; the second line picks up data from Section 5.1.1.

Program 5.1-8: EUCLID Function: Calculating the Non-geodetic Euclidean Norm

```
data EuclidNorm ;
retain SOURCE strt_x strt_y dest_x dest_y EuclidNorm ;
format EuclidNorm 5.2 ;
input SOURCE $ strt_x strt_y dest_x dest_y ;
EuclidNorm=euclid(of strt_x strt_y dest_x dest_y);
cards ;
Ex513  1 2 2  4
Ex511  1 3 9 45
run;
proc print data= EuclidNorm noobs ;
run ;
```

Figure 5.1-5: EUCLID Function Output: Calculating the Non-geodetic Euclidean Norm

SOURCE	strt_x	strt_y	dest_x	dest_y	Euclid Norm
Ex513	1	2	2	4	5.00
Ex5112	1	3	9	45	46.00

The following section presents three SQL queries based on coordinate data.

5.2 SQL Queries for Coordinates

This section introduces typical SQL queries for coordinate data. The examples have in common that they access the DISTANCES data set based on the COORDINATES data set (see Section 5.1.2) using the SPH_DISTANCE macro as well as the GEODIST function for calculating distances. The queries focus on retrieving maximum or minimum distances between departure and destination airports in kilometers, miles, or according to Ivis' spherical definition. The output presents varieties of departure and destination airports and the distances between them.

- **Query 1:** The first query retrieves the most distant destination of all the departure airports. The result is a single row. The approach uses an SQL subquery. The measure queried by the MAX clause is highest distance_sph.
- **Query 2:** This query retrieves the most distant destination for each departure airport. The result is a list. This approach uses a self-join; the MAX clause uses distance_sph.

- **Query 3:** This query retrieves the largest distance value (distance_sph) for each departure airport. The approach uses a simple SQL query and a MAX clause.
- **Query 4:** This query searches for the most distant destination for each departure airport. The query consists of a combination of an SQL query creating a view and a DATA step with IF.FIRST.
- **Query 5:** This query searches for the nearest destination for each departure airport. The only difference is using IF.LAST. For the last two queries, the unit is kilometers.
- **Query 6:** The last query looks for destinations that are more than 10,000 miles away from a departure airport. This query just adds another WHERE clause to the SQL syntax.

The first program creates the data set DISTANCES containing the distances in kilometers, miles, and according to Ivis' spherical definition. Note that the data set DISTANCES is not a symmetrical data set in the sense that a distance value means a symmetric flight movement from point 1 to point 2 and also from point 2 to point 1. Starting airports do not necessarily correspond to destination airports.

Program 5.2-1: Computing the Distances (Data Set DISTANCES)

```
proc sql ;
  create table DISTANCES as
     select S.CITY as START_CITY, D.CITY as DEST_CITY,
        %SPH_DISTANCE
          ( latitude1 = S.Lat_Dec, longitude1 = S.Long_Dec,
            latitude2 = D.Lat_Dec, longitude2 = D.Long_Dec,
              unit = KM) as distance_sph,
 geodist(S.Lat_Dec,S.Long_Dec,D.Lat_Dec,D.Long_Dec, 'K') as distance_km format = 8.2,
 geodist(S.Lat_Dec,S.Long_Dec,D.Lat_Dec,D.Long_Dec, 'M') as distance_m  format = 8.2
        from COORDINATES (where = (SD eq 'S')) as S,
             COORDINATES (where = (SD eq 'D')) as D
        order by START_CITY, Distance_sph ;
  quit ;
```

Figure 5.2-1: Contents of Data Set DISTANCES

START_CITY	DEST_CITY	distance_sph	distance_km	distance_m
Berlin	Lucerne	709.6994	710.42	441.43
Berlin	Stockholm	813.61	814.89	506.35
Berlin	Rome	1183.472	1183.26	735.25
Berlin	Miami	7995.997	8010.70	4977.62
Berlin	Haikou	8705.453	8718.36	5417.34
Berlin	Johannesburg	8866.636	8834.38	5489.43
Frankfurt	Lucerne	341.6523	341.68	212.31
Frankfurt	Rome	963.7325	963.67	598.79
Frankfurt	Stockholm	1189.139	1191.13	740.13
Frankfurt	Miami	7764.86	7779.11	4833.71
Frankfurt	(abbreviated)			

Query 1 looks for the departure airport and its most distant destination. The approach uses an SQL query plus subquery. The result is a single row that contains the combination of departure airport and destination airport with the highest distance_sph value (MAX clause), as well as distances in kilometers and miles.

Program 5.2-2: Query 1: Most Distant Destination (Max)

```
  proc sql ;
  title "Query 1: Most distant destination (max)";
     select *
        from DISTANCES
     where distance_sph ge
          (select max(distance_sph)
           from DISTANCES) ;
  quit ;
```

Figure 5.2-2: Query 1: Result of Query

```
Query 1: Most distant destination (max)
START_CITY      DEST_CITY       distance_sph  distance_km  distance_m
-------------------------------------------------------------------
Sydney          Lucerne             16595.63     16593.24    10310.56
```

Query 2 looks for the most distant destination for each departure airport. The resulting list contains every departure airport together with the most distant destination airport. The approach uses a self-join; the MAX clause uses distance_sph.

Program 5.2-3: Query 2: Most Distant Destination by START_CITY (Max, Groupwise)

```
proc sql ;
title "Query 2: Most distant destination by START_CITY";
   select a.START_CITY, a.DEST_CITY, a.distance_sph, a.distance_km, a.distance_m
     from DISTANCES a inner join
         (select START_CITY, max(distance_sph) as distance_sph
            from  DISTANCES
          group by START_CITY) b
     on a.START_CITY = b.START_CITY and a.distance_sph = b.distance_sph ;
quit;
```

Figure 5.2-3: Query 2: Result of Query

```
Query 2: Most distant destination by START_CITY
START_CITY      DEST_CITY       distance_sph  distance_km  distance_m
-------------------------------------------------------------------
Berlin          Johannesburg        8866.636      8834.38     5489.43
Frankfurt       Haikou              9104.586      9118.75     5666.13
Guangzhou       Miami               14385.36     14402.03     8949.01
London          Haikou              9600.345      9615.59     5974.85
Nanjing         Miami               13275.76     13295.65     8261.53
Shanghai        Miami               13268.71     13288.51     8257.10
Sydney          Lucerne             16595.63     16593.24    10310.56
Zurich          Haikou              9227.704      9242.35     5742.93
```

Query 3 retrieves the largest distance value (distance_sph) for each departure airport. If you want to query just these data, this query is easier to program than the self-join of Query 2. The MAX clause uses distance_sph.

Program 5.2-4: Query 3: Largest Distances by START_CITY

```
proc sql;
title "Query 3: Largest distances by START_CITY";
   select START_CITY as START_CITY,
         max(distance_sph) as distance_sph
      from DISTANCES
   group by START_CITY ;
quit ;
```

Figure 5.2-4: Query 3: Result of Query

```
Query 3: Largest distances by START_CITY
START_CITY      distance_sph
--------------------------
Berlin              8866.636
Frankfurt           9104.586
Guangzhou           14385.36
London              9600.345
Nanjing             13275.76
Shanghai            13268.71
Sydney              16595.63
Zurich              9227.704
```

Query 4 looks for the most distant destination for each departure airport. This query does not contain a MAX clause but a combination of an SQL Query and a DATA step using IF.FIRST. The IF.FIRST searches for the most distant airport and corresponds to a MAX clause.

Program 5.2-5: Query 4: Most Distant Destination by START_CITY (Max, Groupwise) Using a View

```
proc sql ;
create view _distances as
    select S.CITY as START_CITY, Z.CITY as DEST_CITY,
        %SPH_DISTANCE
          ( latitude1 = S.Lat_Dec, longitude1 = S.Long_Dec,
            latitude2 = Z.Lat_Dec, longitude2 = Z.Long_Dec,
               unit = KM) as Distance
        from COORDINATES (where = (SD eq 'S')) as S,
             COORDINATES (where = (SD eq 'D')) as D
        order by START_CITY, Distance ;
quit ;

title "Query 4: Most distant destination for each departure airport (in km)";
data max_distance
   / view = max_distance;
set _distances ;
by START_CITY ;
if last.START_CITY ;
run ;
proc print data = max_distance noobs ;
run ;
```

Figure 5.2-5: Query 4: Result of Query

```
Query 4: Most distant destination for each departure airport (in km)
   START_
   CITY          DEST_CITY       Distance
   Berlin        Johannesburg     8866.64
   Frankfurt     Haikou           9104.59
   Guangzhou     Miami           14385.36
   London        Haikou           9600.35
   Nanjing       Miami           13275.76
   Shanghai      Miami           13268.71
   Sydney        Lucerne         16595.63
   Zurich        Haikou           9227.70
```

The output shows departure and destination airports and the distance between them. According to this, Miami is the most distant destination airport from Nanjing, at around 13.276 km.

Query 5 searches for the nearest destination for each departure airport. In this quite similar query, the main difference is the IF.LAST in the DATA step. The IF.LAST searches for the nearest airport and corresponds to a MIN clause.

Program 5.2-6: Query 5: Closest Destination for Each Departure Airport (in Km) (Min, Grouped, View)

```
title "Query 5: Closest destination for each departure airport (in km)";
data min_distance
   / view = min_distance ;
set _distances ;
by START_CITY ;
if first.START_CITY ;
run ;
proc print data = min_distance noobs ;
run ;
```

The output shows departure and destination airports and the distance between them. According to this, Lucerne is the closest destination airport to London with a distance of about 786 km, and to Zurich with a distance of about 41 km.

Figure 5.2-6: Query 5: SAS Output

```
Query 5: Nearest destination for each departure airport (in km)
    START_CITY    DEST_CITY    Distance
    Berlin        Lucerne        709.70
    Frankfurt     Lucerne        341.65
    Guangzhou     Haikou         456.84
    London        Lucerne        785.79
    Nanjing       Haikou        1576.80
    Shanghai      Haikou        1668.75
    Sydney        Haikou        7395.91
    Zurich        Lucerne         40.88
```

Query 6 searches for destination airports that are more than 10,000 miles away from a departure airport. This query basically consists of the already known SQL query, only supplemented by a WHERE clause.

Program 5.2-7: Query 6: Destinations over 10,000 Miles from Departure

```
title "Query 6: Destinations over 10.000 miles from departure:";
proc sql ;
   select S.CITY as START_CITY, D.CITY as DEST_CITY,
      %SPH_DISTANCE
        (latitude1 = S.Lat_Dec, longitude1 = S.Long_Dec,
         latitude2 = D.Lat_Dec, longitude2 = D.Long_Dec,
         unit = MI) as Distance format = 8.1
   from COORDINATES (where = (SD eq 'S')) as S,
        COORDINATES (where = (SD eq 'D')) as D
   where calculated Distance > 10000
      order by S.CITY, Distance ;
quit ;
```

The output shows departure and destination airports and the distance between them. Rome and Lucerne are each over 10,000 miles away from Sydney.

Figure 5.2-7: Query 6: SAS Output

```
Query 6: Destinations over 10.000 miles from departure:
    START_CITY       DEST_CITY          Distance
    -----------------------------------------------
    Sydney           Rome               10136.2
    Sydney           Lucerne            10312.7
```

5.3 SQL and Maps

Visualizations like maps are highly recommended when the efficient communication of complex geo information has priority over the mathematical precision of individual parameters. This section introduces how SQL can help to handle response and map data. The main focus of this section is the SAS procedures SGMAP and GEOCODE. Starting with SAS 9.4M6, the SGMAP, GINSIDE, GPROJECT, GREDUCE, and GREMOVE procedures were moved from SAS/GRAPH to Base SAS. Technical or mathematical processes are not the subject of this section, nor is the design of maps and diagrams. (See Schendera, 2004, 717-821.)

The example is based on the SAS file FREQUENTFLYERS. If this file has not been delivered with SAS and created under your installation folder (for example, "My SAS Files" or "SASHome"), it can be generated by a program available on the SAS Support page. FREQUENTFLYERS contains fictitious information about participants in a frequent flyer program. Each participant in this program is identified by name, state, city, and US ZIP code.

The goal of the visualization is to display the number of frequent flyers living in major cities on a map of the USA. The amount of raw data is manageable (N=206) and results in a maximum of 10 flyers per city (Chicago IL). The final generated map looks like Figure 5.3-1.

Figure 5.3-1: Map Created by PROC SGMAP using SQL

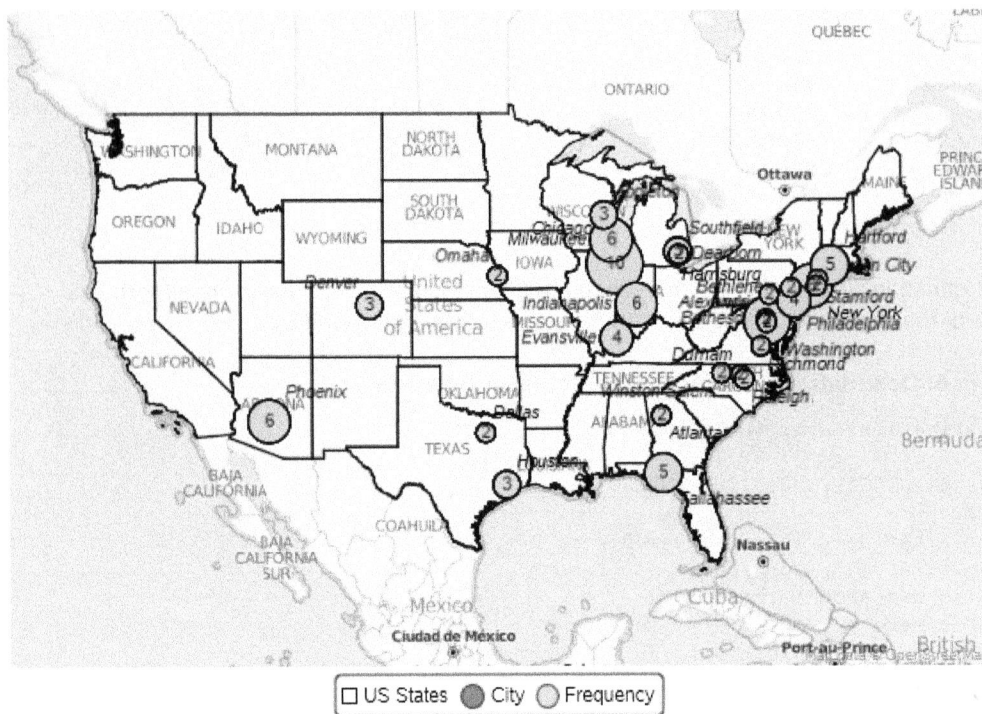

Frequent Flyers (SQL and SGMAP)

5.3.1 SAS Program (Five Steps)

The program required to create this map passes through the following steps:

1. Settings
2. Creating input data set for PLOTDATA
3. Creating input data set for MAPDATA
4. PROC SGMAP: Bringing it together
5. Closing it

Program 5.3-1: SAS Program (Five Steps)

```
*--------------------*
| 1. Settings        |
*--------------------*;
goptions reset=global ;
goptions xpixels=1000 ypixels=750 keymap=winansi;
libname MY_PATH 'D:\';


*---------------------------------------------*
| 2. Creating input data set for PLOTDATA     |
*---------------------------------------------*;
* Aggregate per geographical entity (CITY) *;

proc sql;
  create table CITY_FLYERS as
    select STATE, CITY, count(zipcode) as N_FLYERS
    from MY_PATH.FREQUENTFLYERS
```

```
      where verify(zipcode,'1234567890') EQ 0
    group by STATE, CITY ;
quit;
* Add lat / long per city (keep only cities where lat/long available)*;
proc geocode method=city
    data=CITY_FLYERS
    out=CITY_FLYERS_GEO (where=(_MATCHED_='City' and STATE not in ('AK','HI','PR') ) );
run;


*----------------------------------------*
| 3. Creating input data set for PLOTDATA    |
*----------------------------------------*;

proc greduce data=MAPSGFK.US_STATES out= STATESALL ;
    id STATE ;
run ;
proc sql undo_policy=none ;
    create table STATESRED as select * from STATESALL
      where fipstate(STATE) not in ('AK' 'HI' 'PR')  ;
    drop table STATESALL ;
    drop table CITY_FLYERS ;
quit;


*----------------------------------------*
| 4. PROC SGMAP: Bringing it together       |
*----------------------------------------*;

title 'Frequent Flyers (SQL and SGMAP)';
proc sgmap
      mapdata=STATESRED                               /* Map boundaries          */
      plotdata=CITY_FLYERS_GEO (where=(N_FLYERS>1)) /* Response and location data */
      des='Frequent Flyers (SAS File FREQUENTFLYERS';
    openstreetmap;
    styleattrs
          datacolors=(gray lightblue)
          datacontrastcolors=(black)
          datalinepatterns=(solid) ;
    choromap / mapid=state lineattrs=(color=black) legendlabel='US States';
    bubble x=long y=lat size=N_FLYERS /  fillattrs=(color=gray)
          datalabel=city
          legendlabel='City'
          datalabelattrs=(color=black size=8 style=italic);
    bubble x=long y=lat size=N_FLYERS /  fillattrs=(color=lightblue)
          datalabel=N_FLYERS
          legendlabel='Frequency'
          datalabelpos=center
          datalabelattrs=(color=black size=8 style=normal);
run;
quit;


*------------------------*
| 5. Closing it...          |
*------------------------*;

proc sql ;
    drop table STATESRED ;
    drop table CITY_FLYERS_GEO ;
quit;
ODS HTML CLOSE;
ODS LISTING;
```

5.3.2 Notes and Explanations

1. Settings

The "Settings" step defines the access path to the required SAS table and the desired resolution of the generated map. The keymap=WINANSI statement allows the output of umlauts, for example, ä, ö, and so on.

2. Creating Input Data Set for PLOTDATA

The SGMAP procedure (see 4.) expects at least two input data sets, one for PLOTDATA and one for MAPDATA. (See 3.) The PLOTDATA statement expects a data set that contains response data for the plot overlay. SQL aggregates the data to the total number of flyers per city and keeps the state and city information. The VERIFY condition removes all Canadian ZIP codes. Using a process of geocoding, GEOCODE converts address data like cities into map locations in latitudes and longitudes. The _MATCHED_ condition keeps all cities where latitudes and longitudes were available. For aesthetic reasons, Alaska, Hawaii, and Puerto Rico are removed from the response data. If you are interested, you can include these states and observe the effect on the generated map.

3. Creating Input Data Set for MAPDATA

This step prepares the second input data set for the SGMAP procedure (see 4.), for MAPDATA. PROC GREDUCE simplifies map data sets so that SAS can draw less detailed maps by using fewer boundary points. ID specifies the geographical unit whose borders are simplified. SQL removes Alaska, Hawaii, and Puerto Rico from the map data as well. If you are interested, you can include these states and observe the effect on the generated map. Two separate steps also drop temporary tables.

4. PROC SGMAP: Bringing It Together

In this step, ODS-based PROC SGMAP picks up the map boundaries (MAPDATA, STAESRED file) and response data (PLOTDATA, CITY_FLYERS_GEO). CITY_FLYERS_GEO keeps only frequencies above 1; otherwise the map appears overloaded. DES ("description") describes the generated map. OPENSTREETMAP requests to create a map based on OpenStreetMap. PROC SGMAP overlays an unprojected choropleth map (CHOROMAP) on an OpenStreetMap map, and then overlays a bubble plot (BUBBLE). BUBBLE specifies the coordinates (LONG, LAT) and the response data (N_FLYERS). This map uses two BUBBLE statements. The first BUBBLE statement defines the size of the bubbles; the second BUBBLE statement overlays the values of the response variable.

5. Closing It

SQL drops some more temporary tables. ODS closes the ODS output destinations.

Worth mentioning is the SASHELP.ZIPCODE lookup data set, which contains ZIP code information for the USA, including ZIPCODE centroids (x, y coordinates), so-called area codes, city names, FIPS codes, and two normalized variables (CITY2, STATENAME2) that support city geocoding containing only uppercase alphanumeric characters. SASHELP.ZIPCODE is useful if you want to format maps of the USA or calculate distances between US ZIP codes (Hadden and Zdeb, 2010). Some SAS procedures reference SASHELP.ZIPCODE too.

Chapter 6: Hash Programming as an Alternative to SQL

Hash programming is any programming using hash objects. The hash object and hash iterator enable you to store, search, and query data from lookup tables. Hash objects are nothing new in SAS; they have been around for some time in arrays, SQL (including the Internal Optimizer), and indexes. PROC SQL uses hashing for some joins. However, it does not use the Hash Object facility that is described in this chapter. Instead, PROC SQL provides its own hashing capabilities. Since SAS 9.1, it is possible to program hash objects specifically. You can currently execute hash programming in the DATA step or in the DS2 programming language (PROC DS2 and PROC HPDS2). DS2 is included with Base SAS; its syntax is similar to the DATA step.

Generally speaking, DS2 provides more advanced data manipulation techniques. You can use SQL statements in DS2. It offers higher performance when multithreading in Base SAS or CAS, and it offers a higher precision due to its exact numeric data types like BIGINT, DECIMAL, INTEGER, NUMERIC, or TINYINT. In Section 7.3 of the FedSQL chapter about DS2, you will see many of the same terms and language elements of hash programming presented. For an overview, see Section 6.4. One of the most important features of a hash object is that it resides completely in the physical memory of the DATA step. This is the reason why a separate chapter is dedicated to hash programming and performance, especially with very large data volumes. (See also Theuwissen, 2011, 2010, 2009; Warner-Freeman, 2007.)

Section 6.1 looks at hash programming from different angles, including programming, functional, and entity. From Section 6.2 on, numerous counterparts to SQL and other SAS procedures are presented, as well as variants of hash programming with a DATA step with an initialized data set, a LENGTH variant as a counterpart to the common if 0 then condition, and many more. Section 6.3 presents various applications for two tables (one-to-one scenario). Finally, Section 6.4 presents a list of elements for hash programming.

6.1 What Is Hash Programming?

This question is answered by explaining hash programming from a programming, functional, and entity angle. An easy introductory example introduces basic elements of hash programming, as well as step-by-step variations of this example.

6.1.1 Hash Programming from Three Angles

Hash programming can be best described from three angles: the programming angle, the functional angle, and the entity angle.

- **Programming angle:** Hash programming can best be described as an extended DATA step. In fact, hash objects are generally programmed using a DATA step, the DS2 and HPDS2 procedures, or a DATA _Null_ Step. Hash programming these objects might seem a bit more demanding or at least more time-consuming in the beginning, especially when compared to PROC SQL. (See the next section on aggregation using PROC SQL, PROC SUMMARY, and hash programming.)
- **Entity angle:** Hash programming is based on hash objects. A hash object usually reads a data set into memory. Its structure is that of a table, but it is actually made of an array of items that maps keys to associated values. These "tables" are only temporary; they are not physically present. These hash objects are also not available to other SAS procedures except when saved as a SAS file. At the end of the processing of the hash programming, the temporary hash object disappears. It does not need to be deleted explicitly.

- **Functional angle:** Since memory-based processes are typically faster than disk-based processes, users generally experience faster table lookup operations. In short, a hash object is a fast in-memory method for storing and retrieving data using so-called lookup keys. A hash object enables faster table lookups, storing and retrieving data, replacing and removing data, joining, merging, aggregating, sorting, sampling, and other operations. Hash programming is especially useful for big data, computationally complex applications, or threaded processing. Hash programming is also effective when there is little disk space available for tables or swap files or when time-consuming read/write processes do not allow processing with SQL within a critical time window or even not at all.

Compared with SQL, there is an overlap of hash programming functionalities. Hash programming enables you to create, define, or delete tables, data, or keys. You can merge tables and perform aggregations, queries, and counts. There are also a few differences between hash programming and PROC SQL. Many regard SQL syntax as more intuitive compared to the syntax of hash programming. SQL statements are processed on disk; hash objects are processed in memory. If SQL tables (not views) are created, these are physically existing tables. If objects are created using hash programming, these "tables" do not physically exist. Hash programming can therefore do without the time-consuming read/write processes of SQL. For very large amounts of data, the processing of SAS tables as hash objects is considered to be significantly more performant than PROC SQL.

The background concepts of PROC SQL and hash programming are completely different. So different that there are also differences in terminology. The most important difference is the term for the data to be processed. In SQL, this is the SAS table. In hashing, this is a hash object. A *hash object* is an object provided for inserting or retrieving data entries provided with a key. For readability, this section will also refer to hash objects as tables, although the logic and physics are completely different. Interested users are referred to various SAS Global Forum presentations by Paul Dorfman (2010, 2009, 2008) for more information. The programming languages and logics are accordingly completely different. PROC SQL corresponds (with restrictions) to ANSI SQL.

The advantage of fast processing tables as hash objects in memory might turn into a disadvantage when there is not enough memory to process these tables. If the tables turn out to be too large to be loaded or processed in memory, the processing process generally terminates with an error message. (See the example at the end of Section 6.3.1.) As a countermeasure, you could try to adjust the MEMSIZE option.

Table 6.1-1 presents the advantages and disadvantages of PROC SQL and hash programming, but it is an admittedly subjective selection.

Table 6.1-1: Advantages and Disadvantages of PROC SQL and Hash Programming

+ / -	PROC SQL	Hash Programming
+	• Intuitive logic and programming language. • Does not require large memory for processing tables. • No sorting required (indexing according to requirements). • Contains hash joins (see sqxjhsh). • Interface to SAS Macro Facility (INTO).	• Processing in memory. • Fast processing technique (maybe except for temporary arrays, among others). • No sorting or indexing required. • Modern and flexible programming language extendable via DATA step. • Temporary hash tables no longer exist after DATA Step.
-	• Processing on disk. • Relatively limited programming language. • Generally, slower than hash processing (I/O) and sometimes also slower than DATA step or SAS procedures.	• More abstract programming logic and language (for example, "methods" versus "statements"). • Large memory required for processing tables. • Hash objects are not available to other SAS procedures except when saved as a SAS file. • No interface to SAS Macro Facility.

Both approaches are sensitive to non-distinct key variables. PROC SQL is especially sensitive when joining two or more tables. In hash programming, key variables had to be distinct in SAS 9.1 and later. Starting with SAS 9.2,

MULTIDATA and the option "Y" can be used to indicate that key variables might contain multiple entries. If you are working with an earlier SAS version, you could try to include a second, additional key variable to control the multiple entries in the first key variable.

6.1.2 Rowwise Explanation of an Easy Introductory Example

This example tries to give a precise description of how hash programming works. The terminology might seem complicated such as DECLARE being used for declaring and instantiating as an instance. For better readability, explanations in later sections will be simpler, such as DECLARE just as "declaring" or "creating."

Program 6.1-1: Introductory Example

```
data _null_;
if 0 then set SASHELP.CLASS;
if _n_ = 1 then do;
declare Hash SortMe (ordered:'a');
 SortMe.DefineKey  ('Name');
 SortMe.DefineData ('Name', 'Height', 'Weight', 'Age', 'Sex');
 SortMe.DefineDone ();
end;
set SASHELP.CLASS end=eof;
    SortMe.add ();
 if eof then SortMe.output(dataset:'CLASS_SORTED');
run;

proc print data=WORK.CLASS_SORTED;
run ;
NOTE: The data set WORK.CLASS_SORTED has 19 observations and 5 variables.
NOTE: There were 19 observations read from the data set SASHELP.CLASS.
```

In a Nutshell

The SASHELP.CLASS data is read into the hash object "SortMe" in a modified column sequence, sorted ascending by Name, and then written to the temporary data set CLASS_SORTED. The PROC SQL equivalent is shown in Program 6.1-2.

Program 6.1-2: PROC SQL Equivalent of Program 6.1-1

```
proc sql;
   create table CLASS_SORTED as
      select Name, Height, Weight, Age, Sex
         from SASHELP.CLASS
   order by Name ;
quit;
```

Annotations

The following annotations explain row-by-row every SAS functionality plus some SAS background. Hash programming is a combination of a DATA step and a hash object. This "marriage" begins with:

data _null_;

DATA begins a DATA step; the keyword **_NULL_** as the data set's "name" requests SAS to execute the DATA step but not to create a SAS data set. No output of the DATA step is stored as a SAS data set! Later sections present programming variants using a data set name instead of _NULL_. (See Section 6.3.)

if 0 then set SASHELP.CLASS;

A DATA step consists of two phases: compilation and execution. During compilation, the DATA step sets up the Program Data Vector (PDV) and reads in variables from the data set(s) in the SET statement. During execution, SAS reads in data from the input data set(s) sequentially into memory and the PDV. "if 0 then" is a condition that always fails and thus has two effects. During compilation, the DATA step sets up the PDV with the same variables as in SASHELP.CLASS. During execution, SAS concludes that this condition is not met, the execution is aborted, and consequently no data are read from SASHELP.CLASS, and the DATA step terminates. Instead of "if 0 then", you could use any other always failing variant like ("1=2"), ("a=b"), and so on.

Hash programming is a combination of a DATA step and a hash object. Therefore, it has an important, though rather implicit requirement: variables that are defined in a hash object must also exist in the PDV. The "if 0 then" condition makes sure to add exactly the variables with the exact same attributes to the PDV that will be also added later to the hash object, and thus exist both in the hash object and in the PDV. Later examples demonstrate variants using LENGTH. (See Section 6.2.)

if _n_ = 1 then do;

This IF statement tells SAS to declare the hash object and define the key and data variables as soon as it is reading the first observation. The generated temporary data set WORK.CLASS_SORTED contains 19 data rows. The IF/DO statement is completed by an END. (See a few rows below.) If you drop the "bracket" made of IF/DO and END, then the generated temporary data set will contain only one row, the last data row for "William". Still, SAS processed all 19 data rows and declared and populated the hash object 19 times! If you drop just the IF _1_ = 1 THEN DO clause, then SAS will issue an error because of a non-matching DO statement.

declare Hash SortMe (ordered:'a');
  SortMe.DefineKey ('Name');
  SortMe.DefineData ('Name', 'Height', 'Weight', 'Age', 'Sex');
  SortMe.DefineDone ();

The DECLARE statement creates the hash object "SortMe". In hash programming terminology, the DECLARE statement declares and instantiates the hash object in one step. If you use a _NEW_ operator after a DECLARE statement, then declaring and instantiating is done in two steps. Note when you create a hash iterator, you need to specify a hash object.

One step: Declaring *and* instantiating	**Two steps:** Hash object HT	**Two steps:** Hash iterator HI
declare hash HT( );	*declare hash HT;*	*declare hiter HI;*
declare hiter HI('HT');	*HT = _new_ hash( );*	*HI = _new_ hiter('HT');*

Any other name for hash object or hash iterator is possible. This chapter generally uses HT for hash object and HI for hash iterator, and names indicating the respective join type. The ORDERED method specifies that the data is sorted in ascending ("a") order of the key variable ("Name", see DEFINEKEY) in case you use an OUTPUT method. The DECLARE statement in the example contains the DEFINEKEY, DEFINEDATA, and the DEFINEDONE methods.

- The DEFINEKEY method defines the "Name" column as the key variable for the hash object "SortMe". This method is required.
- The DEFINEDATA method defines the selection and sequence of the selected variables. Please note that the columns sequence differs from the original sequence in SASHELP.CLASS. If you drop "Name" from the list, "Name" is not retained but the data are still sorted by "Name".
- The DEFINEDONE method and the empty parentheses indicate that all definitions are complete. (This expression is abbreviated to "DEFINEDONE indicates..." later.) This method is required.

After DEFINEDONE, variables that were defined in the hash object also exist with the exact same attributes in the PDV. This is necessary because the hash object and the PDV exchange values from each other's variables.

set SASHELP.CLASS end=eof;

The SET statement reads data from one or more data sets. The variables that are read using the SET statement are retained in the PDV. The specified end=eof (eof stands for "end of file") is a binary variable (0/1) with the value 1 for the last observation in the data set and 0 otherwise. Any other name after END= is possible; eof is automatically dropped. In total, the SET statement requests SAS to read in data "until further notice".

SortMe.add ();

The ADD method adds the data of SASHELP.CLASS that is associated with the given key ("Name") to the hash object "SortMe".

if eof then SortMe.output(dataset:'CLASS_SORTED');

The IF statement writes the data from the hash object into a data set created by the OUTPUT method as soon as the last observation of the data set is read. The OUTPUT method creates one or more data sets, each of which contains the data in the hash object. The data set created is named CLASS_SORTED, and the data items are written in ascending key-value order. Instead of just eof, the expression eof=1 would have the same effect.

6.1.3 Step-by-step Variations of the Introductory Example

The following table shows step-by-step variations of the introductory example. The left column shows code variations of the original SAS syntax. The right column explains the consequences of these variations. It also shows relevant information from the SAS log, especially comments in case you cannot find this information in the SAS log.

Table 6.1-2: Step-by-step Variations of the Introductory Example

What if...	... then this happens

... IF EOF THEN is dropped?

Program 6.1-3: Variation 1	**SAS Log 6.1-1: Variation 1**
| ```
data _null_;
if 0 then set SASHELP.CLASS;
if _n_ = 1 then do;
declare Hash SortMe (ordered:'a');
 SortMe.DefineKey  ('Name');
 SortMe.DefineData ('Name', 'Height',
'Weight', 'Age', 'Sex');
 SortMe.DefineDone ();
end;
set SASHELP.CLASS end=eof;
SortMe.add ();
run;
``` | NOTE: There were 19 observations read from the data set SASHELP.CLASS.<br><br>Data (five variables) are still read into the hash object "SortMe", but not written into CLASS_SORTED anymore. |

... IF EOF THEN, END=EOF, and SORTME.ADD (); are dropped?

Program 6.1-4: Variation 2	**SAS Log 6.1-2: Variation 2**
| ```
data _null_;
if 0 then set SASHELP.CLASS;
if _n_ = 1 then do;
declare Hash SortMe (ordered:'a');
 SortMe.DefineKey  ('Name');
 SortMe.DefineData ('Name', 'Height',
'Weight', 'Age', 'Sex');
 SortMe.DefineDone ();
end;
set SASHELP.CLASS ;
run;
``` | NOTE: There were 19 observations read from the data set SASHELP.CLASS.<br><br>Data (five variables) are still read into the hash object, but not written into CLASS_SORTED anymore. |

... SORTME.DEFINEDATA, IF EOF THEN, END=EOF, and SORTME.ADD (); are dropped?

(Continued)

Table 6.1-2: (Continued)

What if...	... then this happens

Program 6.1-5: Variation 3

```
data _null_;
if 0 then set SASHELP.CLASS;
if _n_ = 1 then do;
declare Hash SortMe (ordered:'a');
 SortMe.DefineKey  ('Name');
 SortMe.DefineDone ();
end;
set SASHELP.CLASS ;
run;
```

SAS Log 6.1-3: Variation 3

```
NOTE: There were 19 observations read from
the data set SASHELP.CLASS.
```

Data are still read into the hash object. But only one variable ("Name"), and not into CLASS_SORTED anymore. You cannot find this information in the SAS log.

... SORTME.DEFINEKEY is dropped?

Program 6.1-6: Variation 4

```
data _null_;
if 0 then set SASHELP.CLASS;
if _n_ = 1 then do;
declare Hash SortMe (ordered:'a');
 SortMe.DefineData ('Name', 'Height',
'Weight', 'Age', 'Sex');
 SortMe.DefineDone ();
end;
set SASHELP.CLASS end=eof;
    SortMe.add ();
 if eof then SortMe.output(dataset:'CLASS_
SORTED');
run;
```

SAS Log 6.1-4: Variation 4

```
ERROR: Uninitialized keys for hash object at
line nnn column 2.
ERROR: DATA STEP Component Object failure.
Aborted during the EXECUTION phase.
NOTE: The SAS System stopped processing this
step because of errors.
```

The hash object "SortMe" is not created. The DATA step is aborted during the execution phase. A key variable is required.

... the IF/DO and END "bracket" is dropped?

Program 6.1-7: Variation 5

```
data _null_;
if 0 then set SASHELP.CLASS;
declare Hash SortMe (ordered:'a');
 SortMe.DefineKey  ('Name');
 SortMe.DefineData ('Name', 'Height',
'Weight', 'Age', 'Sex');
 SortMe.DefineDone ();
set SASHELP.CLASS end=eof;
    SortMe.add ();
 if eof then SortMe.output(dataset:'CLASS_
SORTED');
run;
```

SAS Log 6.1-5: Variation 5

```
NOTE: The data set WORK.CLASS_SORTED has 1
observations and 5 variables.
NOTE: There were 19 observations read from
the data set SASHELP.CLASS.
```

SAS processes all 19 data rows, and declares and populates the hash object 19 times, but keeps only the last data row for "William".

... the DEFINEDONE method is dropped?

Program 6.1-8: Variation 6

```
data _null_;
if 0 then set SASHELP.CLASS;
if _n_ = 1 then do;
declare Hash SortMe (ordered:'a');
 SortMe.DefineKey  ('Name');
 SortMe.DefineData ('Name', 'Height',
'Weight', 'Age', 'Sex');
end;
set SASHELP.CLASS end=eof;
    SortMe.add ();
 if eof then SortMe.output(dataset:'CLASS_
SORTED');
run;
```

SAS Log 6.1-6: Variation 6

```
ERROR: Method defineDone must be called to
complete initialization of hash object
before line nnn column 5.
ERROR: DATA STEP Component Object failure.
Aborted during the EXECUTION phase.
NOTE: The SAS System stopped processing this
step because of errors.
NOTE: There were 2 observations read from
the data set SASHELP.CLASS.
```

The hash object "SortMe" is not created. The DATA step is aborted during the execution phase. The DEFINEDONE method is required.

(Continued)

Table 6.1-2: (*Continued***)**

What if...	... then this happens
... the DEFINEDATA method is dropped and the variables are listed after DEFINEKEY instead?	

Program 6.1-9: Variation 7	SAS Log 6.1-7: Variation 7

<table>
<tr><td>

```
data _null_;
if 0 then set SASHELP.CLASS;
if _n_ = 1 then do;
declare Hash SortMe (ordered:'a');
 SortMe.DefineKey ('Name', 'Height',
'Weight', 'Age', 'Sex');
 SortMe.DefineDone ();
end;
set SASHELP.CLASS end=eof;
    SortMe.add ();
 if eof then SortMe.output(dataset:'CLASS_
SORTED');
run;
```

</td>
<td>

NOTE: The data set WORK.CLASS_SORTED has 19 observations and 5 variables.
NOTE: There were 19 observations read from the data set SASHELP.CLASS.

Data (five variables) are still read into the hash object and output into CLASS_SORTED. The data are sorted according to every column after DEFINEKEY and their sequence in ascending order.

</td></tr>
</table>

6.1.4 Sample Data for Benchmark Tests

To illustrate the performance of hash programming in comparison with other approaches, the results of various benchmark tests are presented. The test results are each based on the mean value of three runs to compensate for outliers.

Creating the Sample Data for the Benchmark Tests

The following DATA steps create the SAS files BIG_ONE and BIG_THREE for various **benchmark** tests.

Program 6.1-10: Creating Sample Data for Benchmark Tests

```
data BIG_ONE (drop=i) ;
   retain ID 1 GROUPS PRODUCTS VALUE STRING ;
   do i=1 to 8000000 ;
     VALUE = round(ranuni(12345)*10000) ;
     STRING = strip(translate(round(VALUE), "ABCDEFGHIK", "1234567890"));
     GROUPS =   1 + (i>2000000) + (i>4000000) + (i>6000000) ;
     PRODUCTS = 1 + (i>1000000) + (i>2000000) + (i>3000000)
                  + (i>4000000) + (i>5000000) + (i>6000000)
                  + (i>7000000) ;
        output ;
     ID + 1 ;
   end ;
run ;
proc sort data=BNCHMARK ;
by VALUE ;
run ;

data BIG_THREE (drop=i id2);
   retain ID 10 GROUPS2 PRODUCTS2 VALUE2 STRING2 ;
   do i=1 to 80000000 ;
     VALUE2 = round(ranuni(12345)*10000) ;
     STRING2 = strip(translate(round(VALUE2), "ABCDEFGHIK", "1234567890"));
     GROUPS2 =   1 + (i>20000000) + (i>40000000) + (i>60000000) ;
     PRODUCTS2 = 1 + (i>10000000) + (i>20000000) + (i>30000000)
                   + (i>40000000) + (i>50000000) + (i>60000000)
                   + (i>70000000) ;
        output ;
     ID2 + 1 ;
        ID = abs(ID2-80000000) ;
   end ;
run ;
```

The code for these SAS tables differs in three aspects: the name of the generated table, the column names, and its number of rows. The temporary SAS file BIG_ONE is created with 8,000,000 rows and five fields: ID, GROUPS, PRODUCTS, VALUE, and STRING. The created field GROUPS contains four values 1 to 4 with N=2,000,000 each. PRODUCTS contains eight values 1 to 8 with N=1,000,000 each. VALUE contains a random integer value. STRING is a sequence of alphanumeric characters based on VALUE. The final sorting according to the random variable VALUE ensures that no implicit presorting during the creation of BNCHMARK can affect the test results. BIG_THREE with 80,000,000 rows has a few different column names. In an outer right join, these tables are merged using the key ID. One table should be previously sorted randomly using ID to compensate for a possible uniform presorting. For other table sizes (12,000,000, 40,000,000), you vary the limit values accordingly. To benchmark test an outer right join, tests usually use two tables of the same size. (See BIG_ONE, BIG_THREE.)

If you want to replicate some of the benchmark tests, just replace names of tables and columns. For example, if you want to replicate SAS code that used SASHELP.SHOES, you just replace SASHELP.SHOES by BIG_ONE, SUBSIDIARY by GROUPS, PRODUCT by PRODUCTS, and SALES by VALUE. STRING and STRING2 are added for a bigger data load.

6.2 Working with One Table

This section presents various applications of hash programming within one table as comparative examples:

- Aggregating with PROC SQL, PROC SUMMARY, and hash (Section 6.2.1)
- Sorting with PROC SORT, PROC SQL, and hash (Section 6.2.2)
- Subsetting: Filtering and random sampling (Section 6.2.3)
- Elimination of duplicate keys (Section 6.2.4)
- Querying values (retrieval) (Section 6.2.5)

6.2.1 Aggregating with SQL, SUMMARY, and Hash

This section covers two examples and introduces aggregating with PROC SQL, PROC SUMMARY, and hash programming. The explanations of PROC SQL and PROC SUMMARY are limited to an absolute minimum. Hash programming is covered in detail.

Example 1

Example 1 deals with obtaining the sum of SALES for each level of REGION from SASHELP.SHOES.

SQL Procedure

Accessing SASHELP.SHOES, PROC SQL calculates the sum of SALES per level of REGION and stores it as Total in the table SQL_sum.

Program 6.2-1: Aggregating: Example 1: SQL

```
proc sql ;
    create table SQL_sum as
select REGION, sum(SALES) as Total format=dollar10.
from SASHELP.SHOES
    group by REGION ;
quit ;
```

SUMMARY Procedure

Accessing SASHELP.SHOES, PROC SUMMARY calculates the sum of SALES per level of REGION and stores it as Total in the table SMRY_sum.

Program 6.2-2: Aggregating: Example 1: SUMMARY

```
proc summary data=SASHELP.SHOES nway ;
class REGION / missing groupinternal ;
var SALES ;
output out=SMRY_sum(drop=_:) sum=Total ;
run ;
```

Hash Object

Accessing SASHELP.SHOES, the sum of SALES per level of REGION is calculated using hash programming and printed directly to the SAS output.

Program 6.2-3: Aggregating: Example 1: Hash Object

```
data _null_ ;
if _N_ = 1 then do ;
declare hash HT(suminc:"SALES", dataset:"SASHELP.SHOES") ;
HT.defineKey("REGION") ;
HT.defineDone() ;
do until (eof) ;
set SASHELP.SHOES end=eof ;
HT.ref() ;
end ;
end ;
file print ;
set SASHELP.SHOES ;
by REGION ;
if first.REGION then do ;
HT.sum(sum:Total) ;
put REGION @30 Total dollar10. ;
end ;
run ;
```

SAS Log 6.2-1: Aggregating: Example 1: Hash Object

```
NOTE: 10 lines were written to file PRINT.
NOTE: There were 395 observations read from the data set SASHELP.SHOES.
NOTE: There were 395 observations read from the data set SASHELP.SHOES.
```

Figure 6.2-1: Output: Aggregating: Example 1: Hash Object (Complete):

```
Region                          Total
Africa                     $2,342,588
Asia                         $460,231
Canada                     $4,255,712
Central America/Caribbean  $3,657,753
Eastern Europe             $2,394,940
Middle East                $5,631,779
Pacific                    $2,296,794
South America              $2,434,783
United States              $5,503,986
Western Europe             $4,873,000
```

Explanations

After IF _N_=1, the DECLARE statement creates the hash object HT. Because of the DATASET argument, the hash object is filled with the contents of the SAS table SASHELP.SHOES. The **SUMINC** argument identifies the field whose values are to be used to calculate groupwise totals. The DEFINEKEY method defines the REGION column as the key variable. This approach does not use the DEFINEDATA method. The DEFINEDONE method and the empty parentheses indicate that all definitions are complete. The subsequent DO UNTIL loop loads SASHELP.SHOES repeatedly. The **REF** method adds found SALES values to the provided table grouped by REGION. Using IF FIRST and the SUM method, a sum of SALES is calculated for each level of REGION and stored as TOTAL. PUT assigns the requested formats to the output.

Example 2

Example 2 deals with calculating the sum of SALES for each combination of the SUBSIDIARY and PRODUCT levels of the SASHELP file SHOES.

SQL Procedure

Using the table SASHELP.SHOES, PROC SQL calculates the sum of SALES for each combination of the SUBSIDIARY and PRODUCT levels and stores it as Total in the table SQL_sum.

Program 6.2-4: Aggregating: Example 2: SQL

```
proc sql ;
    create table SQL_sum as
select SUBSIDIARY, PRODUCT, sum(SALES) as Total
from SASHELP.SHOES
    group by SUBSIDIARY, PRODUCT ;
quit ;
```

SUMMARY Procedure

Using the table SASHELP.SHOES, PROC SUMMARY calculates the sum of SALES for each combination of the SUBSIDIARY and PRODUCT levels and stores it as Total in the table SMRY_sum.

Program 6.2-5: Aggregating: Example 2: SUMMARY

```
proc summary data = SASHELP.SHOES nway ;
class SUBSIDIARY PRODUCT ;
var SALES;
output out = SMRY_sum (drop = _:) sum = TOTAL;
run ;
```

Hash Object

Using the table SASHELP.SHOES, hash programming calculates the sum of SALES for each combination of the SUBSIDIARY and PRODUCT levels and stores it as Total in the table HASH_sum. Compared to PROC SQL and PROC SUMMARY, it becomes clear, as in example 1, that more steps and complexity are required to program the same functionality.

Program 6.2-6: Aggregating: Example 2: Hash Object

```
data _null_ ;
  if 0 then set SASHELP.SHOES ;
  declare hash HT (hashexp:18 ) ;
  HT.definekey ("SUBSIDIARY", "PRODUCT" ) ;
  HT.definedata ("SUBSIDIARY", "PRODUCT", "TOTAL") ;
  HT.definedone () ;
  do until (eof) ;
  set SASHELP.SHOES end = eof ;
  if HT.find () ne 0
      then Total= 0 ;
      Total + SALES ;
  HT.replace () ;
```

```
    end ;
  rc = HT.output (dataset: "HASH_sum") ;
run ;
proc print ;
run ;
```

Figure 6.2-2: Output: Aggregating: Example 2: Hash Object (Abbreviated)

```
Subsidiary      Product              Total
-------------------------------------------
Addis Ababa     Boot                 29761
Addis Ababa     Men's Casual         67242
Addis Ababa     Men's Dress          76793
Addis Ababa     Sandal               62819
Addis Ababa     Slipper              68641
Addis Ababa     Sport Shoe            1690
Addis Ababa     Women's Casual       51541
Addis Ababa     Women's Dress       108942
Al-Khobar       Boot                 15062
Al-Khobar       Men's Casual        340201
Al-Khobar       Men's Dress         261445
```

Explanations

The example loads the table SASHELP.SHOES using SET in the DATA step. The DECLARE method defines the hash object HT. By assigning the value 18 to the HASHEXP argument, the hash object is defined to provide 218 (=262,144) slots at the top level. The hash object is then filled with the contents of the SAS table SASHELP.SHOES. The two columns SUBSIDIARY and PRODUCT are defined as key variables (DEFINEKEY method). The DEFINEDATA method defines the selection and sequence of the columns for the hash object, that is, SUBSIDIARY, PRODUCT, and the TOTAL field that is still to be created. The method DEFINEDONE and the empty parentheses indicate that all definitions are complete. The following DO UNTIL loop triggers the calculation of TOTAL; END closes this loop. The REPLACE method causes the slots in the hash object to be overwritten by the calculated values. The return code RC specifies whether the method (OUTPUT) succeeded (0) or failed (1). As long as the RC value is 0, the OUTPUT method stores the complete content of the hash object in the SAS table HASH_sum.

Table 6.2-1: Result Benchmark Test

Test (N rows)	Time in Seconds	PROC SQL	PROC SUMMARY	Hash Object
8,000,000	CPU Time	27.96	8.96	12.76
	Real Time	156.51	18.15	15.06
80,000,000	CPU Time	140.83	45.94	71.87
	Real Time	658.27	251.93	158.09

The benchmark tests show that processing using hash objects is clearly superior to processing PROC SQL and PROC SUMMARY with increasing data volume in real time. Processing using a hash object takes about the same real time as PROC SQL for a 10 times smaller data volume (for example, hash objects: 158.09 secs for 80 million rows versus SQL: 156.51 for 8 million). The performance differences between hash programming, PROC SQL, and PROC SUMMARY would probably become even more apparent with more data columns. However, for data volumes of less than 1,000,000 data rows with possibly even fewer columns, the results could be quite different.

It is interesting to note that PROC SUMMARY always achieves better CPU times than hash programming. The displayed values are based on the average of three runs. If performance is important, you should test the runtimes of the presented hash approaches on your own system with appropriate data volumes. In order to replicate the tests performed, the following substitutions are made in the programs listed above: SASHELP.SHOES by BIG_ONE, SUBSIDIARY by GROUPS, PRODUCT by PRODUCTS, and SALES by VALUE. The more complex a system is and the more relevant the performance, the more throughput data should be included in the calculation of the average performance.

6.2.2 Sorting with SORT, SQL, and Hash

This section introduces sorting with PROC SORT, PROC SQL, and hash programming. Section 6.2.4 presents an example of how to implement the NODUPKEY of PROC SORT. In contrast to all other chapters, no SAS output is given for the examples for sorting.

SORT Procedure

The table SASHELP.PRDSALE is sorted in ascending order by YEAR and MONTH using PROC SORT. The result of the sorting is stored in the SORT_SORT table.

Program 6.2-7: Sorting: SORT

```
proc sort data = SASHELP.PRDSALE
       out = sort_sort ;
       by YEAR MONTH ;
run ;
```

SQL Procedure

The table SASHELP.PRDSALE is sorted in ascending order by YEAR and MONTH using PROC SQL. The result of the sorting is stored in the SQL_SORT table.

Program 6.2-8: Sorting: SQL

```
proc sql;
       create table sql_sort as
               select *
       from SASHELP.PRDSALE
       order by YEAR, MONTH ;
quit ;
```

Hash Object

The table SASHELP.PRDSALE is sorted in ascending order by YEAR and MONTH using hash programming. The result of the sort is stored in the HASH_SORT table.

Program 6.2-9: Sorting: Hash Object

```
data _null_ ;
if _n_ = 0 then set SASHELP.PRDSALE ;
if _n_ = 1 then do ;
declare hash HT (dataset: "SASHELP.PRDSALE", ordered: "A", multidata: "Y") ;
       HT.definekey ("YEAR", "MONTH") ;
       HT.definedata ("ACTUAL","PREDICT","COUNTRY", "REGION", "DIVISION","PRODTYPE",
                      "PRODUCT", "QUARTER","YEAR", "MONTH") ;
HT.definedone ( ) ;
end ;
HT.output (dataset: "hash_sort") ;
run ;
```

The hash object HT is declared after _N_ = 0, _N_ = 1, and the DECLARE statement. The contents of SASHELP. PRDSALE are loaded into the hash object HT. (See DATASET argument.) After the comma, the ORDERED method and the option "A" create HT in ascending order.

Since the topic of this section is sorting, the further ORDERED options shall be introduced at this point: "A" or "a": ascending, "Y" or "y": ascending, "D" or "d": descending, "N" or "n": internal hash sorting (that is, not sorted; an original sorting is not considered).

The MULTIDATA parameter and the "Y" option indicate that key variables might contain multiple entries. The two columns YEAR and MONTH are defined as key variables for HT (DEFINEKEY). The DEFINEDATA method defines the selection and sequence of the columns for HT. The DEFINEDONE method indicates that all definitions are complete. The concluding END closes the DO loop. The OUTPUT method stores the complete content of hash object HT in the SAS table HASH_SORT. The sorting is retained and also stored; key variables are only stored if they were defined using the DEFINEDATA method (as in the example).

6.2.3 Subsetting: Filtering and Random Sampling

This section introduces various approaches to subsetting using hash programming. The following approaches are presented:

- Approach 1: Simple subsetting with WHERE
- Approach 2: Subsetting using a filter table
- Approach 3: Random selection of entries from a SAS table

Approach 1: Simple Subsetting with WHERE

The following comparison between SQL and hash programming will show that programming the same functionality requires more steps and complexity. The aim of the following two examples is to filter out all lines with the level "Africa" in the REGION field and store them in a subset.

From the SASHELP.SHOES table, all rows with the level "Africa" in the REGION column are filtered out and stored in the SAS table SQL_fltrd.

Program 6.2-10: Approach 1: Subsetting with WHERE: SQL

```
proc sql ;
    create table SQL_fltrd as
        select *
        from SASHELP.SHOES
    where REGION="Africa" ;
quit ;
```

SAS Log 6.2-2: Approach 1: Subsetting with WHERE: SQL

```
NOTE: Table WORK.SQL_FLTRD created, with 56 rows and 7 columns.
```

This example illustrates how the result of a query using a condition is *also* stored in a SAS table. All entries that meet the formulated condition are output to the SAS log and stored in parallel in the SAS table.

Program 6.2-11: Approach 1: Subsetting with WHERE: Hash Object

```
data HASH_fltrd (drop=RC) ;
  declare hash HT () ;
  HT.defineKey ( "REGION" ) ;
  HT.defineData ( "SALES" ) ;
  HT.defineDone () ;
  do until (eof) ;
    set SASHELP.SHOES
        (where=( REGION="Africa")) end=eof ;
  rc = HT.add () ;
output ;
  end ;
  stop ;
run ;
```

SAS Log 6.2-3: Approach 1: Subsetting with WHERE: Hash Object

```
NOTE: There were 56 observations read from the data set SASHELP.SHOES.
      WHERE REGION='Africa';
NOTE: The data set WORK.HASH_FLTRD has 56 observations and 7 variables.
```

In this example, the still empty SAS table HASH_fltrd is initialized in the DATA step. After the DECLARE statement, the hash object HT is created. The REGION column is defined as the key variable for HT (DEFINEKEY). SALES is defined as a data column for HT (DEFINEDATA). DEFINEDONE indicates that all definitions are complete. The DO UNTIL loop causes all rows from SASHELP.SHOES to be loaded that contain the level "Africa" in REGION. The ADD method inserts the data into the initialized table. The OUTPUT method stores the data in the initialized SAS table HASH_FLTRD. The STOP statement prevents continuous looping.

Table 6.2-2: Result Benchmark Test

Test (N rows)	Time in Seconds	PROC SQL	Hash Object
8.000.000	CPU Time	1.16	1.17
	Real Time	2.95	1.31
80.000.000	CPU Time	11.43	11.76
	Real Time	213.41	214.43

The benchmark tests show that processing using hash objects and PROC SQL does not achieve significant performance differences in real and CPU time despite substantial data volumes. The differences in the smaller data volume are caused by an outlier in the PROC SQL runs for N=8,000,000. What is striking is the approximately 100 times higher real time at only 10 times higher data load. To be on the safe side, this result was checked and confirmed in a second run of the benchmark tests. To replicate the tests performed, only the following replacements need to be made in the programs listed above: SASHELP.SHOES by BIG_ONE, SUBSIDIARY by GROUPS, PRODUCT by PRODUCTS and SALES by VALUE; the WHERE clause filtered for the character string "FIFI".

Approach 2: Subsetting Using a Filter Table

Hash Object

The following subsetting example uses a filter table, "filter_list". All rows with the "Africa" level in the REGION column are filtered out from SASHELP.SHOES and stored in the SAS table HASH_fltrd. In PROC SQL, users usually apply filter tables in subqueries (not presented).

Program 6.2-12: Approach 2: Subsetting Using a Filter Table: Hash Object

```
data filter_list ;
length REGION $25 ;
REGION="Africa" ;
Output ;
run ;
data HASH_fltrd ;
if _N_ =1 then do ;
declare hash HT (dataset: "filter_list") ;
    HT.defineKey ("REGION" ) ;
    HT.defineDone () ;
end ;
set SASHELP.SHOES ;
    rc = HT.find();
if (rc = 0) then output;
        /* can be abbreviated to: if HT.find()=0 then output ;  */
run ;

proc print data=HASH_fltrd ;
run ;
```

SAS Log 6.2-4: Approach 2: Subsetting Using a Filter Table: Hash Object

```
NOTE: There were 1 observations read from the data set WORK.FILTER_LIST.
NOTE: There were 395 observations read from the data set SASHELP.SHOES.
NOTE: The data set WORK.HASH_FLTRD has 56 observations and 8 variables.
NOTE: There were 56 observations read from the data set WORK.HASH_FLTRD.
```

This approach is based on two steps. In the first step, a filter file is created. In the second step, the hash method FIND between DATA and SET is used to keep all cases that match the criteria of the filter file.

In the DATA step, a filter file ("filter_list") is created. This filter file contains columns and value combinations. In the example, "filter_list" contains one observation "Africa" in the field REGION. In a second step, this filter file ("filter_list") is passed to the hash programming approach. The SAS table HASH_fltrd is initialized in the DATA step. The hash object HT is created after IF _N_ = 1 and the DECLARE statement. In the parentheses, DATASET loads the content of FILTER_LIST into HT. REGION is defined as key variable for HT (DEFINEKEY). DEFINEDONE indicates that all definitions are complete.

The following SET statement loads SASHELP.SHOES. The FIND method checks for a match between the key values from SASHELP.SHOES and the REGION level "Africa" of "filter_list". If a match is found, RC is 0 and the data is stored in the initialized SAS table. So, if a REGION value from SASHELP.SHOES matches the REGION entry of FILTER_LIST, the corresponding rows are stored in HASH_fltrd.

Figure 6.2-3: Output: Approach 2: Subsetting Using a Filter Table: Hash Object (Abbreviated)

Region	Product	Subsidiary	Stores	Sales	Inventory	Returns
Africa	Boot	Addis Ababa	12	$29,761	$191,821	$769
Africa	Men's Casual	Addis Ababa	4	$67,242	$118,036	$2,284
Africa	Men's Dress	Addis Ababa	7	$76,793	$136,273	$2,433

Approach 3: Random Sampling from a SAS Table

The following example shows an application of PROC SQL and hash programming that has the goal to randomly draw a certain number of rows from a SAS table.

SQL Procedure

PROC SQL randomly draws five rows from the table SASHELP.SHOES and stores them in the table RANDOM_SQL.

Program 6.2-13: Approach 3: Random Sampling: SQL

```
proc sql outobs=5 ;
    create table RANDOM_SQL as
select *
from SASHELP.SHOES
    where ranuni(0) between .10 and .20 ;
quit ;
proc print data= RANDOM_SQL noobs ;
run ;
```

PROC SQL draws five rows (see OUTOBS) from the table SASHELP.SHOES, whose RANUNI values are between 0.1 and 0.2. RANUNI is a SAS function that returns a number of a random variable with a uniform distribution between 0 and 1. A seed equal to zero causes the system time to be used as the seed; repeated submitting of the program with a seed equal to zero will result in different sampling results. However, submitting the program repeatedly with a seed greater than zero will generate the same result with unchanged seed (for example, 1234). An additional variation of the range would produce different drawing results though.

Hash Object

The following hash programming randomly draws five rows from the SASHELP.SHOES table and stores them in the RANDOM_HASH table. This very elegant approach is based on a double loop and also excludes double hits. (See Ray & Secosky, 2008, 13-15.) This approach also uses the RANUNI function with a seed equal to zero.

Program 6.2-14: Approach 3: Random Sampling: Hash Object

```
data random_hash (drop=RC I) ;
/* Declaring the Hash Object */
declare hash HT() ;
   HT.defineKey("R") ;
   HT.defineDone() ;
/* 1. Loop. Creating five random values R       */
do I = N-5+1 to N ;
R = int(ranuni(0) * I) + 1 ;
if HT.find() = 0 then R = I ;
   HT.add() ;
end ;
/* Declaring the Hash Iterator                  */
declare hiter HI("HT") ;
rc = HI.first() ;
/* 2. Loop: Compiling lines with R  */
do while (rc = 0) ;
set SASHELP.SHOES nobs=N point=R ;
output ;
rc = HI.next() ;
end ;
stop ;
run ;
proc print data= RANDOM_HASH noobs ;
run ;
```

This approach is basically based on two loops and the creation of a hash object and a hash iterator. The hash iterator object enables you to retrieve the hash object's data in forward or reverse key order. The DATA step initializes a SAS table named RANDOM_HASH, from which the two columns RC and I are to be excluded later. The DECLARE statement creates the hash object HT as an instance. The empty parenthesis means that the hash object is not yet to be filled. DEFINEKEY defines the column R that is still to be created as the key variable for HT. The method DEFINEDONE and the empty parentheses indicate that all definitions are complete. The first loop draws 5 (because a total of five rows are to be drawn) from N random numbers. The loop between DO and END is run five times in the example. In the process, integer random numbers R are drawn from the range between 1 and N. N is initialized by NOBS in the SET statement and corresponds to the value 395 in the example. The FIND method searches for a match; if a match is found, the return code is 0 and R is also I. The ADD method inserts the data into the provided hash object, provided that the key does not yet exist in the table. The second DECLARE creates the hash iterator HI to retrieve entries from the hash object in the brackets ("HT"). The second loop assigns the random numbers R to the SAS file, causing the data rows to be drawn randomly: RC with HI.*first* copies the data of the first entry in HT to the data variables of the hash object as long as data are available. The SET statement loads the data set SASHELP.SHOES to be filtered with the options NOBS and POINT. NOBS= initializes a temporary variable N, which contains the total number of all rows of the respective SAS table (SASHELP.SHOES contains, for example, 395 rows; N is therefore 395). POINT= initializes another temporary variable, R, whose numeric value determines which observation is read. Via R, POINT= causes a random access to the SAS table to be read. The later STOP terminates POINT. OUTPUT causes the output of the read data. The NEXT method copies the data of the next entry in the hash object to the data variables of HT. The concluding END closes the DO loop.

Figure 6.2-4: Output: Approach 3: Random Sampling: Hash Object

```
Region                      Product          Subsidiary     Stores
Central America/Caribbean   Men's Dress      Mexico City         3
United States               Men's Casual     Los Angeles         9
Canada                      Women's Casual   Ottawa              1
Pacific                     Men's Dress      Singapore           3
Western Europe              Sport Shoe       Copenhagen         13
```

6.2.4 Eliminating Duplicate Keys

This section compares approaches using PROC SORT, PROC SQL, and hash programming to filter out multiple occurring keys (NODUPKEY functionality).

SORT Procedure

The SORT procedure proceeds in one step. The table SASHELP.PRDSALE is sorted in ascending ACTUAL order using PROC SORT. In this example, ACTUAL is therefore interpreted as a key variable. The NODUPKEY option deletes all rows that occur more than once in the ID. The result of the sorting is stored in the table SORT_SORT. The EQUALS option ensures that the relative sorting within the BY groups is retained. The NODUPKEY option is very convenient; as result, ten columns are stored in NDPKEY_SORT.

Program 6.2-15: Elimination of Duplicate Keys: SORT

```
proc sort data = SASHELP.PRDSALE out = ndpkey_sort equals nodupkey ;
by ACTUAL ;
run ;
```

SAS Log 6.2-5: Elimination of Duplicate Keys: SORT

```
NOTE: There were 1440 observations read from the data set SASHELP.PRDSALE.
NOTE: 679 observations with duplicate key values were deleted.
NOTE: The data set WORK.NDPKEY_SORT has 761 observations and 10 variables.
```

SQL Procedure

The SQL approach proceeds also in a single step. First, it keeps all the columns required; you could list them individually or just use an asterisk. In case there is no unique ID, you could create an ID using MONOTONIC. Then SQL groups all data by ACTUAL and finally keeps only the row with the smallest ID. Alternatively, you could use the MAX function. The number of selected rows will be the same, but maybe not the selection of rows.

Program 6.2-16: Elimination of Duplicate Keys: SQL (Long and Short Versions)

Long version: ID to be created (MONOTONIC)	Short version: ID available
``proc sql ;
create table ndpkey_sql (drop = ID) as
select *, monotonic() as ID
from SASHELP.PRDSALE
group by ACTUAL
having ID = min(ID) ;
quit;
proc print data=ndpkey_sql ;
run;```	```proc sql ;
create table ndpkey_sql as
select * from SASHELP.PRDSALE
group by ACTUAL
having ID = min(ID) ;
quit;
proc print data=ndpkey_sql ;
run;```

The SQL version on the left creates an ID variable using MONOTONIC. (SASHELP.PRDSALE doesn't have one, actually.) The SQL version on the right uses an existing ID variable. For illustrative reasons, this example simply pretends SASHELP.PRDSALE has an ID. This SQL approach just wants to demonstrate how easily you can convert PROC SORT's NODUPKEY functionality to PROC SQL if only rows with distinct ACTUAL values are needed for further analysis.

SAS Log 6.2-6: Elimination of Duplicate Keys: SQL

```
NOTE: Table WORK.NDPKEY_SQL created, with 761 rows and 2 columns.
```

Hash Object

The hash programming approach also proceeds in one step and is rather manageable. The table SASHELP.PRDSALE is loaded into a hash object. All rows with multiple values in the key variable ACTUAL are automatically excluded. The result is stored in ten columns as a temporary NDPKEY_HASH table. In tables with numerous fields, hardcoding the other fields can be considered unnecessarily labor-intensive. Inserting values into a macro variable after DEFINEDATA could be a remedy at this point; however, you should check whether this additional operation might reduce processing efficiency.

Program 6.2-17: Elimination of Duplicate Keys: Hash Object

```
data _null_;
if _n_ = 0 then set SASHELP.PRDSALE ;
if _n_ = 1 then do ;
declare hash HT (dataset: "SASHELP.PRDSALE", ordered: "A") ;
    HT.definekey ("ACTUAL");
    HT.definedata ("ACTUAL","PREDICT","COUNTRY","REGION","DIVISION","PRODTYPE",
                    "PRODUCT", "QUARTER","YEAR", "MONTH") ;
HT.definedone ( ) ;
end ;
HT.output (dataset: "ndpkey_hash") ;
run ;
```

The hash object HT is created in the DECLARE method after the two _N_s. The contents of SASHELP.PRDSALE are loaded into HT. ORDERED with "A" causes HT to be sorted in ascending order. (See above for further sorting options.) DEFINEKEY defines the ACTUAL column as the key variable for HT. When loading into HT, however, only the first of multiple entries in a key variable is kept. All other values and rows with identical values in the key variable are excluded. DEFINEDATA defines the selection and sequence of columns for HT. DEFINEDONE indicates that all definitions are complete. OUTPUT stores the complete contents of HT in the SAS table NDPKEY_HASH. The sorting order is also stored, as is the key variable ACTUAL, since it was also listed after DEFINEDATA.

SAS Log 6.2-7: Elimination of Duplicate Keys: Hash Object

```
NOTE: There were 1440 observations read from the data set SASHELP.PRDSALE.
NOTE: The data set WORK.NDPKEY_HASH has 761 observations and 10 variables.
```

6.2.5 Querying Values (Retrieval)

This section introduces different approaches for retrieving single values or lists of values by means of hash programming. In contrast to the previous sections, no SQL programming is presented for comparison.

Example 1: Query the Complete Content of Data Fields (NAME, AGE)

The following two approaches illustrate how to query and output complete value lists. The contents of two fields, if filled, are output completely in the SAS log. The hash programming queries the contents of the fields NAME and AGE from the SAS table SASHELP.CLASS and outputs them completely in the SAS log. The format z3. places a zero in front of numerical values with only two digits.

Approach 1

Program 6.2-18: Example 1: Approach 1: Querying Values

```
data _NULL_ ;
  declare hash HT () ;
  HT.defineKey ( "AGE" ) ;
  HT.defineData ( "NAME" ) ;
  HT.defineDone () ;
  do until (eof) ;
     set SASHELP.CLASS end=eof ;
   rc = HT.add () ;
put NAME 10. AGE z3. ;
  end ;
  stop ;
run ;
```

DECLARE creates the hash object HT. DEFINEKEY defines AGE as the key variable for HT. DEFINEDATA defines NAME as a column for HT. DEFINEDONE and empty parentheses indicate that all definitions are complete. The DO UNTIL loop loads all rows from SASHELP.CLASS. The ADD method inserts the data into the initialized table. PUT assigns the requested formats to the output. The STOP statement prevents continuous looping.

Approach 2

Program 6.2-19: Example 1: Approach 2: Querying Values

```
data _null_ ;
  if 0 then set SASHELP.CLASS ;
  declare hash HT (dataset: "SASHELP.CLASS", ordered: "a") ;
  declare hiter HI ("HT") ;
  HT.defineKey ("NAME") ;
  HT.defineData ("NAME", "AGE" ) ;
  HT.defineDone () ;
  call missing(NAME, AGE);
do rc = HI.first () by 0 while (rc = 0) ;
  put NAME 10. AGE z3. ;
  rc = HI.next () ;
end ;
  stop ;
run ;
```

DECLARE creates the hash object HT and loads the data of SASHELP.CLASS, sorted by NAME in ascending order. You could also insert a HASHEXP argument into the parenthesis expression. Its value determines how many slots HT should provide at the top level. The second DECLARE creates the hash iterator HI to retrieve entries from HT. DEFINEKEY defines NAME as the key variable for HT. DEFINEDATA defines the columns for HT, NAME, and AGE. DEFINEDONE indicates that all definitions are complete. Using CALL MISSING is a trick to turn off the "Variable *xyz* is not initialized" messages in the SAS log. CALL MISSING first sets keys and data in the PDV to missing. Within the DO/WHILE loop, RC uses HI.*first* to copy entries in NAME and AGE from HI to the SAS log as long as data are available. Except for *z3.*, the output in the SAS log is not formatted.

Figure 6.2-5: Output: Approach 1 and 2 in SAS Log (Abbreviated)

```
Alfred     014
Alice      013
Barbara    013
Carol      014
Henry      014
James      ...
```

Example 2: **Query of Values Using a Condition (AGE=14): Output of One Hit**

This example presents querying a field using a condition. The content of a field, if filled, is only output in the SAS log by means of one representative hit. The hash programming queries the contents of the field AGE from SASHELP. CLASS for the condition 14 and, if several values are available, outputs one hit only in the SAS log. This approach therefore outputs only one hit in the SAS log, even if several hits occur. In contrast to example 1 earlier, the output in the SAS log is formatted.

Program 6.2-20: Example 2: Querying Values Using a Condition

```
data _null_;
length AGE HEIGHT WEIGHT 8 NAME $8 SEX $1 ;
declare hash HT(dataset:"SASHELP.CLASS") ;
HT.defineKey("AGE") ;
HT.defineData("AGE", "NAME", "HEIGHT", "WEIGHT");
HT.defineDone() ;
call missing (of _all_) ;
age = 14 ;
rc = HT.find() ;
if rc = 0
   then put "Data available for AGE " AGE ": " NAME
                                      HEIGHT WEIGHT ;
   else put "Data not available for: " "Age: " AGE ;
run ;
```

In contrast to the **if 0 then set** mechanism introduced at the beginning of this chapter, this DATA _NULL_ step uses length. **LENGTH** defines the length of the fields to be output later. DECLARE creates the hash object as HT and fills it with the contents of SASHELP.CLASS. DEFINEKEY and DEFINEDATA define key variables and data columns for HT. It is important that the field for the condition is defined using **DEFINEKEY** and not DEFINEDATA. DEFINEDONE indicates that all definitions are complete. The condition is specified after CALL MISSING. The system searches for at least one AGE value with the level 14. FIND checks for a match between the condition 14 and the AGE values from SASHELP.CLASS. If a match is found, a hit is output in the SAS log. PUT assigns the requested formats to the output. IF/THEN determines which formatting should be output to the SAS log in which situation (hit or not). Caution: Keys and data fields to be output in the SAS log must be defined in DEFINEKEY and DEFINEDATA before.

Figure 6.2-6: Output: Example 2: Querying Values Using a Condition

Output in the event of a hit:

```
Data available for AGE 14: Alfred 69 112.5
```

Output if data is not available:
```
Data not available for: Age: 19
```

The following sections present two examples of querying a field using a condition. In example 3, all entries are output to the SAS log for which the condition applies. The differences between examples 3 and 4 are not so much in the hash programming as in the formatting of the output and the number of fields to be output to the SAS log.

Example 3: Querying Values Using a Condition (AGE=14): Outputting Multiple Hits

This example illustrates querying a field using a condition. The content of the respective field, if filled, is completely output to the SAS log. The hash programming queries the AGE column of SASHELP.CLASS if it meets the WHERE condition. For all hits, data entries in the fields AGE and NAME are output to the SAS log. This example can easily be extended to more complex conditions.

Program 6.2-21: Example 3: Querying Values Using a Condition

```
data _NULL_  ;
  declare hash HT () ;
  HT.defineKey ("AGE") ;
  HT.defineData ("NAME") ;
  HT.defineDone () ;
  do until (eof) ;
     set SASHELP.CLASS (keep=NAME AGE WEIGHT HEIGHT SEX
                        where=(AGE=14))
            end=eof ;
   rc = HT.add () ;
put AGE = z2. NAME = 10. ;
  end ;
  stop ;
run ;
```

Figure 6.2-7: Output: Example 3: Querying Values Using a Condition

```
Age=14 Name=Carol
Age=14 Name=Judy
Age=14 Name=Alfred
Age=14 Name=Henry
```

DEFINEKEY and DEFINEDATA define key variables and data columns for HT. The DO UNTIL loop loads rows from SASHELP.CLASS that contain the value 14 in AGE. The ADD method inserts the data into the initialized table. PUT formats the output in the SAS log.

Example 4: Querying Several Values (NAME="Joyce", AGE=11): One Hit

This example illustrates a query using two conditions. There are subtle differences from the previous example. In addition to a numerical condition (AGE), a second condition of the type String ("Joyce" in NAME) must now be met. The conditions are specified after CALL MISSING, and not using a WHERE clause (although the effect is the same). All variables in DEFINEKEY need to be used in the conditions. The number of output hits differs too. For all data entries that match both conditions, only one hit is output in the SAS log. For example, if there were duplicate rows, only one copy would be displayed in the SAS log.

Program 6.2-22: Example 4: Querying Several Values

```
data _null_  ;
length AGE HEIGHT WEIGHT 8 NAME $8 ;
declare hash HT(dataset:"SASHELP.CLASS") ;
HT.defineKey("NAME", "AGE") ;
HT.defineData("HEIGHT", "WEIGHT") ;
```

```
HT.defineDone() ;
call missing (of _all_) ;
name = "Joyce" ;
age = 11 ;
rc = HT.find() ;
if rc = 0
   then put "Data available for:          "
                              NAME AGE HEIGHT WEIGHT ;
   else put "Data not available for: "
                              NAME AGE ;
run;
```

The FIND method checks if there is a match between the keys in HT and the specified conditions. Depending on whether a match was found or not, PUT prints the first or the second result text line to the log.

Figure 6.2-8: Output: Example 4: Querying Several Values

Output in the event of a hit:

```
Data available for:        Joyce 11 51.3 50.5
```

Output if data is not available:

```
Data not available for: Joyce 12
```

Example 5: Querying Several Values (AGE=14, SEX="M"): Several Hits

This example illustrates a query using two conditions. All hits are output in the SAS log for which both conditions apply. If there were duplicate rows, all rows would be displayed in the SAS log.

Program 6.2-23 Example 5: Querying Several Values

```
data _null_ ;
if _n_ = 1  then do ;
declare hash HT() ;
HT.definekey("AGE", "SEX") ;
HT.definedata("AGE") ;
HT.definedone() ;
end ;
do until(eof) ;
set SASHELP.CLASS (keep=NAME AGE WEIGHT HEIGHT SEX
   where=(AGE=14 & SEX="M")) end=eof ;
  rc = HT.add () ;
  put "Data available for: " NAME AGE HEIGHT WEIGHT ;
end ;
stop ;
run ;
```

Figure 6.2-9: Output: Example 5: Querying Several Values

Output in the event of one hit (or more):
```
Data available for: Alfred 14 69 112.5
Data available for: Henry 14 63.5 102.5
```

Example 6: Querying Several Values (AGE=14, SEX="M"): Save Query Output in SAS Table

This example illustrates how the result for a query using two conditions is also stored in a SAS table. All entries for which both conditions apply are output to the SAS log and stored parallel in the SAS table SUBSET. Note that the log displays only two columns (AGE, NAME); the SUBSET table has six variables, however. The number of rows is the same.

Program 6.2-24: Example 6: Querying Several Values: Output in a SAS table

```
data SUSBET ;
  declare hash HT () ;
  HT.defineKey ( "AGE" ) ;
  HT.defineData ( "NAME" ) ;
  HT.defineDone () ;
  do until (eof) ;
     set SASHELP.CLASS (keep=NAME AGE WEIGHT HEIGHT SEX
          where=(AGE=14)) end=eof ;
   rc = HT.add () ;
put AGE = z2. NAME = 10. ;
output ;
  end ;
  stop ;
run ;
```

6.3 Working with Multiple Tables

Combining tables is one of the basic tasks of data management. The "combining" of different variables for the same cases is called one-to-one merging or joining. The "appending" of different cases with the same variables is known as concatenating or interleaving. The following section presents various join approaches using hash programming. Special features of joining using PROC SQL are highlighted as they are to be replicated using hash programming. The join examples are restricted to the one-to-one scenario. In a one-to-one relationship, a case in a first table references exactly one case in a second table based on one or more key variables. A one-to-one relationship exists if, in the case of a single key variable, the values of the key variables occur only once in all tables to be merged or, in the case of multiple key variables, combinations of the values of all key variables occur only once in all tables to be merged. (For one-to-many, many-to-one, and many-to-many scenarios, see Schendera, 2011, Chapter 7.) The following examples of hash programming with two tables are presented:

- Inner and outer joins with two tables (Section 6.3.1)
- Fuzzy Join with two tables (Section 6.3.2)
- Splitting of an unsorted SAS table (Section 6.3.3)

6.3.1 Inner and Outer Joins with Two Tables

PROC SQL uses join operators for joining different variables from several tables. Variants of this form of combining tables include inner joins (InnerJoin1, InnerJoin2) and outer joins (Left, Right, Full). Hash objects can also be used to program the inner and outer joins of tables in an uncomplicated way. First, a join approach with PROC SQL is presented. Then, the respective join functionality is implemented using hash programming. The data and examples are taken from Schendera (2011, Chapter 6.2).

Program 6.3-1: Demo Data for JOIN Operators

```
/**********************************/
/*  Demo data for JOIN Operators   */
/**********************************/;

data ONE ;                              data THREE ;
   input ID A B C ;                        input ID E F G ;
datalines;                              datalines;
01 1    2   3                           01 4    4   4
02 1    2   3                           02 4    4   4
03 99 99 99                             03 99 99 99
04 1    2   3                           05 4    4   4
05 1    2   3                           06 4    4   4
;                                       ;
                                        run ;
```

Table ONE contains the IDs 1 to 5 and the variables A, B, and C. Table THREE contains the variables E, F, G, and also the variable ID. However, the two tables differ in their values in ID. In ONE, the value 6 is missing (instead, the value 4 occurs). In THREE, the value 4 is missing (the value 6 appears instead). You use join operators to combine these tables.

Inner Join

An inner join can be defined with PROC SQL in two variants. (See Schendera, 2011, 195ff.). The Inner Join1 variant is programmed using commas in the FROM clause and returns the cross product of all tables in the FROM clause as the result table. The Inner Join2 variant is programmed using the keywords INNER, JOIN, and ON and returns as a result table only all of the rows with matching IDs of the tables in the FROM clause. The Inner Join2 variant thus introduces an additional condition: the equality in the rows in the ON expression. (See the WHERE clause for equi-joins.) In the FROM clause, an inner join can include up to 32 tables at the same time; all other join variants always have only two tables (apart from subqueries).

Program 6.3-2: Inner Join: SQL

```
proc sql ;
create table IJ_SQL
as select ONE.ID, A,B,C,E,F,G
from THREE inner join
ONE on THREE.ID=ONE.ID ;
quit ;
proc print data=IJ_SQL noobs ;
run ;
```

The tables ONE and THREE are joined and linked using the key variable ID, which occurs in both tables. ONE provides the structure and specifies the ID. Because of ON and in it ONE.ID = THREE.ID, all rows whose IDs match are kept. (See Figure 6.3-1.)

Program 6.3-3: Inner Join: Hash Object

```
data IJ_HASH(drop=rc) ;
if 0 then set ONE ;
declare hash HT(dataset:"ONE") ;
rc=HT.defineKey("ID") ;
rc=HT.defineData("A", "B", "C") ;
rc=HT.defineDone() ;
do until(eof) ;
set THREE end=eof ;
call missing(A, B, C) ;
rc=HT.find() ;
if rc=0 then output ;
end ;
stop ;
run ;
proc print data=IJ_HASH noobs ;
run ;
```

Figure 6.3-1: Output: Inner Join: Hash Object

ID	A	B	C	E	F	G
1	1	2	3	4	4	4
2	1	2	3	4	4	4
3	99	99	99	99	99	99
5	1	2	3	4	4	4

The resulting tables IJ_SQL and IJ_HASH match; they contain 4 rows and 5 variables. The row of ID 4 from ONE is completely missing, as is the row of ID 6 from THREE.

Outer Joins

Outer joins are basically inner joins of type Inner Join2, which also contain the rows that do not match the other tables in the resulting table. Outer joins are usually extended inner joins of the Inner Join2 type. After an outer join, a resulting table has at least as many, but usually more values than after an inner join of type Inner Join2. All three outer join variants contain all rows of the Cartesian product of the two tables restricted by the ON expression for which the SQL expression is true. The left outer join also contains the rows from the left table that do not match any rows in the second table. In contrast, the right outer join also has the rows from the right table that do not match a row in the first table. The full outer join also has all the rows from both tables that do not match any row in the other table.

Outer Left Join

Program 6.3-4: Outer Left Join: SQL

```
proc sql ;
create table OLJ_SQL as
select *
from ONE a left join THREE b
on a.ID=b.ID ;
quit ;

proc print data=OLJ_SQL noobs ;
run ;
```

The tables ONE and THREE are joined and linked using the key variable ID, which occurs in both tables. ONE provides the structure and specifies the ID. (See Figure 6.3-2.)

Program 6.3-5: Outer Left Join: Hash Object

```
data OLJ_HASH (drop=RC) ;
length ID A B C E F G 4. ;
if _n_=0 then set ONE ;
if _n_=1 then do ;
declare hash HT (dataset: "THREE") ;
HT.definekey("ID") ;
HT.definedata("E", "F", "G") ;
HT.definedone() ;
call missing(of _ALL_) ;
end ;
set ONE ;
RC=HT.find() ;
/* RC eq 0 then output ; Decomment if necessary */
run ;
proc print data=OLJ_HASH noobs ;
run ;
```

After DECLARE, the contents of table THREE are loaded. If you remove the comment, the condition "if RC ne 0" will also exclude the row with ID 4 from OLJ_HASH.

Figure 6.3-2: Output: Outer Left Join: Hash Object

```
ID     A     B     C     E     F     G
 1     1     2     3     4     4     4
 2     1     2     3     4     4     4
 3    99    99    99    99    99    99
 4     1     2     3     .     .     .
 5     1     2     3     4     4     4
```

The resulting OLJ_SQL and OLJ_HASH tables match; they contain 5 rows and 6 variables. The row of ID 6 from THREE is completely missing. In PROC SQL, the length of the "first" structuring table ONE dominates the presence of IDs in THREE.

Outer Right Join

Program 6.3-6: Outer Right Join: SQL

```
proc sql ;
create table
ORJ_SQL as select
b.ID,A,B,C,E,F,G
from ONE a right join THREE b
on a.ID=b.ID
order by ID ;
quit ;

proc print data=ORJ_SQL noobs ;
run ;
```

The tables ONE and THREE are joined and linked using the key variable ID, which occurs in both tables. ONE provides the structure and specifies the ID. Because of ON and ONE.ID = THREE.ID, all rows are kept whose IDs match. (See Figure 6.3-3.)

Program 6.3-7: Outer Right Join: Hash Object

```
data ORJ_HASH(drop=RC) ;
if 0 then set ONE ;
declare hash HT(dataset:"ONE") ;
rc=HT.defineKey("ID") ;
rc=HT.defineData("A", "B", "C") ;
rc=HT.defineDone() ;
do until(eof) ;
set THREE end=eof ;
call missing(of A, B, C) ;
rc=HT.find() ;
output ;
end ;
stop ;
run ;
proc print data=ORJ_HASH noobs ;
run ;
```

After DECLARE, the contents of table ONE are loaded. The FIND method checks if the ID of THREE matches the ID in one of the data rows in HT. If there is a match, the observation from THREE "merged" to the ONE data set.

Figure 6.3-3: Output: Outer Right Join: Hash Object

ID	A	B	C	E	F	G
1	1	2	3	4	4	4
2	1	2	3	4	4	4
3	99	99	99	99	99	99
5	1	2	3	4	4	4
6	.	.	.	4	4	4

The resulting tables IJ_SQL and IJ_HASH match; they contain 5 rows and 6 variables. The line of ID 4 from ONE is completely missing.

Table 6.3-1: Result Benchmark Test

Test (N rows)	Time in Seconds	PROC SQL	Hash Object
8.000.000	CPU Time	47.14	61.00
	Real Time	318.65	236.12
12.000.000	CPU Time	68.07	61.74
	Real Time	587.88	224.40

When processing outer right joins using hash objects and PROC SQL, significant performance differences in real time become apparent. Higher limit values such as 80,000,000 or 40,000,000 caused the benchmark testing of the hash object to terminate due to lack of memory.

SAS Log 6.3-1: Hash Object Insufficient Memory Error

```
FATAL: Insufficient memory to execute DATA step program. Aborted during the EXECUTION phase.
ERROR: The SAS System stopped processing this step because of insufficient memory.
WARNING: The data set WORK.ORJ_HASH may be incomplete.  When this step was stopped there
were 0 observations and 9 variables.
```

6.3.2 Fuzzy Join with Two Tables

All examples for merging tables using PROC SQL and hash programming so far have covered the exact combining of tables. In exact combining, rows from different tables are only joined if they have exactly the same values in the key variables. If there are no identical values in the common key variable, key-based joining of SAS tables is logically not possible.

With fuzzy joins, on the other hand, rows from different tables are joined even if they only have approximately the same value in the key variable. This means that tables can be linked together using fuzzy joins even if they do not have exactly the same values in the key variables.

Program 6.3-8: Fuzzy Join with Two Tables

```
data REFERENCE ;                         data DATA_ADDED ;
input                                    input
VAL_SLOT TIMESTMP time5. ;               VAL_ADDED TIME_ADDED time5. ;
datalines ;                              Datalines ;
105 08:00                                102 08:02
110 09:59                                110 10:03
115 12:51                                110 13:00
120 14:30                                130 14:38
125 16:59                                135 16:52
;                                        ;
```

The REFERENCE file contains the lines to which the data from DATA_ADDED are added. The table REFERENCE contains the column VAL_SLOT as the key and TIMESTMP as another field. (See example 2 for joining using a timestamp.) The table DATA_ADDED contains the column VAL_ADDED as the key and TIME_ADDED as another field. The REFERENCE and DATA_ADDED tables contain only one matching key value, namely 110. A successful join or match-merge based on common entries in a key variable cannot be assumed. The fuzzy join approach provides a remedy.

A fuzzy join uses a condition to define the extent to which the values in the two key variables are allowed differ from each other. The *permissible tolerance* or *fuzziness* defined by the user allows tables to be joined together even if the entries in the key variable do not correspond exactly, but only approximately.

The tolerance is only one aspect that should be specified before a fuzzy match; as a further point, you should specify whether you want to keep

- all Cartesian hits (see Program 6.3-9)
- only hits with an optimal=minimal tolerance including multiple hits (see Program 6.3-11)
- only hits that match the grid of a leading table ideally with perfect matches only (see Program 6.3-11)
- hits where meeting the tolerance condition might have priority over a perfect match (tolerance = 0) (see Program 6.3-10).

These considerations also extend to time and string variables. Fuzzy joins with string variables differ from numeric or date variables only in an additional step of quantifying the distance between strings by using SAS functions like COMPLEV or COMPGED (dissimilarity between two strings), SPEDIS (spelling distance between two strings), SOUNDEX (phonetic matching, preferably of English strings), or user-written SAS functions using PROC FCMP. The following SQL and hash programming examples illustrate these points.

SQL 1: Target: All possible combinations of fuzzy matching rows

Program 6.3-9 FUZZY_SQL1: All Possible Combinations of Fuzzy Matching Rows

```
proc sql ;
    create table FUZZY_SQL_1 as select
           VAL_SLOT, VAL_ADDED, TIMESTMP, TIME_ADDED,
        abs(VAL_ADDED - VAL_SLOT) as TOLERANCE,
        monotonic() as ID
    from REFERENCE, DATA_ADDED
           where calculated TOLERANCE lt 15  ;
quit;
```

Result: Cartesian Product (N=14). The use of MONOTONIC() is for the SQL examples 2 and 3. The TOLERANCE condition is met, but maybe you only want unique, but not multiple (Cartesian) hits.

Figure 6.3-4: Output: FUZZY_SQL_1

VAL_SLOT	VAL_ADDED	TIMESTMP	TIME_ADDED	ID	TOLERANCE
105	102	8:00	8:02	1	3
105	110	8:00	10:03	2	5
105	110	8:00	13:00	3	5
110	102	9:59	8:02	6	8
110	110	9:59	10:03	7	0
110	110	9:59	13:00	(abbreviated)	

SQL 2: Target: Selection of fuzzy matches according to "grid" of leading REFERENCE table.

Program 6.3-10: FUZZY_SQL2: Selection of Fuzzy Matches According to "Grid" of Leading REFERENCE Table

```
proc sql ;
    create table FUZZY_SQL_2 as
    select a.VAL_SLOT, a.VAL_ADDED, TIMESTMP, TIME_ADDED, a.ID, a.TOLERANCE
    from FUZZY_SQL a inner join
        (select min(VAL_SLOT) as VAL_SLOT,
               min(VAL_ADDED) as VAL_ADDED,
               max(ID) as ID
        from FUZZY_SQL_1
        group by VAL_SLOT) b
    on a.VAL_SLOT = b.VAL_SLOT and
        a.ID = b.ID ;
 quit ;
```

Result: N=5 unique hits according to the "grid" of the leading REFERENCE table. The TOLERANCE condition is met, but not minimal=optimal though (see 105, 125). The ON clause includes VAL_SLOT and ID.

Figure 6.3-5: Output: FUZZY_SQL_2

VAL_SLOT	VAL_ ADDED	TIMESTMP	TIME_ ADDED	ID	TOLERANCE
105	110	8:00	13:00	3	**5**
110	110	9:59	13:00	8	0
115	110	12:51	13:00	13	5
120	130	14:30	14:38	19	10
125	135	16:59	16:52	25	**10**

SQL 3: Target: Fuzzy matches with minimal tolerance. Multiple matches (see I., result not shown), unique fuzzy matches with REFERENCE "grid" only (see II.).

Program 6.3-11: FUZZY_SQL3 and FUZZY_SQL4: Fuzzy Matches with Minimal Tolerance

```
proc sql ;
/*(I) Optimized matches: Multiple (correct) hits per VAL_SLOT*/
    create table FUZZY_SQL_3 as
    select a.VAL_SLOT, a.VAL_ADDED, TIMESTMP, TIME_ADDED, a.ID, a.TOLERANCE
    from FUZZY_SQL_1 a inner join
        (select max(VAL_SLOT) as VAL_SLOT,
                min(VAL_ADDED) as VAL_ADDED,
                min(TOLERANCE) as TOLERANCE
        from FUZZY_SQL_1
        group by VAL_SLOT) b
    on a.VAL_SLOT = b.VAL_SLOT and
        a.TOLERANCE = b.TOLERANCE ;
/*(II) Only unique (correct) hits per VAL_SLOT                  */
    create table FUZZY_SQL_4 as
    select *
    from FUZZY_SQL_3
    group by VAL_SLOT
            having ID=min(ID) ;
quit;
```

Result of FUZZY_SQL4: N=5 hits according to the "grid" of the leading REFERENCE table. Now, TOLERANCE is minimal=optimal (see 105, 125). This time, the ON clause includes VAL_SLOT and TOLERANCE.

Figure 6.3-6: Output: FUZZY_SQL_4

VAL_SLOT	VAL_ ADDED	TIMESTMP	TIME_ ADDED	ID	TOLERANCE
105	102	8:00	8:02	1	**3**
110	110	9:59	10:03	7	0
115	110	12:51	10:03	12	5
120	110	14:30	10:03	17	10
125	130	16:59	14:38	24	**5**

If the condition for fuzziness, *abs(VAL_ADDED - VAL_SLOT)*, is zero, this corresponds to a perfect match, a match-merge, if you will. The greater the absolute deviation, that is, "fuzziness," the greater the tolerance when matching entries in a key variable. Depending on your definition, a fuzzy join might not distinguish between an exact and an approximate match. Therefore, an approximate match might take precedence over an exact match.

The following hash programming example produces another result variant.

Example 1: Fuzzy Join for Numeric Key Variables

The tables REFERENCE and DATA_ADDED are to be combined. The numeric values in the key variables VAL_SLOT and VAL_ADDED are to establish the connection. The REFERENCE table is the reference. The entries from DATA_ADDED are assigned to the structure of REFERENCE using the extent of the approximate fit of VAL_SLOT and VAL_ADDED.

Program 6.3-12: FUZZY_NUM: Fuzzy Join of Numerical Values (Hash Object)

```
*--------------------------*
| Fuzzy join of num. values |
*--------------------------*;
data FUZZY_NUM ;
   length VAL_SLOT TIMESTMP 8 ;
   if _n_=1 then do ;
     declare hash HT(dataset:"DATA_ADDED") ;
     HT.definekey("VAL_ADDED") ;
     HT.definedata("VAL_ADDED", "TIME_ADDED") ;
     declare hiter HI("HT") ;
     HT.definedone() ;
     call missing(VAL_SLOT, TIMESTMP) ;
   end ;
   set REFERENCE ;
     RC1=HI.first() ;
   do while (RC1=0) ;
     if abs(VAL_ADDED - VAL_SLOT) lt 15 then leave ;
     RC2=HI.next() ;
     if RC2 ne 0 then do ;
      VAL_ADDED=. ; TIME_ADDED=. ;
       leave ;
     end ;
   end ;
run ;

proc print data= FUZZY_NUM noobs ;
var VAL_SLOT VAL_ADDED TIMESTMP TIME_ADDED ;
  format TIMESTMP  TIME_ADDED time5. ;
run ;
```

VAL_SLOT specifies the grid 105, 110, 115, 120, and 125. The "fit" of VAL_SLOT and VAL_ADDED assigns the value 102 from DATA_ADDED to 105 from REFERENCE; 102 from DATA_ADDED is also assigned to 110 and 115 from REFERENCE, respectively. 130 from DATA_ADDED is assigned to 120 and 125 from REFERENCE, respectively. 110 from REFERENCE is not assigned the value 110 from DATA_ADDED.

Figure 6.3-7: Output: FUZZY_NUM

VAL_SLOT	VAL_ ADDED	TIMESTMP	TIME_ ADDED
105	102	8:00	8:02
110	102	9:59	8:02
115	102	12:51	8:02
120	130	14:30	14:38
125	130	16:59	14:38

The tables REFERENCE and DATA_ADDED were combined by means of an approximate correspondence of the entries in the key variables VAL_SLOT and VAL_ADDED. A possible exact match between exactly matching values in the key variables is not taken into account in the fuzzy join.

Example 2: Fuzzy Join for Time Values

Example 2 differs only in the format of the key variables from example 1. The key variables TIMESTMP and TIME_ ADDED are now in the SAS time format TIME5. The REFERENCE table also is the reference here. The entries from DATA_ADDED are assigned to the structure of REFERENCE by the extent of the approximate fit of TIMESTMP and TIME_ADDED.

Program 6.3-13: FUZZY_TMS: Fuzzy Join for Time Values (Hash Object)

```
*---------------------*
| Fuzzy join of times |
*---------------------*;
data FUZZY_TMS ;
   length TIMESTMP VAL_SLOT 8 ;
   if _n_=1 then do ;
     declare hash HT(dataset:"DATA_ADDED") ;
     HT.definekey("TIME_ADDED") ;
     HT.definedata("TIME_ADDED", "VAL_ADDED") ;
     declare hiter HI("HT") ;
     HT.definedone() ;
     call missing(TIMESTMP, VAL_SLOT) ;
   end ;
   set REFERENCE ;
     RC1=HI.first() ;
   do while (RC1=0) ;
     if abs(TIME_ADDED-TIMESTMP) lt 600 then leave ;
     RC2=HI.next() ;
     if RC2 ne 0 then do ;
       VAL_ADDED=. ; TIME_ADDED=. ;
       Leave ;
     end ;
   end ;
run ;


proc print data= FUZZY_TMS noobs ;
var TIMESTMP TIME_ADDED VAL_SLOT VAL_ADDED ;
  format TIMESTMP  TIME_ADDED time5. ;
run ;
```

For example, TIMESTMP specifies the grid 8:00, 9:59, 12:51, 14:30, and 16:59. The "fit" of TIMESTMP and TIME_ ADDED assigns the value 8:02 from DATA_ADDED to the value 8:00 from REFERENCE; 10:03 from DATA_ADDED is assigned 9:59 from REFERENCE, and so on.

Figure 6.3-8: Output: FUZZY_TMS

TIMESTMP	TIME_ ADDED	VAL_SLOT	VAL_ ADDED
8:00	8:02	105	102
9:59	10:03	110	110
12:51	13:00	115	110
14:30	14:38	120	130
16:59	16:52	125	135

Although the REFERENCE and DATA_ADDED tables do not have exactly the same values in the key variables, it was possible to link them together using an approximate correspondence of the entries in the TIMESTMP and TIME_ ADDED key variables. If the result of a fuzzy join is not satisfactory, a user could adjust the conditions for fuzziness in the expression *abs(VAL_ADDED - VAL_SLOT)* until a better result is obtained. Exactly matching values in the key variables did not occur in this example.

6.3.3 Splitting of an Unsorted SAS Table

This section introduces the splitting of an unsorted SAS table using PROC SQL and hash programming, each in conjunction with the DATA step and in the case of the SQL approach also with macro variables. The task of PROC SQL and hash programming is to segment the SAS table SASHELP.CLASS according to the values in the AGE column and to store the contents of SASHELP.CLASS in accordingly filtered subsets. As a further possibility, reference is made to macro approaches elsewhere in this volume.

SQL Procedure

Program 6.3-14 SQL: Splitting of an Unsorted SAS Table

```
proc sql noprint ;
select distinct "AGE" || put (AGE, best.-l)
into :name_list
        separated by " "
from SASHELP.CLASS ;
select "when (" || put (AGE, best.-l) || ")
        output AGE" || put (AGE, best.-l)
into :renamelist
    separated by ";"
from SASHELP.CLASS ;
quit ;

proc sort data = SASHELP.CLASS force ;
by AGE ;
run ;
data &name_list. ;
set SASHELP.CLASS ;
select (AGE) ;
&renamelist. ;
otherwise ;
end ;
run ;
```

The SQL approach consists of two steps. In the first step, the future names of the SAS tables to be generated are created in the form "AGE*nn*" (for example, AGE11) and stored in the macro variable NAME_LIST. In a second step, the renaming statements are created in the form "when (14) output AGE14" and stored in the macro variable RENAME_LIST. To improve performance, SASHELP.CLASS is also sorted by AGE in a third step. Finally, the DATA step accesses the entries in the macro variables NAME_LIST and RENAME_LIST, thereby segmenting SASHELP.CLASS according to the values in AGE and storing the correspondingly filtered contents in subsets with the names AGE11 and so on.

SAS Log 6.3-2 SQL: Splitting of an Unsorted SAS Table

```
NOTE: There were 19 observations read from the data set SASHELP.CLASS.
NOTE: The data set WORK.AGE11 has 2 observations and 5 variables.
NOTE: The data set WORK.AGE12 has 5 observations and 5 variables.
NOTE: The data set WORK.AGE13 has 3 observations and 5 variables.
NOTE: The data set WORK.AGE14 has 4 observations and 5 variables.
NOTE: The data set WORK.AGE15 has 4 observations and 5 variables.
NOTE: The data set WORK.AGE16 has 1 observations and 5 variables.
```

SAS Macro SPLIT_DS2

I would also like to point out the two macros SPLIT_DS1 (alphanumeric split variable) and SPLIT_DS2 (numeric split variable; see Program 6.3-15). The syntax of the SPLIT_DS2 macro is reproduced for illustration. With the exception of the numbering of the SAS tables created, the result corresponds to the task formulated at the beginning. An explanation of this macro is in Chapter 4.

Program 6.3-15: SAS Macro SPLIT_DS2: Splitting of an Unsorted SAS Table

```
%macro split_ds2 (SPLIT_DS=, SPLIT_BY=);
  proc sql ;
     select count(distinct &SPLIT_BY.)
        into :n
         from  &SPLIT_DS. ;
     select distinct &SPLIT_BY.
        into :&SPLIT_BY.1 - :&SPLIT_BY.%left(&n)
        from  &SPLIT_DS. ;
  quit ;
  %do i=1 %to &n;
       data WORK.&SPLIT_BY._&i ;
          set &SPLIT_DS. ;
          if &SPLIT_BY.=&&&SPLIT_BY.&i ;
          run;
     %end;
  %mend split_ds2 ;

%split_ds2  (SPLIT_DS=SASHELP.CLASS, SPLIT_BY=AGE);
```

SAS Log 6.3-3: SAS Macro SPLIT_DS2: Splitting of an Unsorted SAS Table
```
NOTE: The data set WORK.AGE_1 has 2 observations and 5 variables.
NOTE: The data set WORK.AGE_2 has 5 observations and 5 variables.
NOTE: The data set WORK.AGE_3 has 3 observations and 5 variables.
NOTE: The data set WORK.AGE_4 has 4 observations and 5 variables.
NOTE: The data set WORK.AGE_5 has 4 observations and 5 variables.
NOTE: The data set WORK.AGE_6 has 1 observations and 5 variables.
```

Hash Object

In contrast to the SQL and macro approaches, the hash programming approach essentially proceeds in one single step.

Program 6.3-16: Hash Object: Splitting of an Unsorted SAS Table

```
data _null_ ;
declare hash HT (ordered: "a") ;
    HT.definekey  ("Name", "Sex", "Age", "Height", "Weight", "_n_") ;
    HT.definedata ("Name", "Sex", "Age", "Height", "Weight") ;
    HT.definedone ( ) ;
do _n_ = 1 by 1 until (last.age) ;
   set  SASHELP.CLASS  ;
   by age ;
   HT.add() ;
end ;
    HT.output (dataset: "AGE_" || put (age, best.-l)) ;
run ;
```

DECLARE creates the hash object HT as an ascending sorted instance. DEFINEKEY defines the key columns for HT. The "_n_" option in the DEFINEKEY line keeps rows even if they contain duplicate entries in rows across the columns from NAME to WEIGHT. The SAS log does not tell you explicitly that there might be duplicate values in the key variable. If you want to keep unique rows only, you just need to remove _ n _. The SAS log informs you by an "ERROR" note that it found duplicate keys. DEFINEDATA defines the selection and sequence of columns for HT. DEFINEDONE indicates that all definitions are complete. The subsequent DO/END loop loads a first distinct value of AGE, uses this value to add each row from the table SASHELP.CLASS to the hash object, and finally uses the OUTPUT method to store the content under the name AGE_*nn*. This DO/END loop is processed until no further levels of AGE need to be processed. Because of the BY statement after the SET statement, the input data set needs to be sorted by the column after BY.

SAS Log 6.3-4: Hash Object: Splitting of an Unsorted SAS Table

```
NOTE: The data set WORK.AGE_11 has 2 observations and 5 variables.
NOTE: The data set WORK.AGE_12 has 5 observations and 5 variables.
NOTE: The data set WORK.AGE_13 has 3 observations and 5 variables.
NOTE: The data set WORK.AGE_14 has 4 observations and 5 variables.
NOTE: The data set WORK.AGE_15 has 4 observations and 5 variables.
NOTE: The data set WORK.AGE_16 has 1 observations and 5 variables.
```

6.4 Overview: Elements of Hash Programming

The following overview compiles elements of hash programming (as of SAS 9.4). The elements are sorted by attributes, operators, statements, and methods. For more information, please refer to the SAS technical documentation.

Attributes:

> **ITEM_SIZE:** Returns the size (in bytes) of an item in a hash object.
> **NUM_ITEMS:** Returns the number of items in the hash object.

Operators:

> **_NEW_:** Creates an instance of a hash or hash iterator object.

Statements:

> **DECLARE:** Declares a hash or hash iterator object; creates an instance of and initializes data for a hash or hash iterator object.

Methods:

> **ADD:** Adds the specified data that is associated with the given key to the hash object.
> **CHECK:** Checks whether the specified key is stored in the hash object. If found (success), RC=0 is returned, otherwise (failure) a value unequal to 0.
> **CLEAR:** Removes all items from the hash object without deleting the hash object instance (see also DELETE).
> **DEFINEDATA:** Defines data, associated with the specified data variables, to be stored in the hash object. You define data by passing the DATA step data variable name to the DEFINEDATA method. You also use these variables to initialize the hash object.
> **DEFINEDONE:** Indicates that all key and data definitions are complete.
> **DEFINEKEY:** Defines key variables for the hash object.
> **DELETE:** Deletes the hash or hash iterator object.
> **DO_OVER:** Traverses a list of duplicate keys in the hash object.
> **EQUALS:** Determines whether two hash objects are equal.
> **FIND:** Determines whether the specified key is stored in the hash object.
> **FIND_NEXT:** Sets the current list item to the next item in the current key's multiple item list and sets the data for the corresponding data variables.
> **FIND_PREV:** Sets the current list item to the previous item in the current key's multiple item list and sets the data for the corresponding data variables.
> **FIRST:** Returns the first value in the underlying hash object. If the hash object sorted, then the data item that is returned is the one with the 'least' key (smallest numeric value or first alphabetic character).
> **HAS_NEXT:** Determines whether there is a next item in the current key's multiple data item list. The FIND method determines whether the key exists in the hash object. The HAS_NEXT method determines whether the key has multiple data items associated with it.
> **HAS_PREV:** Determines whether there is a previous item in the current key's multiple data item list. The FIND method determines whether the key exists in the hash object. The HAS_PREV method determines whether the key has multiple data items associated with it.
> **LAST:** Returns the last value in the underlying hash object. If the hash object sorted, then the data item that is returned is the one with the 'highest' key (largest numeric value or last alphabetic character).

NEXT: Returns the next value in the underlying hash object.

OUTPUT: Creates one or more data sets each of which contain the data in the hash object. If the data in the hash object is sorted, it is also stored sorted. To store key variables, they must be defined in DEFINEMETHOD.

PREV: Returns the previous value in the underlying hash object.

REF: Consolidates the CHECK and ADD methods into a single method call. The REF method is useful for counting the number of occurrences of each key in a hash object.

REMOVE: Removes the data that is associated with the specified key from the hash object.

REMOVEDUP: Removes the data that is associated with the specified key's current data item from the hash object.

REPLACE: Replaces the data that is associated with the specified key with new data.

REPLACEDUP: Replaces the data that is associated with the current key's current data item with new data.

RESET_DUP: Resets the pointer to the beginning of a duplicate list of keys when you use the DO_OVER method.

SETCUR: Specifies a starting key item for iteration.

SUM: Retrieves the summary value for a given key (when there is only one data element per key) from the hash table and stores the value in a DATA step variable.

SUMDUP: Retrieves the summary value for the current data item of the current key (when there is more than one data element per key) and stores the value in a DATA step variable.

Chapter 7: FedSQL

SQL is a success story. Originally developed by IBM in the SEQUEL, XRM, and System R projects, this database language soon turned into a quasi-industry standard for relational databases. At the heart of this success is the American National Standards Institute (ANSI) standard that made communication and interaction between manufacturers and databases easier. The SAS version of SQL, PROC SQL, does *not* fully comply with the ANSI-1992 standard. Deviating PROC SQL functionalities include the support of SAS macro language (see Chapter 4), how PROC SQL handles missing values (see Chapter 2), and legacy SAS data types. Other database manufacturers followed the ANSI-1992 and later ANSI standards in a similar fashion with proprietary enhancements.

Over time, the standards and the database technologies evolved, leading to FedSQL. As the interaction of PROC SQL with these ANSI-compliant databases became unnecessarily difficult, SAS 9.4 introduced FedSQL as of ANSI-1999. FedSQL is a vendor-neutral SQL dialect that means you can write programs that are independent of the database system. PROC FEDSQL provides a data access technology in SAS that offers a scalable, threaded, high-performance way to access, manage, and share nonrelational data in multiple data sources.

Let's start with a bit of ANSI history. SAS Institute provides two SQL dialects and philosophies: PROC SQL and PROC FEDSQL. PROC SQL is partially compliant with ANSI-1992 plus its many SAS additions. PROC FEDSQL is fully ANSI-1999-compliant, and it is vendor-neutral.

Figure 7.1-1: SAS Institute's Two SQL Languages and Their ANSI standards

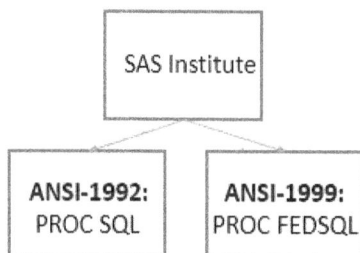

Note: For the sake of precision, the box for PROC SQL should probably read, "Partial ANSI-1992 plus SAS Additions". For the sake of space, I would like to keep it short though. When I am referring to the FEDSQL procedure or PROC FEDSQL, I use capital letters, not when I am referring to the FedSQL concept or language. It is not always easy to keep them apart.

Figure 7.1-2 makes a few points clear:

- the differences between the platforms
- the apparent differences between processing environments
- where SQL and FEDSQL can be used and where not
- the similarities between SQL and FEDSQL syntax (indicated by overlapping boxes)

The second row of Figure 7.1-2 illustrates that SAS offers two platforms. SAS 9.4 is for the client/server. SAS Viya is for cloud-computing. SAS Viya 3.5 includes the Compute Server (formerly known as SPRE, SAS Programming Run-Time Environment), which provides the same capabilities as SAS 9.4 using cloud-native technology. But it

Figure 7.1-2: SAS Platforms, Processing Environments, and SQL and FEDSQL Variants

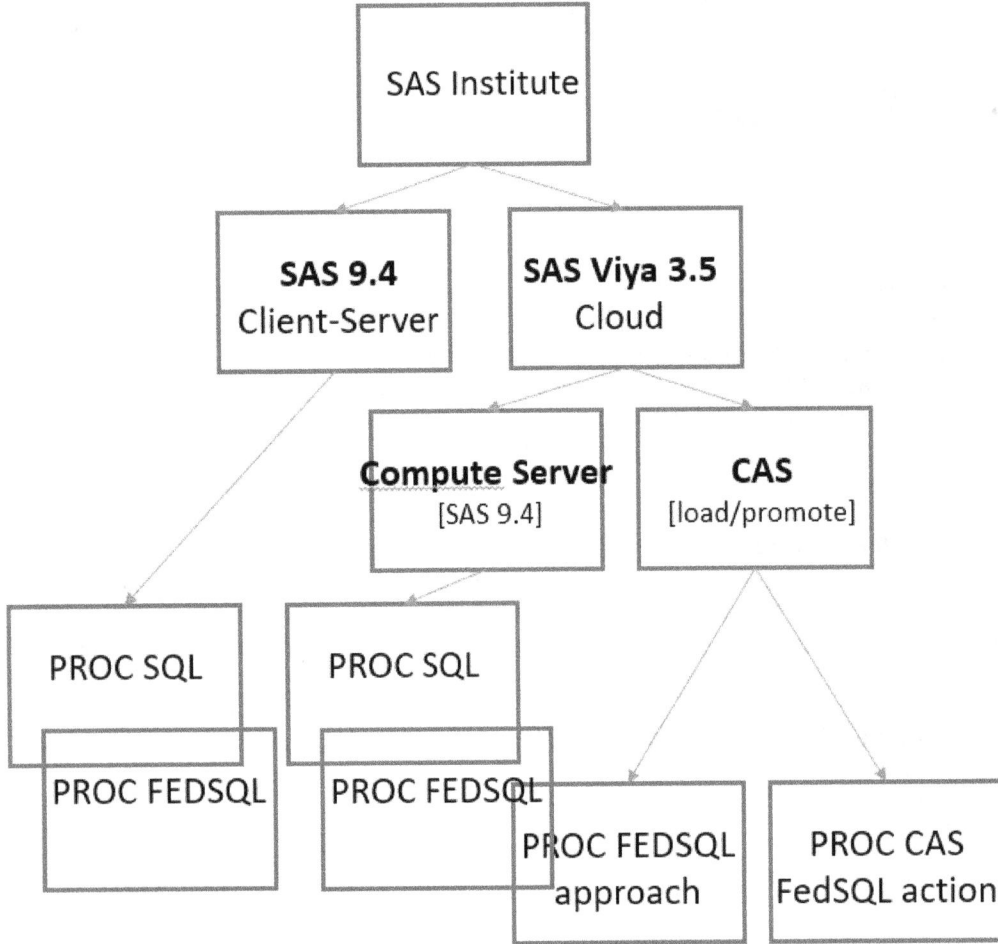

also includes the Cloud Analytic Services server (CAS). So with SAS Viya, you get "SAS" and "CAS" all together. For the sake of completeness, there is also the SAS Micro Analytic Service (MAS). You can use MAS for the repeated processing of data running in streams (SAS Event Stream Processing). MAS is outside the scope of the following introduction to FedSQL, however.

The bottom rows of Figure 7.1-2 attempt to illustrate on which platforms you can use PROC SQL and PROC FEDSQL. PROC SQL is easy: you can program and process PROC SQL code on SAS and the Compute Server, not on CAS.

PROC FEDSQL is a little more complicated. SAS provides at least three functional different varieties of PROC FEDSQL. You can program and process PROC FEDSQL code on SAS, the Compute Server, *and* CAS. In CAS there are two varieties: a FEDSQL approach and a FEDSQL action. Unfortunately, SAS and CAS do not support the same PROC FEDSQL statements and the same SAS functions and function calls, whereas programs written in SAS run on the Compute Server without any problems. This means that you could rewrite your SQL into FEDSQL code in SAS 9.4, then move to SAS Viya 3.5's Compute Server and use its advantages. This chapter tested all Compute Server programs on SAS and vice versa; they all work fine on both platforms.

Even if the syntax were equally supported in all platforms, the processing would be completely different. On SAS and the Compute Server, PROC SQL is single-threaded on-premises; in contrast, PROC FedSQL is fully threaded in SAS 9.4, on the Compute Server, and in CAS. From a data view, the significant difference in CAS processing is that data are already memory-resident and distributed, and that there are usually a significantly larger number of threads available.

CAS technology runs in-memory and involves at least one additional process of loading into memory. If necessary, you can share in-memory tables globally ("promote" in the figure) and also save in-memory tables permanently on disk. *Promoted* tables are not *permanently stored* tables. This means that PROC SQL and PROC FEDSQL are conceptually and technically different. PROC FEDSQL and PROC SQL share many, though not all statements and options. The subtleties begin with the fact that in FEDSQL you will not find the PROC SQL "SAS extras" that you maybe have found rather useful in the past and might now need to refactor. FEDSQL lacks the former SQL features that were beyond ANSI-1992, including Boolean expressions or macro functionalities like INTO. PROC FEDSQL also includes ANSI-1999 options for simultaneously connecting to numerous ANSI-compliant data sources like Google, Hive etc., and facilities to load the data in-memory for processing. With PROC SQL and FEDSQL, SAS gives you two programming languages that enable you to get the best of both ANSI worlds.

PROC FEDSQL runs on SAS 9.4 (local, on servers, Compute Server) and in the cloud. (See Sections 7.1 and 7.2.) These FEDSQL versions for SAS and CAS share many statements but not all. (See Subsection 7.2.2.) Besides this hopefully temporary restriction, FEDSQL is a programming language that helps you get going in the SAS and CAS worlds. Even in the CAS environment, you can use FEDSQL in at least two varieties: as PROC FEDSQL (procedure approach) and in PROC CAS (as an action approach) using fedsql.execdirect. (See Subsection 7.2.3.) You can also use FEDSQL statements in the DS2 language. (See Section 7.3.) Taking all that together, this means that your PROC SQL expertise as a starting point offers you a springboard into the myriad of possibilities and nuances of the CAS technology by putting FEDSQL to practice.

Chapter 7 introduces the following aspects of FedSQL (the language and concept) and PROC FEDSQL (the procedure):

- **Benefits and specifics of FedSQL language**, especially in comparison with its implementation as PROC FEDSQL versus PROC SQL and when used in SAS versus CAS (Section 7.1).
- **Using FEDSQL syntax in the SAS and in the CAS environment** (Section 7.2) in three different scenarios: replacing "SQL" by "FEDSQL" (Subsection 7.2.1), when the simple replacing "SQL" by "FEDSQL" needs a bit of modifying (Subsection 7.2.2), and using FEDSQL on CAS (Subsection 7.2.3).
- You can also use FEDSQL and SQL syntax in **DS2** programs (Section 7.3). This section introduces the building blocks of the DS2 language and presents numerous applications of DS2 from simple FEDSQL and SQL in DS2 to threaded processing in DS2. With DS2, you can take advantage of its built-in multi-threading capabilities, new high-precision numeric data types, and SAS web apps that run on DS2 only.
- Overviews of the PROC **FEDSQL syntax**, systematic and complete (Section 7.4): An outline (Subsection 7.4.1), FEDSQL statements used in SAS and CAS (Subsection 7.4.2), and details of connection and processing options (Subsections 7.4.3-7.4.6).
- Section 7.5 compiles questions that might help you choose between PROC FEDSQL and PROC SQL.

For those not so familiar with the SAS product range, I suggest you browse the overviews in Subsection 7.1.4 before we dive into the sophisticated processing worlds of FedSQL and SQL. Subsection 7.1.4 describes components and functionalities related to the CAS environment (some might be already mentioned above) with a few keywords. CAS is an essential component of SAS Viya's high-performance in-memory analytics. Please see also the preface that compares several programming languages and their advantages especially in the CAS environment.

7.1 Benefits and Specifics of FedSQL Language

Subsection 7.1.1 highlights benefits and specifics of the FedSQL language as it is implemented in the FEDSQL procedure. (See Tables 7.1-1 to 7.1-3.) Subsection 7.1.2 discusses differences in FedSQL functionalities on SAS versus CAS. (See Tables 7.1-4 and 7.1-5.) Subsection 7.1.3 provides a quick overview of the FEDSQL syntax of version SAS 9.4. Section 7.4 provides a more detailed description of the FEDSQL syntax, especially when it comes to the processing and connection options. Subsection 7.1.4 introduces selected components and functionalities related to the CAS environment. (See Table 7.1-6.)

7.1.1 FedSQL Compared to PROC SQL

This subsection consists of three tables and a note about precision of PROC SQL:

- Comparing SQL and FedSQL, **Table 7.1-1** highlights benefits of FedSQL over PROC SQL.
- **Table 7.1-2** compiles differences in writing programs in PROC SQL and PROC FEDSQL.
- **Table 7.1-3** lists the FEDSQL data type support for SAS data sets on CAS.

The last section discusses the precision of PROC SQL or PROC FEDSQL when it comes to big integer and possible factors involved in procedures, hardware, or operating environment.

Benefits of FedSQL Compared to PROC SQL

Table 7.1-1 highlights the benefits of FedSQL compared to PROC SQL.

Table 7.1-1: Comparing PROC SQL and FedSQL

Keyword	PROC SQL	FedSQL
Introduction in SAS	SAS 6.06	SAS 9.4
In a nutshell	Requires less code and processing time in almost every case and is more suitable than the DATA step for large data volumes. Less flexible in dealing with data sets in more complicated structures.	FedSQL is the next level SQL. This vendor-neutral SQL dialect provides a technology to process relational and nonrelational data in multiple data sources.
Benefit of FedSQL: **SQL language elements.** FedSQL can execute many SQL statements and language elements in SAS and CAS. PROC SQL as a procedure is not supported on CAS (see below).		
Support in	PROC SQL programs run in SAS 9.4 and the Compute Server, but not in CAS.	FedSQL programs run in SAS 9.4, the Compute Server, and in CAS.
Benefit of FedSQL: **CAS technology and performance.** PROC SQL is not supported on CAS. Note: FedSQL has different functionality on the CAS server than it has in a SAS library.		
ANSI 1999 compliant	**No.** ANSI SQL-1992. Not fully compliant with the current ANSI standard for SQL, contains many non-ANSI SAS enhancements such as the support of SAS macro language, the CALCULATED keyword, remerging, mnemonics, or the handling of missing values. According to the ANSI standard, these expressions become NULL. In PROC SQL, they become false.	**Yes.** ANSI SQL-1999. ANSI 1999 conformance allows FedSQL to process queries in its own language as well as the native languages of other DBMSs that conform to the ANSI 1999 standard. Only a few non-ANSI SAS enhancements.
Benefit of FedSQL: **Communication.** When connecting to a DBMS with a libref using LIBNAME, the language matches or translates the target data source's definition to the FedSQL data types (64 bit floating point) which might result in a loss of numeric precision. When connecting with a caslib, the language translates the data source's definition to native CAS data types. Using the drive enables FedSQL to retrieve the data while preserving the original precision of the data.		
Accesses	Tables and RDBMS on-premises. Only one type of data source per access.	Tables and RDBMS, and non-relational data sources (April 2019), on-premises or in the cloud. Different types of data source per access.
Benefit of FedSQL: **Non-relational data sources.** Access to and handling of data types of non-relational data sources like MongoDB 3.6 and later.		

(*Continued*)

Table 7.1-1: (*Continued*)

Keyword	PROC SQL	FedSQL
Data types	Processes only SAS character (ANSI equivalent: CHAR) and numeric (ANSI equivalent: DOUBLE) data types. SAS numeric data types (integer, decimal, and date) are DOUBLE. INTEGER, DECIMAL, and DATE data types are not supported.	Processes 17 ANSI data types. Data source's definition, Time(p), Timestamp(p), Double, NChar, Varchar, BIGINT. Please see the overview below.

Benefit of FedSQL: **Precision.** Using librefs or caslibs, translating or matching the data source's definition to the FEDSQL data types allows greater precision. The DS2 procedure also allows you to take advantage of the new high-precision numeric integer data types.

Keyword	PROC SQL	FedSQL
Number of DB connections per query	One	Multiple

Benefit of FedSQL: **Multiple numbers of DB connections** per query are possible because databases, connections, languages, and queries operate under the same ANSI standard (see above).

Keyword	PROC SQL	FedSQL
Federated queries	**Possible.** SQL is able to query data from more than two (or more) different data sources.	**Possible.** FEDSQL is able to query data from more than two (or more) different data sources. I'll go out on a limb and say that FEDSQL tends to require less code for the same purpose.

Benefit of FedSQL: **Federated queries** access data from multiple different data sources such as Oracle and Google and return a single result set. Easier programming, logic with fewer passes, and faster processing. FedSQL can also split a single SQL query connected to multiple databases and send the parts to the individual databases.

Keyword	PROC SQL	FedSQL
Multi-threading	Only for sorting and indexing.	Fully multi-threaded.

Benefit of FedSQL: **Multi-threading.** PROC FEDSQL is able to take full advantage of the multi-threading technology. The DS2 procedure is multi-threaded by nature, too.

Keyword	PROC SQL	FedSQL
Implicit and explicit SQL pass-through	**Yes.** Implicit: PROC SQL statements are translated into data source-specific code internally. Explicit: Requires a CONNECT and a DISCONNECT respectively when submitting code.	**Yes.** Implicit: FEDSQL statements are translated into data source-specific code internally. Explicit: **Native** SELECT statements are submitted using a specially formed SELECT statement.

Benefit of FedSQL: **Performance.** Passing data source-specific query code directly to the data source for processing, improves query response time, reduces the data volume being transferred to FedSQL, thus improving the overall query processing time. Implicit SQL pass-through is available for SAS libraries and for caslibs, although support in CAS is currently limited.

Writing SAS Programs in PROC SQL and PROC FEDSQL

Table 7.1-2 compiles differences in writing programs in PROC SQL and PROC FEDSQL.

Table 7.1-2: Writing SAS Programs Using PROC SQL and PROC FEDSQL

Keyword	PROC SQL	PROC FEDSQL
Reference to newly created column	**CALCULATED** keyword	Expression of a calculation must be repeated.
Assigning formats	**FORMAT=**	Use **PUT**(field,format.).
Mnemonics	Can use **mnemonics** (EQ, NE, and so on).	Cannot use mnemonics (use =, <>, and so on).

(*Continued*)

Table 7.1-2: (*Continued*)

Keyword	PROC SQL	PROC FEDSQL
Handling missings: where NAME **ne** ' '	Equivalents: NAME <> ' ' NAME < > null NAME is not null NAME is not missing	Equivalents: NAME <> ' ' NAME is not null NAME is not missing
Operators	**CONTAINS:** where NAME contains 'PET' ; Note that PROC SQL also supports the LIKE operator.	**LIKE:** where NAME like ' %PET %' ;
Remerge Functionality	**Supports** the remerge functionality (not an ANSI concept). Remerge queries perform a statistical analysis and integrate the results of the analysis with other columns from the data. You can identify a remerge by syntax specifics or a message to the log.	Does not support the remerge functionality. To produce the same results as PROC SQL, a remerge query in FEDSQL requires writing a summary (sub)query and an explicit join with the other column values.
Overwrite an existing table (CREATE TABLE)	Repeated submitting CREATE TABLE overwrites an existing table.	Repeated submitting CREATE TABLE does not overwrite an existing table. In SAS, you must first delete the existing table with DROP TABLE. You can use the FORCE option to suppress errors if the table does not exist. In CAS, you need to specify as table option REPLACE=TRUE.
SAS Log (terminology), for example, removing one row	"NOTE: 1 row was deleted from WORK.ONE."	"NOTE: Execution succeeded. One row affected."
Access concatenated librefs, for example, SASHELP	No issue.	PROC FEDSQL cannot access concatenated librefs. To access these preset SAS libraries, you need to specify an individual LIBNAME statement for each individual folder in the concatenated library or use a DS2 approach (see examples later).
Table without libref / caslibs	No issue. PROC SQL handles [WORK.]MY_TABLE as temporary.	Avoid. FEDSQL handles MY_TABLE as temporary on SAS and CAS. However, processing is different. (See Subsection 7.2.4.) Recommendation: always change to tables with librefs or caslibs.

FEDSQL Data Type Support for SAS Data Sets and CAS Tables

PROC FEDSQL supports the data types for reading and creating SAS data sets and CAS tables shown in Table 7.1-3. SAS data sets support only character (CHAR) and numeric (DOUBLE) no matter where you process them. FEDSQL supports the CAS data types only in CAS. Once SAS data sets are loaded into CAS, they are CAS tables. This table highlights that some of the data types listed cannot be saved to either SAS data sets or CAS tables. The short description column illustrates how they are translated to and from SAS and CAS data types. For additional details on the new character data types, please consult the SAS documentation. (See, for example, SAS Institute, 2016c.)

Table 7.1-3: FEDSQL Data Type Support for SAS Data Sets and CAS Tables

	SAS Data Type	FEDSQL Data Type	CAS Data Type	Short description ["CAS:" applies to CAS data types only]
character	character	BINARY(n)		**FEDSQL:** Applies the SAS format $n. N: Maximum n bytes wide.
	character	CHAR(n)	CHAR	**FEDSQL:** Applies the SAS format $n. N: Maximum n characters wide [1 byte per character]. **CAS:** Fixed-length character string.
	character	NCHAR(n)		**FEDSQL:** Applies the SAS format $n. N: Maximum n characters wide [2-4 bytes per character].
	character	NVARCHAR(n)		**FEDSQL:** Applies the SAS format $n. N: Maximum n multi-byte chars.
	character	VARBINARY(n)	VARBINARY	**FEDSQL:** Applies the SAS format $n. N: Maximum n bytes. **CAS:** Varying-length binary opaque data. This data type is used to store image data, audio, documents, and other unstructured data.
	character	VARCHAR(n)		**FEDSQL:** Applies the SAS format $n. N: Maximum n multi-byte chars. **CAS:** Varying-length character string.
date and time	numeric	DATE		**FEDSQL:** Applies the SAS format DATE9. Valid SAS date values range from 1582-01-01 to 9999-12-31. Dates outside that date range are treated as invalid dates finally.
	numeric	TIME(p)		**FEDSQL:** Applies the SAS format **TIME8.** P: Precision.
	numeric	TIMESTAMP(p)		**FEDSQL:** Applies the SAS format **DATETIME19.2.** P: Precision.
numeric	numeric	BIGINT	INT64	**CAS:** Large signed, exact whole number. **FEDSQL: Please note:** Potential loss of precision. A SAS numeric is a DOUBLE, which is an approximate, not an exact numeric data type. See the notes about the precision of PROC SQL below. Up to 19 significant digits. Integer.
	numeric	DECIMAL		**FEDSQL:** Fractional. Up to 52 significant digits.
	numeric	DOUBLE	DOUBLE	**FEDSQL:** Fractional. Up to 16 significant digits. Compare with NUMERIC. **CAS:** Signed, approximate, double-precision, floating-point number.
	numeric	FLOAT(p)		**FEDSQL:** Fractional. Up to 16 significant digits. P: Precision.
	numeric	INTEGER	INT32	**FEDSQL:** Integer. Up to 10 significant digits. **CAS:** Regular signed, exact whole number.
	numeric	NUMERIC (p,s)		**FEDSQL:** Compare with DOUBLE. P: Precision, S: Scale.
	numeric	SMALLINT		**FEDSQL:** Integer. Up to 5 significant digits.
	numeric	TINYINT		**FEDSQL:** Integers from -128 to 127. Up to 3 significant digits.

NVARCHAR, VARBINARY, and VARCHAR store data of varying length, not fixed-length data. For date and time values, use DATE, TIME, or TIMESTAMP; do not assign date and time SAS formats to a numeric data type. Once you start working with the new data types, you need to be extra careful when it comes to handling several of them, especially in DS2. Many things are no longer self-evident. Some questions might arise such as is it possible to add DOUBLE and INTEGER values? REAL is not listed because it is not supported on CAS. (Under certain circumstances, you might use FLOAT, though.) Do not confuse NUMERIC (p,s; a fixed-decimal data type) with DECIMAL (which is a floating point numeric type), or the SAS data type "numeric", which is DOUBLE.

16 Digits or the Precision of PROC SQL: A Clarification

In Base SAS, all numeric values are fractional, have a maximum of eight bytes, and a maximum of 16 or 17 digits precision, depending on the operating system. A SUGI/Regional User Group paper by Mohammed and colleagues (2015, p. 3-4) pointed out that PROC SQL might calculate inaccurate results for big integers. For SAS and Compute Server, big integers are values with more than 16 digits. The conclusion "PROC FEDSQL on the other hand returned correct values from the table and correct sum value as the result set" needs a bit of clarification though. The root cause is not the procedures used, but the way the data is stored. (See Section 8.3.) The column BIGINTEGER might be truncated when you store it in an environment that does not support high-precision numerics. As soon as SAS or any other data source does not hold values correctly, neither PROC SQL nor PROC FEDSQL, or any other procedure, is able to calculate correct values from truncated data. As you can see in the PROC PRINT output, the numeric version (BIGINTEGER_NUM) does not perfectly represent the original digit sequence in character format (BIGINTEGER). Try the code below on CAS with minor modifications; please also see the DS2 utility in Example 8 in Subsection 7.3.2. The digit sequences were taken from Mohammed et al. (2015).

Program 7.1-1: Creating the Test Data

```
data TEST_SQL ;
input ID 2. BIGINTEGER $19. ;
          BIGINTEGER_NUM
          =input(BIGINTEGER,19.) ;
datalines ;
4611686018427387904
2222222222222222222
2312346890765434687
 ;
run;
```

Program 7.1-2: Printing the Test Data (PROC PRINT)

```
proc print data=TEST_SQL noobs ;
   var ID BIGINTEGER BIGINTEGER_NUM ;
   format BIGINTEGER_NUM 19. ;
run ;
```

Figure 7.1-2: Printing the Test Data (PROC PRINT)

```
ID         BIGINTEGER         BIGINTEGER_NUM
1    4611686018427387904    4611686018427387904
2    2222222222222222222    2222222222222222336
3    2312346890765434687    2312346890765434368
```

Program 7.1-3: Computing a Sum (SQL)

```
proc sql ;
   select sum(BIGINTEGER_NUM) as
      TOTAL_SQL format=19.
   from TEST_SQL ;
quit ;
```

Program 7.1-4: Computing a Sum (FEDSQL)

```
proc fedsql ;
   select put(sum(BIGINTEGER_NUM),19.) as
      TOTAL_SQL
   from TEST_SQL ;
quit ;
```

Figure 7.1-3 Computing a Sum (SQL)

```
TOTAL_SQL
-------------------
9146255131415044096
```

Figure 7.1-4 Computing a sum (FEDSQL)

```
TOTAL_SQL
-------------------
9146255131415044096
```

Program 7.1-5: Computing a Sum: Template for FEDSQL on CAS

```
proc fedsql sessref=my_CAS_session ;
   select put(inputn(BIGINTEGER, '19.'),19.) as BIGINTEGER_NUM
         from CASUSER.TEST_SQL ;
   select put(sum(inputn(BIGINTEGER, '19.')),19.) as TOTAL_SQL
         from CASUSER.TEST_SQL ;
quit ;
```

Adjust this code template to your specific CAS environment. The lesson learned is to be extra careful with integers that have more than 15 digits. Check computations, procedures, and data storage specifics.

7.1.2 Specifics of FedSQL: SAS versus CAS

Table 7.1-4 highlights different FedSQL functionalities in SAS compared to CAS. This table also answers the questions that you should know when you want to write SAS programs (see also subsection 7.2.4):

- Will your program run in a SAS environment (local, on a server on the premises, or in the Compute Server) or will it run in the cloud, in the CAS environment?
- Will your program run single- or multi-threaded? Note that you might get undesired results if a formerly single-thread program is not properly adjusted to run in multiple threads. Example: Sum (+) statement. Use the DATA step option SINGLE= YES to run a program single-threaded on CAS.
- Will your program process legacy SAS data types or use the data source's definitions? This innocent question has two important implications. First, you need to know exactly which data type you need to use. Second, if you choose an inappropriate data type, your results might be incorrect.

Table 7.1-4: Specifics of FedSQL: SAS versus CAS

Keyword	SAS (incl. Compute Server)	CAS
To execute FedSQL statements	Specify the PROC FEDSQL statement followed by FedSQL language statements.	Specify connection options (SESSREF= or SESSUUID=) with a CAS session name in the procedure statement.
Fundamental difference in technology	Basic steps: Set LIBNAME, access data, process data.	CAS is an in-memory processing environment. Therefore, CAS requires an extra step before data processing: load data in-memory using PROC CASUTIL.
Load tables into your CAS session	It's not a cloud environment; this step is not necessary.	Either load tables into your CAS session using CASUTIL or other tools, or set a caslib for your CAS session and query tables from the caslib by name.
Functions / routines	Table 10.2-2 in Subsection 10.2.2 introduces selected SAS functions and routines in categories. Not all functions and routines are supported for running on a CAS server. The column "CAS" indicates which do.	
Writing SAS programs **(1) Functions** **(2) BY-group processing** (also first/last, merge) **(3) DATA Step:** Filtering tables SAS Formats Data types	Check the **sum (+)** statement if multi-threaded. **BY** statement (requires sorting). You need to decide: Legacy or ...	Check the **sum (+)** statement if multi-threaded. No **BY** statement and **no** sorting required (remove, if necessary). Use **subsetting if** instead of filtering rows on output table, *data PERM.TABL (where=(VALU eq ''))* ; Remove WORDS and WORDF formats; replace WORDDATE with NLDATE. ... data source definition?
Table Output	FEDSQL output tables in SAS are permanent or temporary tables.	In the CAS environment, FEDSQL output tables are temporary in-memory tables by default, even if promoted. To save CAS tables permanently use CASUTIL.
WORK tables	Can handle WORK tables.	CAS cannot handle WORK tables; rename them to permanent tables.
Pass-through	Full and partial FedSQL pass-through available for FEDSQL in SAS.	Partial FedSQL pass-through is not yet available for FEDSQL in CAS.

(Continued)

Table 7.1-4: (*Continued***)**

Keyword	SAS (incl. Compute Server)	CAS
FEDSQL functionality	Not limited.	Limited.
Default behavior when not using librefs/caslibs	Creates a temporary table.	Creates temporary table in SAS. Creates a temporary table in CAS (see below).

If you do not use a LIBNAME in SAS (SAS 9.4, Compute Server), SAS creates a temporary data set by default. If you do not use a caslib, CAS still creates a temporary data set by default. The libref in the log might be misleading. The following overview in Table 7.1-5 highlights the different default behavior in SAS and CAS when you do not use a libref or caslib.

Table 7.1-5: Default Difference When You Do Not Use a Libref or Caslib in SAS and CAS

Environment?	PROC SQL	PROC FEDSQL

SAS 9.4	**Program 7.1-6: SQL/SAS: No libref**
	
	``proc sql;
	    create table MY_CLASS as
	        select *
	    from SASHELP.CLASS
	        order by AGE ;
	quit ;```
		Program 7.1-7: FEDSQL/SAS: No libref
		
		``proc fedsql;
		    create table MY_CLASS as
		        select *
		    from mybase.CLASS
		        order by AGE ;
		quit ;```

What happens?	A missing LIBNAME makes MY_CLASS automatically a temporary table: **SAS Log 7.1-1 SQL/SAS: No libref** ```NOTE: Table WORK. MY_CLASS created, with 19 rows and 5 columns.```	A missing LIBNAME makes MY_CLASS automatically a temporary table: **SAS Log 7.1-2 FEDSQL/SAS: No libref** ```NOTE: Execution succeeded. 19 rows affected.```
SAS Viya (Compute Server)	Use SAS Code from Program 7.1-6 'as is'. See Program 7.1-6.	Use SAS Code from Program 7.1-7 'as is'. See Program 7.1-7.
What happens?	A missing libref makes MY_CLASS automatically a *temporary* table.	A missing libref makes MY_CLASS automatically a *temporary* table.

SAS Viya (CAS)	**Program 7.1-8: SQL/CAS: No caslib**
	
	``proc sql sessref=my_CAS_session ;
	    create table MY_CLASS as
	        select *
	    from CASUSER.CAS_CLASS ;
	quit ;```
		Program 7.1-9: FEDSQL/CAS: No caslib
		
		``proc fedsql sessref=my_CAS_session ;
		    create table MY_CLASS as
		        select *
		    from CASUSER.CAS_CLASS ;
		quit ;```

What happens?	Nothing. A missing caslib in PROC SQL does not make any difference on CAS, because it runs only on Compute Server. CAS does not process PROC SQL code. The log contains several error notes. **SAS Log 7.1-3 SQL/CAS: No caslib** ```NOTE: The SAS System stopped processing this step because of errors.``` This example was deliberately constructed to illustrate that SQL does not run on CAS. For example, if SQL was written instead of FEDSQL out of sheer habit.	A missing caslib makes MY_CLASS **not** a permanent table even if the libref in the log seems to say so. Global-scope is temporary. **SAS Log 7.1-4 FEDSQL/SAS: No caslib** ```NOTE: Table MY_CLASS was created in caslib CASUSER(abc@xyz.com) with 19 rows returned.``` Use Program 7.2-70 to generate the demo data.

7.1.3 FEDSQL Syntax: Outline and Supported Statements

PROC SQL as a procedure is not supported on the CAS server. However, you can use many SQL statements and language elements as FEDSQL expressions for processing in SAS or CAS. This subsection gives you a quick overview of the FEDSQL syntax of version SAS 9.4 and the subset available for CAS. Section 7.4 provides a more detailed description of the FEDSQL syntax, especially when it comes to the processing and connection options.

Subsections 7.2.1 and 7.2.2 provide FEDSQL syntax examples for processing in SAS. Subsection 7.2.3 provides FEDSQL syntax examples for processing in CAS.

PROC FEDSQL Outline

This outline gives a simplified overview of the FEDSQL syntax. The numbers in brackets refer to chapters where you can find more detailed information.

PROC FEDSQL

                    <Connection options> [see Subsections 7.4.3 and 7.4.5]
                    <Processing options> [see Subsections 7.4.4 and 7.4.6] ;

    [FEDSQL statements, see Subsection 7.4.2] ;

QUIT ;

FEDSQL Statements Supported in SAS 9.4

This list gives you an overview of the FEDSQL statements that are supported in SAS 9.4.

    ALTER TABLE

    BEGIN

    COMMIT

    CREATE INDEX | TABLE| VIEW

    DELETE

    DESCRIBE TABLE | VIEW

    DROP INDEX | TABLE | VIEW

    EXECUTE

    INSERT

    ROLLBACK

    SELECT

    UPDATE

The FEDSQL statements available for CAS server are a subset of those available for SAS 9.4.

CREATE TABLE with the **AS** query expression

SELECT
    UNION (SAS Viya 3.5 and later)

DROP TABLE

Statements Not Supported in CAS

At the moment, the following SELECT statement features are not supported in CAS:

EXCEPT
INTERSECT
correlated subqueries (however, see Example 6 in Subsection 7.2.3)
dictionary queries
views
DS2 user-defined functions (package expressions).

7.1.4 The FedSQL Environment: Components of SAS Viya: SAS and CAS

For those not familiar with the SAS product range, I will describe components and functionalities related to the CAS environment with a few keywords before we dive into the sophisticated processing worlds of FedSQL and SQL. (See Table 7.1-6.) CAS is one essential component of the new high-performance in-memory analytics architecture, SAS Viya. The other component you actually already know: it's SAS. In the SAS Viya environment, it's called Compute Server.

Table 7.1-6: Components of SAS Viya and CAS

SAS Solution	In a nutshell
SAS Viya	SAS Viya (pronounced like "vaya con dios" without the "con dios" part) is a new architectural approach that provides a cloud/grid model offering high-performance in-memory analytics. Its interfaces are public and accessed through third-party clients, open-source clients, and Java (Pendergrass, 2017). SAS Viya contains CAS and Compute Server (see below). Data sources can be relational, unstructured, on-premises, in the cloud, SAS data sets, SASHDAT files, XML, JSON, and so on.
SAS Cloud Analytic Services (CAS)	CAS is a shared server that provides SAS Viya's in-memory run-time environment for data management and analytics. CAS consists of 1 to N hosts. These hosts can be a part of a cluster, a single system, in the Amazon cloud, or OpenStack. Data is processed in parallel. For multi-threaded processing, CAS offers symmetric multi-processing (SMP) on one node and massive parallel processing (MPP) on multiple nodes. Data is retained in-memory for multiple users or steps, and can be saved permanently on disk. CAS supports only UTF-8 encoding.
Compute Server (aka SPRE, SAS Programming Run-Time Environment)	Compute Server (aka SPRE, SAS Programming Run-Time Environment; including SAS Studio as the programming interface) is the familiar SAS Foundation session and provides client sessions for individual users. Users interact with CAS from these sessions. SAS 9.4M5 or later can be used instead of SPRE. If you start SAS Studio, you can submit and process SAS code on SPRE only without having to start a CAS session. For SPRE, you use librefs.

(Continued)

Table 7.1-6: (*Continued*)

SAS Solution	In a nutshell
SAS Studio	SAS Viya's integrated, browser-based programming environment. You can program using the point-and-click interface or enter code directly on the CODE tab. From SAS Studio, you can submit SAS code to the Compute Server without having to establish a CAS session (see above; single-threaded SAS Foundation) or send to CAS (fully-threaded) using the CAS statement. For CAS, you use caslibs. If you do not start a CAS session, you work in the SPRE (SAS Foundation) environment. The connection with the CAS server is automatically created when you use SAS Studio 5.2 in Linux or Windows. The connection information, the host name, and port are set during the installation.
Enterprise Guide	Starting with EG 8.2, a desktop client, you can directly connect to SAS Viya and submit SAS code to CAS without having to install a full SAS 9 environment.
SAS Visual Analytics (VA)	Programming tools with baseline functionality, including reporting and basic analytics.
SAS Visual Statistics (VS)	Advanced analytic functionality that builds on SAS VA, including comparing predictive models.
SAS Visual Data Mining and Machine Learning	Advanced analytic functionality that builds on SAS VS, including tuning machine learning algorithms, performing advanced statistical operations, and analyzing complex data.
SAS Visual Text Analytics	A text analytics framework that combines text mining, contextual extraction, categorization, sentiment analysis, and search capabilities.

There are more components; this selection is just the ones mentioned in the context of PROC SQL and PROC FEDSQL. The following Subsections 7.2.1 and 7.2.2 provide SQL and FEDSQL syntax examples for processing in SAS. Subsection 7.2.3 provides examples for processing in CAS.

7.2 Using FEDSQL Syntax in SAS and CAS

This section focuses on writing programs with PROC FEDSQL. One of the maxims of programming is: Use the code that you know. If you are already familiar with PROC SQL, then the obvious next step is to try the simplest approach: just find and replace "SQL" with "FEDSQL" in your already existing PROC SQL program code. (See Subsection 7.2.1.) Sometimes it might be all you need to do, and your FEDSQL code will still work fine.

Not all replacements might be successful though, and not all seemingly successful were successful indeed. Subsection 7.2.2 demonstrates that several SQL programs just need a bit of modifying. This subsection will highlights a pitfall of the find-and-replace approach. The FEDSQL programs that you write in these two subsections will work in a SAS environment whether it is local, on a server, or on SAS Viya's Compute Server. All SAS programs were tested on SPRE, and vice versa; they all work fine on both platforms. A side-effect is you can still use most of the SQL manuals and adjust your SQL programs to work as FEDSQL programs from now on.

Subsection 7.2.3 introduces FEDSQL programs for a CAS environment. The SQL code itself will not differ much; you just need to consider additional CAS-specific options like connecting to CAS, loading data to CAS (CAS processes in-memory), explicitly stating required precision, and maybe multi-threading. You will also find an example how to write the same SAS program as a FEDSQL procedure approach and as an action approach using fedsql.execdirect in PROC CAS. For programming on CAS with DS2, the DATA step, or FEDSQL, you might need to turn to more specialized manuals.

Subsection 7.2.4 summarizes how to run programs in CAS not only using FEDSQL, but also the DATA step, and others. It explains when a program is processed in CAS, under which circumstances possibly in SAS (for example, input, language elements, and so on), and when to expect unexpected results. Some tables introduce CAS procedures and CAS-enabled SAS procedures.

7.2.1 FEDSQL on SAS: Replacing SQL with FEDSQL Successfully

This subsection presents SAS code where you replace SQL (left column of Table 7.2-1) with FEDSQL (right column of Table 7.2-1) and it still works. This subsection also highlights an important pitfall of the find-and-replace approach. For safety and performance reasons, I recommend that you prefix temporary table names with WORK after FROM in FEDSQL.

Table 7.2-1: FEDSQL on SAS: Replacing PROC SQL with FEDSQL

Before: SQL	After: FEDSQL
Printing a SAS table	
Program 7.2-1: SQL: Printing a SAS Table	**Program 7.2-2: FEDSQL: Printing a SAS Table**

<pre>proc sql ;
   select *
   from ONE ;
quit ;</pre>
	<pre>proc fedsql ;
	   select *
	   from WORK.ONE ;
	quit ;</pre>

Subsetting a SAS table row-wise using DELETE	
Program 7.2-3: SQL: Using DELETE	**Program 7.2-4: FEDSQL: Using DELETE**

<pre>proc sql ;
   delete from ONE
   where B= 99 ;
quit;</pre>
	<pre>proc fedsql ;
	   delete from WORK.ONE
	   where B= 99 ;
	quit;</pre>

Creating a SAS view	
Program 7.2-5: SQL: Creating a SAS View	**Program 7.2-6: FEDSQL: Creating a SAS View**

<pre>proc sql;
   create view TWO as
      select *
   from ONE ;
quit;</pre>
	<pre>proc fedsql;
	   create view TWO as
	      select *
	   from WORK.ONE ;
	quit;</pre>

Subsetting a SAS table columnwise using column list ("KEEP")	
Program 7.2-7: SQL: Keeping Columns	**Program 7.2-8: FEDSQL: Keeping Columns**

<pre>proc sql ;
   create table TWO as
   select ID, B from ONE ;
quit ;</pre>
	<pre>proc fedsql ;
	   drop table TWO force /*tip!*/;
	   create table TWO
	         {options replace=true} as
	   select ID, B from WORK.ONE ;
	quit ;</pre>

Comment: Assuming table TWO already exists, you can repeatedly overwrite it by simply repeatedly submitting SQL. With FEDSQL, you cannot overwrite an existing table by default.	Comment: FEDSQL offers you two options: Use either FORCE to explicitly delete an existing table first in the same procedure request. Or, use the table option REPLACE=TRUE for overwriting in CAS.
Sorting a SAS table (ORDER BY)	
Program 7.2-9: SQL: Sorting	**Program 7.2-10: FEDSQL: Sorting**

<pre>proc sql ;
   create table TWO as
   select * from ONE
   order by ID ;
quit;</pre>
	<pre>proc fedsql ;
	   create table TWO as
	   select * from WORK.ONE
	   order by ID ;
	quit;</pre>

(Continued)

Table 7.2-1: (*Continued*)

Before: SQL	After: FEDSQL

Subsetting a SAS table: Removing duplicate values

Program 7.2-11: SQL: Removing Duplicates	**Program 7.2-12: FEDSQL: Removing Duplicates**

<table>
<tr><td>

```
proc sql ;
   create table TWO as
   select distinct *
   from ONE  ;
quit;
```

</td><td>

```
proc fedsql ;
   create table TWO as
   select distinct *
   from WORK.ONE ;
quit;
```

</td></tr>
</table>

Comment: If you use an asterisk, SQL and FEDSQL remove duplicate rows; if you list individual column names after DISTINCT, SQL and FEDSQL remove duplicate value *combinations* within these columns. An ORDER BY clause is only required if the data retrieved should be ordered.

Combining tables: Concatenating (UNION ALL)

Program 7.2-13: SQL: Concatenating Rows	**Program 7.2-14: FEDSQL: Concatenating Rows**

<table>
<tr><td>

```
proc sql ;
   create table TWO as
   select * from ONE
      union all
   select * from ONE ;
quit ;
```

</td><td>

```
proc fedsql ;
   create table TWO as
   select * from WORK.ONE
      union all
   select * from WORK.ONE ;
quit ;
```

</td></tr>
</table>

Combining tables: Merging (FULL JOIN)

Program 7.2-15: SQL: Merging	**Program 7.2-16: FEDSQL: Merging**

<table>
<tr><td>

```
proc sql ;
   create table FULLJOIN as
   select coalesce (a.ID, b.ID)
      as ID,A,B,C,E,F,G
   from ONE a full join THREE b
      on a.ID=b.ID
   order by ID ;
quit ;
```

</td><td>

```
proc fedsql ;
   create table FULLJOIN as
   select coalesce (a.ID, b.ID)
      as ID,A,B,C,E,F,G
   from WORK.ONE a full join
      WORK.THREE b
      on a.ID=b.ID
   order by ID ;
quit ;
```

</td></tr>
</table>

Combining several statements into one procedure call

Program 7.2-17: SQL: Combining Statements	**Program 7.2-18: FEDSQL: Combining Statements**

<table>
<tr><td>

```
proc sql ;
   delete from WORK.LEFT
   where B= 99 ;
   create table FULLJOIN as
   select coalesce (a.ID, b.ID)
     as ID,A,B,C,E,F,G
from WORK.LEFT a full join
   WORK.RIGHT b
   on a.ID=b.ID
   order by ID ;
quit ;
```

</td><td>

```
proc fedsql ;
   delete from WORK.LEFT
   where B= 99 ;
   create table FULLJOIN as
   select coalesce (a.ID, b.ID)
     as ID,A,B,C,E,F,G
from WORK.LEFT a full join
   WORK.RIGHT b
   on a.ID=b.ID
   order by ID ;
quit ;
```

</td></tr>
</table>

(*Continued*)

Table 7.2-1: (*Continued*)

Before: SQL	After: FEDSQL

Aggregating functions: Columnwise (Numeric)

Program 7.2-19: SQL: Aggregating (Numeric)	**Program 7.2-20: FEDSQL: Aggregating (Numeric)**

```
proc sql ;
   create table TWO as
   select count(*) as N_rows,
          sum(A) as A_sum
   from ONE ;
quit ;
```

```
proc fedsql ;
   create table TWO as
   select count(*) as N_rows,
          sum(A) as A_sum
   from WORK.ONE ;
quit ;
```

Comment: The FEDSQL code is executed on SAS 9.4 and generates a SAS table TWO as requested. Using FEDSQL, the SAS log gives an interesting insight into how SAS works:

SAS Log 7.2-1: Feedback Using BASE Driver: BIGINT

```
NOTE: BASE driver, creation of a BIGINT column has been requested, but is not supported
by the BASE driver. A DOUBLE PRECISION column has been created instead.
NOTE: Execution succeeded. One row affected.
```

From Table 7.1-3 we know that SAS data sets support only CHAR and DOUBLE on SAS 9.4 and the Compute Server. FEDSQL supports CAS data types only in CAS. The log tells you that SAS tried to assign BIGINT for CAS first, but as it was executed on SAS 9.4, it changed it to DOUBLE for SAS. If a BIGINT was successfully assigned in the CAS environment, this numeric precision may not be appropriate though. (TINYINT may be sufficient with small integer values; see also the discussion about precision in Subsection 7.1.1.)

Aggregating functions: Columnwise (Character)

Program 7.2-21: SQL: Aggregating (Character)	**Program 7.2-22: FEDSQL: Aggregating (Character)**

```
proc sql ;
   create table TWO as
   select *,
                A / B as QUOT,
     (trim(ID)||'/'||ID) as STRING
   from ONE ;
quit ;
```

```
proc fedsql ;
   create table TWO as
   select *,
                A / B as QUOT,
     (trim(ID)||'/'||ID) as STRING
   from WORK.ONE ;
quit ;
```

Comment: If you just look at the FEDSQL code, the code runs and seems to be OK. This impression might be misleading. The SAS log gives a hint though:

SAS Log 7.2-2: Feedback Using BASE driver: VARCHAR

```
NOTE: BASE driver, creation of a VARCHAR column has been requested, but is not supported
by the BASE driver. A CHAR column has been created instead.
```

Comment: The FEDSQL version works fine. The NOTE was generated by executing FEDSQL on the Compute Server. The cause of the note is the BASE LIBNAME engine, which automatically converts VARCHAR to CHAR for storing as a SAS data set. If this code was executed in CAS, the output would have been a CAS table with a new VARCHAR column, and no note would be generated.

7.2.2 FEDSQL on SAS: Replacing PROC SQL with FEDSQL with Tuning

This subsection presents SAS code where replacing SQL with FEDSQL needs a bit of adjusting afterward. These adaptations are necessary because SQL and FEDSQL are based on different ANSI standards. Some SAS-specific extensions of PROC SQL are not supported by the vendor-neutral FEDSQL. On the left of Table 7.2-2, you see the original SQL code, on the right the *adjusted* FEDSQL code. Sometimes the relevant code parts are highlighted in bold. Below each SQL-FEDSQL pair, you find some keywords that comment on the necessary adjustments.

Table 7.2-2: FEDSQL on SAS: Replacing PROC SQL with FEDSQL with Tuning

Before: SQL	After: FEDSQL

Creating an empty table

Program 7.2-23: SQL: Empty Table	Program 7.2-24: FEDSQL: Empty Table
| ```
proc sql ;
 create table EMPTY
(ID num label="ID",
  D char (30) label="D",
  E num label="E",
  F date format=ddmmyy8. label="F") ;
quit ;
``` | ```
proc fedsql ;
 create table EMPTY
(ID double having format 4. label 'ID',
  D char (30) having format $10. label 'D',
  E double having format comma6. label 'E',
  F double having format ddmmyy8. label 'F') ;
quit ;
``` |

What needed to be done?

Single quotation marks (ID), explicit format (double), change FORMAT= to FORMAT, and change LABEL= to LABEL (ID). Replace NUM by DOUBLE or other appropriate formats. However, you should use the new formats in PROC FEDSQL whenever possible, for example, BIGINT in CAS environment.

Reading in data (rowwise)

Program 7.2-25: SQL: Reading Raw Data	Program 7.2-26: FEDSQL: Reading Raw Data
| ```
proc sql ;
create table ONE (ID
 char(2),A num,B num,C num) ;
insert into ONE
 values ("01", 1, 2, 3)
 values ("02", 1, 2, 3)
 values ("03",99,99,99)
 values ("04", 1, 2, 3)
 values ("05", 1, 2, 3) ;
quit ;
``` | ```
proc fedsql ;
create table ONE
        (ID  char(2) having format $2.,
         A double having format 8.,
         B double having format 8.,
         C double having format 8.) ;
insert into WORK.ONE values ('01', 1, 2, 3) ;
insert into WORK.ONE values ('02', 1, 2, 3) ;
insert into WORK.ONE values ('03',99,99,99) ;
insert into WORK.ONE values ('04', 1, 2, 3) ;
insert into WORK.ONE values ('05', 1, 2, 3) ;
quit ;
``` |

What needed to be done?

Proper assignment of FEDSQL formats (having format), repeatedly stating the target table after INSERT, repeatedly closing of INSERT rows with a semicolon. Single quotation marks for character entries. (See ID.)

Assigning formats

Program 7.2-27: SQL: Assigning Formats	Program 7.2-28: FEDSQL: Assigning Formats
| ```
proc sql ;
   select sum(ACTUAL) as
     TOTAL_ACTUAL format=dollar10.2
   from SASHELP.PRDSALE ;
quit ;
``` | ```
proc fedsql ;
   select put(sum(ACTUAL),dollar10.2) as
     TOTAL_ACTUAL
   from mybase.PRDSALE  ;
quit ;
``` |

What needed to be done?

Use a PUT function to add a dollar format to the sum of ACTUAL. In FEDSQL, the resulting TOTAL_ACTUAL is in character format, compared to the numeric format in PROC SQL. Below you will find an example that elaborates the use of PUT and INPUT.

(Continued)

Table 7.2-2: (*Continued*)

PROC FEDSQL cannot access concatenated librefs (SASHELP and SASUSER). Concatenated librefs "pretend" that SAS data files are stored at one place when in fact they are in more than one physical location. To access the SAS files, you need to specify a LIBNAME statement, as shown in the following examples:

- **SAS:** libname MYBASE base 'C:\Program Files\SASHome\SASFoundation\9.4\core\sashelp\';
- **SPRE:** libname MYBASE v9 '/opt/sas/spre/home/SASFoundation/sashelp' ; for example, PRDSALE, CLASSFIT, CLASS.

Use LIBNAME with the LIST option, for example, "libname SASHELP list", to narrow down the required paths.

CASE/WHEN processing I

Program 7.2-29: SQL: CASE/WHEN	**Program 7.2-30: FEDSQL: CASE/WHEN**
| ```
proc sql ;
create table TWO
  as select *,
  case
    when ID = "03"
      then " with 99"
    else "without 99"
  end as CODE
from ONE ;
quit ;
``` | ```
proc fedsql ;
create table TWO
  as select *,
  case
    when ID = '03'
      then ' with 99'
    else 'without 99'
  end as CODE
from WORK.ONE ;
quit ;
``` |

What needed to be done?
Single quotation marks for character entries ('03' in ID) and text strings (' with 99' in CODE).

Boolean expressions in a SUM function (CASE/WHEN processing II)

Program 7.2-31: SQL: Boolean Expressions	**Program 7.2-32: FEDSQL: Boolean Expressions**
| ```
proc sql ;
 create table WEIGHT_GAIN as
 select SEX,
 count(*) as PupilCount,
 mean(WEIGHT) as WEIGHT_mean,
 mean(PREDICT) as PREDICT_mean,
 sum(WEIGHT < PREDICT) as WGT_LT_PRED
    from sashelp.classfit
    group by SEX ;
quit ;
``` | ```
proc fedsql ;
 create table  WEIGHT_GAIN_FED as
 select SEX,
 count(WEIGHT) as PupilCount,
 mean(WEIGHT) as WEIGHT_mean,
 mean(PREDICT) as PREDICT_mean,
 sum(case
    when ( WEIGHT < PREDICT ) is true then 1
    end) as WGT_LT_PRED
    from mybase.classfit
    group by SEX ;
quit ;
``` |

What needed to be done?
In contrast to PROC SQL, PROC FEDSQL does not allow Boolean expressions (simplified 0 for "false", and 1 for "true"). In PROC SQL, the expression after SUM counts how many times a WEIGHT value is smaller than a PREDICT value. In FEDSQL, a Boolean expression in the SUM function would trigger the following error note:

```
ERROR: Function SUM(bool) does not exist
ERROR: No function matches the given name and argument types.
```

In FEDSQL, this Boolean expression could be specified using a CASE/WHEN/END expression.

(*Continued*)

Table 7.2-2: (*Continued*)

& operator

Program 7.2-33: SQL: '&' Operator	**Program 7.2-34: FEDSQL: 'and' Operator**

<pre>proc sql ;
create table TWO
as select * from ONE
   where A >= 1 & C = 3 ;
quit ;</pre>
	<pre>proc fedsql ;
	create table TWO
	as select * from WORK.ONE
	   where A >= 1 and C = 3 ;
	quit ;</pre>

What needed to be done?
Replace '&' with 'and'.

Reference on a calculated column (CALCULATED option)

Program 7.2-35: SQL: CALCULATED Option	**Program 7.2-36: FEDSQL: CALCULATED Workaround**

<pre>proc sql ;
  create table DOPPELTE
  as select ID, A, B, C,
    count(*) as DOPP_CNT
  from ONE
      group by ID, B
  having calculated
        DOPP_CNT > 1 ;
quit ;</pre>
	<pre>proc fedSQL ;
	  create table DOPPELTE
	  as select ID, A, B, C,
	    count(*) as DOPP_CNT
	  from WORK.ONE
	      group by ID, B, A, C
	  having
	            count(*) > 1 ;
	quit ;</pre>

What needed to be done?
Replace HAVING CALCULATED "new column" with the original expression, HAVING "count(*)". Group by all columns in the table, for example, compare 'group by ID, B' versus 'group by ID, B, A, C'.

Handle remerging properly

Program 7.2-37: SQL: Remerging (Example)	**Program 7.2-38: FEDSQL: Remerging Workaround**

<pre>proc sql;
   select *
     from SASHELP.CLASS
     group by SEX
     having AGE=max(AGE);
quit;</pre>
	<pre>proc fedsql;
	   select
	     mybs.*
	   from
	       (select SEX, max(AGE) as max_AGE
	        from mybase.CLASS
	        group by SEX
	       ) as sbqry
	   inner join mybase.CLASS as mybs on
	          mybs.SEX=sbqry.SEX and
	          mybs.AGE=sbqry.max_AGE;
	quit;</pre>

What needed to be done?
Find out whether remerging needs to be tackled in the first place. The easiest way to identify remerging is to check the SAS log for a submitted PROC SQL statement. If you see the following message: *NOTE: The query requires remerging summary statistics back with the original data.*, then you need to rewrite the code in PROC FEDSQL. The above FEDSQL variant replaces the HAVING statement with a subquery and an inner join.

(*Continued*)

Table 7.2-2: (*Continued*)

CONTAINS operator

Program 7.2-39: SQL: CONTAINS Operator

```
proc sql ;
   create table TWO as
      select * from ONE
   where ID contains "0" ;
quit ;
```

Program 7.2-40: FEDSQL: CONTAINS Workaround (LIKE)

```
proc fedsql ;
   create table TWO as
      select * from WORK.ONE
      where ID like '%0%'  ;
   quit ;
```

What needed to be done?

Replace CONTAINS with LIKE. Replace '0' with '%0'. Use single quotation marks.

Correlated Subquery (query plus subquery)

Program 7.2-41: SQL: Correlated Subquery

```
proc sql;
    select *
    from SASHELP.PRDSAL3 A
    where exists
        (select *
        from SASHELP.PRDSALE B
        where A.ACTUAL=B.PREDICT
                and
            COUNTRY="CANADA");
quit;
```

Program 7.2-42: FEDSQL: Correlated Subquery

```
proc fedsql;
    select *
    from mybase.PRDSAL3 A
    where exists
        (select *
        from mybase.PRDSALE B
        where A.ACTUAL=B.PREDICT
                and
            COUNTRY='CANADA');
quit;
```

What needed to be done?

Use single quotation marks. Access preset SAS libraries like SASHELP by using LIBNAME as follows:
SAS: libname MYBASE base 'C:\Program Files\SASHome\SASFoundation\9.4\core\sashelp\';
SPRE: libname MYBASE v9 '/opt/sas/spre/home/SASFoundation/sashelp' ; for example, PRDSALE, CLASSFIT, CLASS.

Query for Grouped mean

Program 7.2-43: SQL: Grouped Mean

```
proc sql;
select DIVISION,
   avg(PREDICT) as PRED_MEAN
from SASHELP.PRDSALE
     group by DIVISION
       order by DIVISION ;
quit;
```

Program 7.2-44: FEDSQL: Grouped Mean

```
proc fedsql;
select DIVISION,
   avg(PREDICT) as PRED_MEAN
from mybase.PRDSALE
     group by DIVISION ;
quit;
```

What needed to be done?

Access preset SAS libraries like SASHELP, by using LIBNAME as follows:
SAS: libname MYBASE base 'C:\Program Files\SASHome\SASFoundation\9.4\core\sashelp\';
SPRE: libname MYBASE v9 '/opt/sas/spre/home/SASFoundation/sashelp' ; for example, PRDSALE, CLASSFIT, CLASS.
ORDER in FEDSQL not necessary. (See the notes about sorting order in CAS in 7.2.4.) QUIT recommended.

(*Continued*)

Table 7.2-2: (*Continued*)

Specifying SAS Formats in a Query Expression

Program 7.2-45: Before: SQL: Specifying SAS Formats in a Query Expression

```
proc sql ;
   create table TWO as
   select count(*) as N_months,
          sum(ACTUAL) as ACTUAL_sum format=dollar10.2,
          mean(ACTUAL) as ACTUAL_mean format=commax10.2
   from SASHELP.PRDSALE;
quit ;
```

Program 7.2-46 After: FEDSQL: Specifying SAS Formats in a Query Expression

```
proc fedsql ;
   create table TWO as
   select count(*) as N_months,
          put(sum(ACTUAL),dollar10.2) as ACTUAL_sum,
          put(mean(ACTUAL),commax10.2) as ACTUAL_mean
   from mybase.PRDSALE;
quit ;
```

What needed to be done?

Honestly? Nothing. This is just another formatting approach; if you would replace FEDSQL with SQL in the FEDSQL version (yes, this time the other way round), the FEDSQL code would also work for SQL. The result of the PUT function is always character.

INTO :column

Program 7.2-47: SQL: INTO :column

```
proc sql ;
   select mean(AGE) into :AGE_mac
   from SASHELP.CLASS;
   select *
   from mybase.CLASS
   where AGE >= &AGE_mac. ;
quit;
```

Program 7.2-48: FEDSQL: INTO Workaround

```
proc fedsql noprint ;
create table WORK.MAC_INPUT as
   select mean(AGE) as AGE_mean
   from mybase.CLASS;
quit;
data _null_;
   set WORK.MAC_INPUT ;
   if _n_=1 then call symput('AGE_
mac',trim(left(AGE_mean)));
run;
proc fedsql ;
   select *
   from mybase.CLASS
   where AGE >= &AGE_mac. ;
quit;
```

What needed to be done?

The INTO facility is not supported in PROC FEDSQL. Use DATA NULL plus SYMPUT as a workaround. Access preset SAS libraries by setting a LIBNAME.

(*Continued*)

Table 7.2-2: (*Continued*)

Using PUT and INPUT functions

Program 7.2-49 SQL: Using PUT and INPUT

```
proc sql ;
 create table TWO as select
  input(A,yymmdd8.)
     as A format=yymmdds10.,
  input(B,comma12.2)
     as B format=dollarx12.2,
  input(C,time8.)
     as C format=hour4.1
 from ONE ;
quit ;
```

Program 7.2-50: FEDSQL: Using PUT and INPUTN (PUTN)

```
proc fedsql ;
/* result in TWO is NUM! */
    create table TWO as
  select
    inputn(A,'yymmdd8.') as A_num,
    inputn(B,'comma12.2') as B_num,
    inputn(C,'time8.') as C_num
        from WORK.ONE ;
/* formatted result in THREE is CHAR! */
  create table THREE as
  select
    put(A_num,yymmdds10.) as A_nfrm,
    put(B_num,dollarx12.2) as B_nfrm,
    put(C_num,hour4.1) as C_nfrm
      from WORK.TWO ;
  drop table TWO ;
  quit ;
```

What needed to be done?

The input of table ONE is date and time values in character format. SQL: The result in table TWO is of format NUMERIC. FEDSQL: The result in table THREE is of format CHAR. INPUTN specifies a numeric informat to a character column at run time. PUT specifies a numeric informat to a character source at compile time. PUT converts the numeric values to character values in the specified format, for example, DOLLARX12.2. PUT is faster than PUTN because you can specify a format at compile time rather than at run time.

Workaround for ALTER TABLE functions

Program 7.2-51: SQL: Using ALTER TABLE

```
proc sql ;
   alter table ONE
   modify B format=dollar12.2 ;
quit ;
```

Program 7.2-52: FEDSQL: ALTER TABLE Workaround

```
proc fedsql ;
   drop table TWO force;
   create table TWO as
     select
       put(B,dollar12.2) as B_CHAR
         from WORK.ONE ;
 * alter table ONE
   alter B set format dollar12.2 ;
quit;
```

What needed to be done?

The SQL output is numeric; the transformation is done without any passes. The FEDSQL output is character; the transformation requires one read and one write pass. ALTER TABLE is supported in PROC FEDSQL on SAS SPRE, but not on CAS.

Explicit Pass-Throughs

Program 7.2-53: SQL: Explicit Pass-through

```
proc sql ;
   connect to oracle as MY_ORA
   (path='My_path' schema='My_Scm'
   user='CFGS'    password='xxx');
 create table MY_TABLE2 as
 select * from connection to MY_ORA
  (select * from MY_ORA.MY_TABLE1);
   disconnect from oracle ;
quit ;
```

Program 7.2-54: FEDSQL: Explicit Pass-through

```
proc fedsql ;

 create table MY_TABLE2 as
 select * from connection to MY_ORA
  (select * from MY_ORA.MY_TABLE1) ;

 quit ;
```

(Continued)

Table 7.2-2: (*Continued*)

What needed to be done?

FedSQL uses librefs or caslibs to establish an explicit pass-through connection to the data source; you do not need to specify CONNECT and DISCONNECT statements anymore, you can remove them from the original SQL code. (See the code lines in bold.) Currently only the SELECT from CONNECTION TO is supported for FEDSQL in CAS.

Heterogeneous Joins (Federated Queries)

Program 7.2-55: SQL: Heterogeneous Join	Program 7.2-56: FEDSQL: Heterogeneous Join

```
proc sql;
/*Pass 1: Separate Queries */
 create table myoracle as
    select product, prodid
       from myoracle.products ;
 create table mybase as
    select totals, prodid, custid
       from mybase.sales ;
 create table myoracle2 as
    select city, custid
       from  myoracle.customers ;
/*Pass 2: Joining the Subsets */
select product, sales, city
where
  myoracle.prodid=mybase.prodid
       and
  maybase.custid=myoracle2.custid ;
quit ;
```

```
proc fedsql;
    select myoracle.products.product,
           mybase.sales.totals,
           myoracle.customers.city
    from  myoracle.products,
          mybase.sales,
          myoracle.customers
    where products.prodid=sales.prodid
       and sales.custid=customers.custid ;
quit ;
```

What needed to be done?

One step in FEDSQL replaces the separate queries in SQL. FEDSQL can do this type of query in one pass, whereas SQL requires at least two passes. First, perform three separate queries, and then join the subsets (not to mention dropping the utility tables). These last two examples hopefully show how elegantly and efficiently you can program queries with FEDSQL.

7.2.3 Using FedSQL in CAS: PROC FEDSQL and PROC CAS

The following three subsections provide you with in-depth guidance in three thematically separate areas that reflect the three basic steps in working on CAS: connect, prepare and process.

- **Connecting** to the CAS server and retrieving connection details (Subsection 7.2.3.1).
- **Accessing, Loading, and Retrieving Data** in the CAS environment (Subsection 7.2.3.2).
- **Data Processing Examples** in CAS with queries, joins, and so on (Subsection 7.2.3.3).

The following subsections used SAS Studio 5.2 on SAS Viya 3.5. Some of the technical output is anonymized, such as User or UUID.

7.2.3.1 Connecting to the CAS Server and Retrieving Connection Details

This subsection introduces how to create a connection with CAS, retrieve and view information about CAS sessions, and disconnect or terminate a CAS session. The most important statements are used in Subsection 7.2.3.3. I recommend looking at Example 1: Preparing the CAS Sessions and Loading SASHELP Files to CAS.

Connecting to the CAS Server

SAS Studio is the web-based programming environment in SAS Viya. As soon as you start SAS Studio 5.2 in Linux or Windows, the connection with the CAS server is automatically created. You do not need to enter in connection information anymore.

View the Information Connecting to the CAS Server

You can retrieve the connection information used by submitting the following PROC OPTIONS code:

```
proc options group=CAS ;
run ;
```

The SAS log shows the installed values system options. Selected relevant information is in bold. The output is self-explanatory.

SAS Log 7.2-3: Output PROC OPTIONS (CAS Group)

```
     SAS (r) Proprietary Software Release V.03.05  TS1M0
Group=CAS
CASAUTHINFO=       Specifies an authinfo or netrc file that includes authentication
information.
CASDATALIMIT=100M Specifies the maximum number of bytes that can be read from a file.
CASHOST=sasserver.demo.sas.com   The CAS server name associated with a CAS session.
CASLIB=            Specify the default CASLIB name.
CASLOPT=           Debugging options for CASL.
CASNCHARMULTIPLIER=1.5   Specifies a multiplication factor to increase the number of
                   bytes when transcoding fixed CHAR data.
CASNWORKERS=ALL    Specify the number of workers to use with a CAS session.
CASPORT=5570       The port associated with a CAS session.
CASSESSOPTS=       Identify CAS server session options.
CASTIMEOUT=60      The CAS session timeout in seconds.
CASUSER=           The userid associated with a CAS session.
SESSREF=CASAUTO    Identify the name to associate with a generated CAS session.
```

Creating a CAS Session

Creating a CAS session depends on the environment you are working in. If you work in SAS Studio, you simply submit the following CAS statement to create a session:

Program 7.2-57: Create a CAS Session

```
cas CASAUTO ;
```

You could successfully create a CAS session by just using CAS without using the session name CASAUTO. If your CAS client is SAS 9.4M5 or later or SAS Enterprise Guide, you connect to the CAS server explicitly using the retrieved CAS connection information as shown in Program 7.2-58.

Program 7.2-58: Create a CAS Session II

```
options cashost="cloud.example.com" casport=5570;
cas CASAUTO;
```

The SAS log was created by SAS Studio.

SAS Log 7.2-4: Creating a CAS Session

```
NOTE: The session CASAUTO connected successfully to Cloud Analytic Services sasserver.demo.
sas.com using port 5570. The UUID is 01-02-03-04-05-06. The user is abc@xyz.com and the
active caslib is CASUSER(abc@xyz.com).
```

```
NOTE: The SAS option SESSREF was updated with the value CASAUTO.
NOTE: The SAS macro _SESSREF_ was updated with the value CASAUTO.
NOTE: The session is using 0 workers.
```

In both scenarios, the CAS statement starts a CAS session named CASAUTO. When you start a CAS session, the server authenticates your identity, listens for a session request, and starts a session between the client process and the session process. Sessions provide some important functionalities. One is the fault tolerance for each session: problems occurring in one session do not impact other sessions or the server.

Retrieving Information about CAS Sessions

If you want to know everything about all your CAS sessions, including CASAUTO, MY_CAS_SESSION (see Program 7.2-61), and others, submit the LISTSESSIONS option.

Program 7.2-59: Retrieving Information about a CAS Session
```
cas CASAUTO listsessions;
```

LISTSESSIONS lists all of the CAS sessions known to the CAS server for the userID associated with "CASAUTO". SAS Studio prints this information to the SAS log.

SAS Log 7.2-5: Output Information about a CAS Session
```
NOTE: LISTSESSIONS lists information about sessions known to Cloud Analytic Services.
Sessions can be created with the CAS statement.
NOTE: SessionName = CASAUTO:Wed Jan 20 01:28:52 2021
        UUID = 01-02-03-04-05-06
        State = Connected
        Authentication = OAuth
        Userid = abc@xyz.com
NOTE: Request to LISTSESSIONS completed for session CASAUTO.
```

If you want to retrieve information about the properties of the specific CAS session, CASAUTO, submit the LISTSESSOPTS option.

Program 7.2-60: Retrieving Information about Properties of a CAS Session
```
cas CASAUTO listsessopts;
```

LISTSESSOPTS lists properties for session CASAUTO, including Name, UsageType, Group, or Description. I do not provide an example output for LISTSESSOPTS at this point.

You need to be connected to an existing session, CASAUTO, in order to use these options.

Create Your Own CAS Sessions

You can even create your own CAS sessions. Using the details of the existing CAS server connection, you can override an active caslib, a time-out (in seconds), and locale defaults.

Program 7.2-61: Creating Your Own CAS Session
```
cas my_CAS_session sessopts=(caslib=casuser timeout=1800 locale="de_CH");
```

Another interesting thing happens in the brackets after SESSOPTS=. CASUSER is now the active caslib no matter which other active caslib was set before. The name of the new CAS session is my_CAS_session, and the locale was set to German (Switzerland). Chinese (China) would require "zh_CN" and English (United States) "en_US".

SAS Log 7.2-6: Output Information about Your Own CAS Session
```
NOTE: The session MY_CAS_SESSION connected successfully to Cloud Analytic Services
sasserver.demo.sas.com using port 5570. The UUID is 01-02-03-04-05-06. The user is abc@xyz.
com and the active caslib is CASUSER(abc@xyz.com).
NOTE: The SAS option SESSREF was updated with the value MY_CAS_SESSION.
NOTE: The SAS macro _SESSREF_ was updated with the value MY_CAS_SESSION.
NOTE: The session is using 0 workers.
NOTE: 'CASUSER(abc@xyz.com)' is now the active caslib.
NOTE: The CAS statement request to update one or more session options for session MY_CAS_
SESSION completed.
```

Disconnect, Reconnect, or Terminate Your CAS Sessions

When you disconnect from a session, you must reconnect to the session before the connection time-out expires (default: 60 seconds), otherwise the session is automatically terminated. Before you disconnect, it is strongly recommended that you adjust the TIMEOUT= value in seconds. The following example requests CAS to terminate the session if nobody returns within 15 minutes.

Program 7.2-62: Disconnect, Reconnect, or Terminate a CAS session
```
cas CASAUTO sessopts=(timeout=900) ;
%if &CASSTMTERR eq 0 %then %do ;
   cas CASAUTO disconnect ;
%end ;

cas CASAUTO reconnect ;

cas CASAUTO terminate ;
```

When you return within the preset time-out interval, in this example 15 minutes, submit a RECONNECT statement using the same CAS session name (CASAUTO). When you want to terminate the CAS session name, submit the TERMINATE statement.

7.2.3.2 Accessing, Loading, and Retrieving Data in the CAS Environment

This subsection introduces concepts like caslibs, the CASLIB statement, utilities such as PROC CASUTIL, and their versatile application in loading, deleting, or saving tables permanently.

As soon as you start SAS Studio, it automatically connects to your SAS and CAS data. By default, the SAS Studio Libraries tree (top left) displays pre-set librefs that point to the location of your SAS data sets, be it SAS default or user-defined librefs. However, the Libraries tree does not show the storage location of your CAS tables yet. To see your CAS tables, you first need to assign SAS librefs to existing CAS libraries (caslibs) that contain your CAS tables. From this angle, you could understand caslibs for CAS tables as a cloud-equivalent combination of librefs *plus* libraries for SAS tables. All data-processing CAS operations use caslibs.

Caslibs contain controls, data, and connection information. The key concept of caslibs is access-control: Permissions define which users and groups are authorized to access and handle the contents of the caslib. Next to these access controls (user IDs, permissions), caslibs contain connection information to data sources (password, schema, session, path), provide access to data from the source and in-memory tables, space for data (one or more CAS tables, either in-memory, from disk, or streams), and provide a secure interface to data providers. There are different types of caslibs, including predefined (for example, MODELS, PUBLIC) or personal (CASUSER). The data containers in caslibs are like tables whose columns contain information like names of variables and labels, data types, and so on (Pendergrass, 2017).

Assigning SAS Librefs for Existing Caslibs

Caslibs make data available to CAS. To make existing caslibs visible in the SAS Studio Libraries tree, you need to assign librefs first. _ALL_ assigns librefs to all existing caslibs by using the name of the caslib for its libref. If you submit the first CASLIB statement, you associate a libref for each of your caslibs. (You could also specify your active SESSREF "my_CAS_session"; see Program 7.2-61 version II.) In both instances, you submitted the CAS statement before. (See above.) If you do not want to or cannot (see below) assign librefs to each and every caslib which, in turn, would add each and every existing caslib to the Libraries tree, you can use the LIBNAME statement to assign a libref to only the caslib that you want to add to the tree. Program 7.2-70 (Example 1) demonstrates how to use a LIBNAME to associate a libref (see "MY_CAS") with a caslib (see "CASUSER"). The name of the libref and the caslib does not have to be the same. Sometimes this is not even possible as the naming rules for caslibs are different from those for librefs.

Program 7.2-63: Assigning Librefs for Existing Caslibs

```
caslib _all_ assign ;                        /* version */
caslib _all_ assign sessref=my_CAS_session; /* version II, with SESSREF */
```

Now the Libraries tree is updated. Using the names of the caslibs, the new librefs make the caslibs *usually* visible. You can recognize them by the small cloud symbol; the drawer symbol indicates SAS libraries. If you cannot see a libref (thus the caslib remains invisible), you might need to check several root causes. Maybe the original caslib name did not comply with the naming rules for librefs and could not get successfully translated to a libref. For example, it might have exceeded eight characters or contained one or more blanks. (See the notes about "QASMartStore" in SAS Log 7.2-7.) In this case, use the above LIBNAME approach to assign an adjusted libref to that caslib. Another reason why you might not see an expected libref is that the caslib did not exist. You can use "CASLIB _ALL_ LIST;" to check whether that caslib had been defined before in the first place. Caslibs might also be *hidden* in the sense of intentionally reduced visibility. Hidden caslibs are listed on the Data page in SAS Environment Manager, though. If you *can* see a libref, but not *all* CAS tables, then check accordingly whether the names of the "invisible" tables did not comply with naming rules, or whether they exist at all. *All* in-memory tables in CAS are *temporary* unless they are explicitly permanently stored.

SAS Log 7.2-7: Output Information about Assigning Librefs to Existing Caslibs

```
NOTE: A SAS Library associated with a caslib can only reference library member names that
conform to SAS Library naming conventions.
NOTE: CASLIB CASUSER(abc@xyz.com) for session MY_CAS_SESSION will be mapped to SAS Library
CASUSER.
NOTE: CASLIB Formats for session MY_CAS_SESSION will be mapped to SAS Library FORMATS.
NOTE: CASLIB ModelPerformanceData for session MY_CAS_SESSION will not be mapped to SAS
Library ModelPerformanceData. The CASLIB name is not valid for use as a libref.
NOTE: CASLIB Models for session MY_CAS_SESSION will be mapped to SAS Library MODELS.
NOTE: CASLIB Public for session MY_CAS_SESSION will be mapped to SAS Library PUBLIC.
NOTE: CASLIB QASMartStore for session MY_CAS_SESSION will not be mapped to SAS Library
QASMartStore. The CASLIB name is not valid for use as a libref.
NOTE: CASLIB Samples for session MY_CAS_SESSION will be mapped to SAS Library SAMPLES.
```

You can add or manage caslibs with the CASLIB statement in SAS Studio only if you are authorized to do so. In one of the next examples ("Loading CSV files to the CAS Server"), you find a manually added caslib.

Managing Tables within Caslibs (PROC CASUTIL)

For processing in the CAS environment, data must be loaded to the CAS server. The following code examples demonstrate how to load data from local to CAS and remove them again.

The first CASUTIL example loads the SAS table CLASS from the physical folder SASHELP on the Compute Server, renames it to CLASS_CAS, and moves it to the in-memory folder CASUSER on the CAS server. The REPLACE option specifies that a new file can overwrite an already existing file with the same name. Alternatively, you could use a DATA step to load a table from SAS to CAS. The note from the SAS log is for PROC CASUTIL. CAS tables loaded into memory are in-memory copies of the associated CAS file. Loaded tables do not have to fit into memory. CAS manages the tables to allow rows to be processed by workers.

PROC CASUTIL offers three LOAD options: DATA= for client-side SAS data sets, FILE= for client-side files like Excel, and CASDATA= for server-side SASHDAT files.

Program 7.2-64: Uploading Data (PROC CASUTIL)
```
proc casutil;
        load data=sashelp.class
        outcaslib="CASUSER" casout="CLASS_CAS" replace ;
run;
```

Program 7.2-65: Uploading Data (DATA Step)
```
data casuser.class_cas ;
    set sashelp.class ;
  run;
```

SAS Log 7.2-8: Feedback after Successfully Uploading Data
```
NOTE: SASHELP.CLASS was successfully added to the "CASUSER(abc@xyz.com)" caslib as
"CLASS_CAS".
```

Using DROPTABLE, the next CASUTIL example removes the table CLASS_CAS from memory of the CAS library CASUSER. The QUIET option suppresses error messages and avoids setting SYSERR when the specified table is not found.

Program 7.2-66: Dropping Data (PROC CASUTIL)
```
proc casutil ;
   droptable casdata="CLASS_CAS" quiet ;
run; quit;
```

Loading CSV Files to the CAS Server

This CASUTIL example loads the CSV file HMEQ.CSV located at the end of the specified path, renames it to CAS_ HMEQ, and moves it to the newly specified CAS folder CAS_CSV. From the CASDATA= you can see that HMEQ.CSV is a server-side file.

Program 7.2-67: Uploading CSV files (PROC CASUTIL)
```
cas CASAUTO;
caslib cas_csv task=add type=dnfs path="/opt/open/data" ;
caslib _all_ assign;

proc casutil ;
  load casdata="hmeq.csv"
      importoptions=(filetype="csv" getnames="true")
      casout="cas_hmeq"
      replace ;
    contents casdata = "cas_hmeq" ;
quit ;
```

SAS Log 7.2-9: Output Information about Uploading CSV Files
```
NOTE: The UUID '01-02-03-04-05-06' is connected using session CASAUTO.
NOTE: Cloud Analytic Services made the DNFS file hmeq.csv available as table CAS_HMEQ in
caslib CAS_CSV.
NOTE: The Cloud Analytic Services server processed the request in 0.02891 seconds.
NOTE: Cloud Analytic Services processed the combined requests in 0.002617 seconds.
```

Promoting versus Storing CAS In-Memory Contents (DATA Step, PROC CASUTIL)

When you're about to save a table, you can choose between a combination of four options: saving a CAS table permanently on disk or temporarily in-memory and promoting its contents or not. For example, the first DATA step creates a local temporary in-memory table; the second DATA step creates a global temporary in-memory table if you want to share that table (see "promote=yes"). Again, both tables are temporary. If you want to create a permanent copy, PROC CASUTIL's SAVE saves the file as CLASS_CAS_F.sashdat.

Program 7.2-68: Promoting versus Storing CAS In-memory Contents

```
data MY_CAS.CLASS_CAS_F ;
   set CASUSER.CLASS_CAS ;
      if SEX="F" ;
run;

data MY_CAS.CLASS_CAS_F
         (promote=yes) / sessref=my_CAS_session;
   set CASUSER.CAS_CLASS ;
      if SEX="F" ;
      by AGE ;
run;
proc casutil;
   save casdata='CLASS_CAS_F'
   outcaslib='CASUSER' replace;
quit;
```

The PROMOTE=YES statement in the second DATA step has important consequences for file attributes and accessibility. By default, a local session-scope table is accessible within the session by the user only. They cannot be shared with other users and are dropped when the CAS session is terminated. Promoting a table to global scope allows other sessions and users to see the table and possibly access the data if they are allowed to. The user can set the respective access rights. Promoting a table does not copy the table itself. It just allows other sessions and users to request access and information about the table. Global-scope tables also persist in-memory; they are not permanent and are in essence temporary copies of the original files. So, if you restart the server, these files are gone. If in doubt, you can use PROC CASUTIL's LIST statement to check the attributes of the files in the respective libraries.

Program 7.2-69: List Files from a Caslib's Data Source or In-memory Tables

```
proc casutil;
    list files  incaslib='CASUSER' ;
*   list tables  incaslib='CASUSER' ;
quit;
```

TABLES lists the in-memory tables in a caslib. FILES lists the files that are available in the caslib's data source. If you submit Program 7.2-69, CASUTIL displays the files including their SASHDAT suffix, CLASS_CAS_F.sashdat.

7.2.3.3 Data Processing Examples in CAS (Queries, Joins, and so on)

This subsection presents several examples of how to process data in CAS. Starting from preparing a CAS session, saving CAS data to SASHDAT, as well as querying and creating one table, creating several tables in a single FEDSQL statement, and joining tables from multiple CAS libraries. The final example does not use FEDSQL statements; however, it requires you to set up your CAS session to be able to perform that visual analysis.

- *Example 1:* Preparing the CAS session and loading SASHELP files to CAS
- *Example 2:* Simple query: FedSQL in CAS: PROC FEDSQL versus PROC CAS
- *Example 3:* Creating a CAS table using PROC FEDSQL and PROC CAS (enhanced options)
- *Example 4:* Aggregating several tables in the same FEDSQL statement
- *Example 5:* Joining multiple tables within the same CAS library
- *Example 6:* Querying data using a correlated subquery
- *Example 7:* Saving CAS in-memory data to SASHDAT (PROC CASUTIL)
- *Example 8*: Visual analysis

The examples use the SASHELP data sets CLASS, PRDSAL2, and PRDSAL3.

Example 1: Preparing the CAS Session and Loading SASHELP Files to CAS

The first example is about entering the CAS environment. In Program 7.2-70, the code lines from CAS to CASLIB prepare the CAS session. PROC CASUTIL and the DATA step approach load the SASHELP.CLASS files from the Compute Server into the memory of the CAS server.

The code lines from CAS to CASLIB start the session, assign CAS engine librefs, and SAS librefs to existing caslibs. The preparation of a CAS session should have at least three elements:

1. **CAS:** Starts a CAS session, for example, my_CAS_session. When you start a session, a personal caslib is available. CASUSER is personal and only your user ID can access the data.
2. **LIBNAME:** By assigning the libref MYCAS to the caslib CASUSER, the CAS engine LIBNAME statement connects a Compute Server session with a CAS session. The caslib has to exist, and in this example, it is a default caslib.
3. **CASLIB** and _ALL_ ASSIGN assign SAS librefs to existing caslibs. This enables you to see caslib names in the Libraries panes or windows of SAS applications.

Program 7.2-70: Example 1: Preparing the CAS Session and Loading SASHELP Files to CAS
```
cas my_CAS_session ;
libname MY_CAS cas caslib=CASUSER ;
caslib _all_ assign sessref=my_CAS_session;
```

This code illustrates how to start a CAS session (see "cas"), how to use a LIBNAME to associate a libref (see "MY_CAS") with a caslib (see "CASUSER"), and how to display all caslibs in the SAS Studio Libraries tree (see "caslib _all_"). Assigning caslibs usually shows five default CAS dictionaries:

1. **Casuser:** Stores personal data.
2. **Formats:** Stores user-defined formats.
3. **Models:** Stores models created by VA and VS.
4. **Public:** Shared and writeable caslib, accessible to all.
5. **Samples:** Stores sample data, supplied by SAS.

You can recognize caslibs by the small cloud symbol. Depending on the implementation and authorization, you might see more than these caslibs. If they do not see caslibs as expected, you might want to refresh the view.

PROC CASUTIL and the DATA step load SASHELP tables from the Compute Server to the caslib CASUSER. In the Compute Server's SASHELP, they are physical tables (on disk); in the CAS server's CASUSER, they are virtual tables (in-memory). If you submit the CASUTIL or DATA step code, CASUSER and MY_CAS should contain three files (CAS_CLASS, CAS_PRDSAL2, and CAS_PRDSAL3).

Using PROC CASUTIL

```
proc casutil;
    load data=sashelp.class
        outcaslib="CASUSER" casout="CAS_CLASS" replace ;
    load data=sashelp.PRDSAL2
        outcaslib="CASUSER" casout="CAS_PRDSAL2" replace ;
    load data=sashelp.PRDSAL3
        outcaslib="CASUSER" casout="CAS_PRDSAL3" replace ;
quit ;
```

Using a DATA Step

```
data CASUSER.CAS_CLASS;          data CASUSER.PRDSAL2;          data CASUSER.PRDSAL3;
set sashelp.class ;              set sashelp.prdsal2 ;          set sashelp.prdsal3 ;
run;                             run;                           run;
```

SAS Studio uses different symbols for these different storage methods: a cloud for in-memory and a drawer for on-disk storage folders in the SAS Studio Libraries tree. You can access the in-memory tables by the caslib CASUSER and the CAS engine libref MY_CAS. (See Example 7.) QUIT terminates the CASUTIL procedure. You will notice in the SAS Studio libraries that as soon as you move a file to a caslib (CASUSER), then it will also appear in the MY_CAS folder. As soon as you delete a file from CASUSER, then it will also disappear in the MY_CAS folder.

Example 2: Simple Query: FedSQL in CAS: PROC FEDSQL versus PROC CAS

In the CAS environment, you can use PROC FEDSQL in at least two varieties: as PROC FEDSQL (Program 7.2-71) and in PROC CAS using fedsql.execdirect (Program 7.2-72).The FEDSQL code used in both programs is identical, as are the names of the CAS table ("CAS_CLASS") and its CASLIB, "CASUSER". The only differences are minor variations programming the query in PROC FEDSQL and PROC CAS.

SESSREF indicates that the FEDSQL code runs in CAS. PROC FEDSQL requires the CAS session (SESSREF=) after the call of the procedure. In PROC CAS, the CAS session is specified in a separate line after SESSION and closed with a semicolon. PROC CAS requires the CAS action fedsql.execdirect. The FEDSQL code is specified as a string after QUERY=; in contrast to PROC FEDSQL, this string is enclosed in single quotation marks. The result of the queries is not displayed.

Program 7.2-71: Example 2: PROC FEDSQL Procedure Approach	**Program 7.2-72: Example 2: PROC CAS Action Approach**
```proc fedsql sessref=my_CAS_session ; select * from CASUSER.CAS_CLASS ; quit;```	```proc cas; session my_CAS_session ; fedsql.execdirect query='select * from CASUSER.CAS_CLASS' ; quit;```

Note: The FEDSQL example on the left tells you by its prefixes and caslib that this code is supposed to run in the CAS environment. If you do not have these express pointers you just need to check whether the code contains a SESSREF= option; this option specifies whether the FEDSQL code is supposed to run in a CAS or a SAS environment. If a FEDSQL does not contain a SESSREF= option, it is supposed to run in the SAS environment. (See Subsections 7.2.1 and 7.2.2.)

Example 3: Creating a CAS Table using PROC FEDSQL and PROC CAS (Enhanced Options)

PROC FEDSQL (Program 7.2-73) and PROC CAS (Program 7.2-74) create the table CASUSER.CAS_SUBSET querying the CAS source CASUSER.CAS_CLASS. The following descriptions focus on the minor differences: the _METHOD option in PROC FEDSQL writes a brief text description of the FedSQL query plan. By default, FedSQL does not overwrite an existing in-memory table of the same name. REPLACE and REPLICATION are CAS table options: REPLACE=TRUE deletes a possibly existing table of the same name; REPLICATION=0 specifies the number of copies of the table to make for fault tolerance. The result of the queries is not displayed. Examples 4 and later will be written as FEDSQL versions only.

Program 7.2-73: Example 3: PROC FEDSQL	**Program 7.2-74: Example 3: PROC CAS**
```proc fedsql sessref=my_CAS_session _method ; create table CASUSER.CAS_SUBSET {option replication=0 replace=true} as select * from CASUSER.CAS_CLASS where AGE >= 14 ; quit;```	```proc cas; session my_CAS_session ; fedsql.execdirect / method=true query='create table CASUSER.CAS_SUBSET {option replication=0 replace=true} as select * from CASUSER.CAS_CLASS where AGE >= 14' ; quit;```

SAS Log 7.2-10: SAS Logs for PROC FEDSQL and PROC CAS

```
Methods for full query plan
---------------------------
        SeqScan with qual from CASUSER.CAS_CLASS

Methods for stage 1
-------------------
        SeqScan from {Push Down}.Child 1

NOTE: Table CAS_SUBSET was created in caslib CASUSER(abc@xyz.com) with 9 rows returned.
```

The _METHOD output provides a high-level query plan and a brief summary of the execution stages. Because the above programs are essentially simple queries, the output begins and ends with the high-level view of the plan nodes in the query plan. FEDSQL programs including joins and subqueries produce more complicated query plans and multi-leveled execution stages.

Example 4: Aggregating Several Tables in the Same FEDSQL Statement

Example 4 accesses two tables in CASUSER, aggregates each by COUNTRY, YEAR, and PRODUCT, and creates the result tables CAS_PRDAGG2 and CAS_PRDAGG3.

Program 7.2-75: Example 4: Aggregating Several Tables in the Same FEDSQL Statement

```
proc fedsql sessref=my_CAS_session ;
   create table CASUSER.CAS_PRDAGG2
                  {option replication=0 replace=true}  as
     select COUNTRY, YEAR as YEAR, PRODUCT,
                  sum(ACTUAL) as SUM_SALES
     from CASUSER.CAS_PRDSAL2
          group by COUNTRY, YEAR, PRODUCT;
   create table  CASUSER.CAS_PRDAGG3
                  {option replication=0 replace=true}  as
     select COUNTRY, YEAR as YEAR, PRODUCT,
                  sum(ACTUAL) as SUM_SALES
     from CASUSER.CAS_PRDSAL3
          group by COUNTRY, YEAR, PRODUCT;
quit;
```

SAS Log 7.2-11: Example 4: Aggregating Several Tables in the Same FEDSQL Statement

```
NOTE: Table CAS_PRDAGG2 was created in caslib CASUSER(abc@xyz.com) with 48 rows returned.
NOTE: Table CAS_PRDAGG3 was created in caslib CASUSER(abc@xyz.com) with 24 rows returned.
```

Example 5: Joining Multiple Tables within the Same CAS Library

Example 5 accesses the result tables CAS_PRDAGG2 and CAS_PRDAGG3 from Example 4, filters them, and creates the result table CAS_PRD_UNION.

Program 7.2-76: Example 5: Joining Multiple Tables within the Same CAS Library

```
proc fedsql sessref=my_CAS_session ;
   create table CASUSER.CAS_PRD_UNION
                  {option replication=0 replace=true} as
     select SUM_SALES, PRODUCT, YEAR, COUNTRY
        from CASUSER.CAS_PRDAGG2
     where   COUNTRY='Canada'
union
     select SUM_SALES, PRODUCT, YEAR, COUNTRY
        from CASUSER.CAS_PRDAGG3
     where   COUNTRY='U.S.A.'  ;
quit;
```

SAS Log 7.2-12: Example 5: Joining Multiple Tables within the Same CAS Library
```
NOTE: Table CAS_PRD_UNION was created in caslib CASUSER(abc@xyz.com) with 24 rows returned.
```

The same principle applies when you join tables from several different CAS libraries. You might just need to specify the different caslibs (see an example in 7.2-56) or other connection options. (For details, please refer to Subsection 7.4.3.)

Example 6: Querying Data Using a Correlated Subquery

Example 6 uses the minimum PREDICT value for "Canada" in CAS_PRDSAL3 and keeps all rows of table CAS_PRDSAL2 where ACTUAL is greater than this PREDICT value. The resulting table is named CAS_QUERY.

Program 7.2-77: Example 6: Querying Data Using a Correlated Subquery
```
proc fedsql sessref=my_CAS_session  ;
   create table PUBLIC.CAS_QUERY as
     select * from CASUSER.CAS_PRDSAL2
     where ACTUAL >
             (select min(PREDICT)
              from CASUSER.CAS_PRDSAL3
              where upcase(COUNTRY)='CANADA') ;
quit;
```

In contrast to other examples, Example 6 stores the generated CAS table in PUBLIC. (See Example 2.)

Example 7: Saving CAS In-memory Data to SASHDAT (PROC CASUTIL)

Even global-scope tables are not permanent; they are temporary copies of the original files that will be gone if you restart the server. SASHDAT files are a SAS propriety format for CAS and represent the in-memory table stored on disk. The interesting part is that when files are stored as a SASHDAT file they are split into pieces and distributed across multiple hosts. When CAS loads that table, the pieces that make up the SASHDAT file are read from storage and put are back together as single table in CAS memory. SASHDAT files are often the fastest way to reload a CAS table from disk.

Program 7.2-78: Saving CAS In-memory Data to SASHDAT (PROC CASUTIL)
```
proc casutil;
   save casdata='CLASS_CAS_F'
   incaslib='CASUSER'
   outcaslib='CASUSER' replace;
   contents casdata='CLASS_CAS_F';
quit;
```

When you're storing in-memory data using PROC CASUTIL, the statements that you use are CASDATA and maybe INCASLIB; all the other statements are usually optional. The CASDATA option names the in-memory file that you want to save. INCASLIB specifies the caslib where the in-memory file lies that you want to save. The default is the current active caslib (so, this example code is redundant). The OUTCASLIB specifies the caslib where you want to store the in-memory file on disk. Using CASOUT, you can give an in-memory file a different name before you save it. When you save to a caslib, a *.sashdat file suffix is added by default, including saving to path-based caslibs. The CONTENTS statement displays table metadata such as column names and data types for the table CLASS_CAS_F. Promoted tables contain a flag. For example, the Table Information for the table CLASS_CAS_F contains the Flag "Yes" in the column "Promoted Table".

SAS Log 7.2-13: Example 7: Joining Tables from Multiple CAS Libraries
```
NOTE: Cloud Analytic Services saved the file CLASS_CAS_F.sashdat in caslib
CASUSER(abc@xyz.com).
NOTE: The Cloud Analytic Services server processed the request in 0.005444 seconds.
```

Example 8 is a SAS program processed in the CAS environment that does not use any FEDSQL statements. However, to be able to run this simple SGPLOT code in CAS, it requires you to set up your CAS session to be able to do so.

Program 7.2-79: Example 8: Visual Analysis

```
proc sgplot data=CASUSER.cas_prdsal2;
  vbar COUNTRY / response=ACTUAL datalabel categoryorder=respdesc;
run;
```

If you have done everything right, you should see output shown in Figure 7.2-1.

Figure 7.2-1: Result of Example 8: Visual Analysis

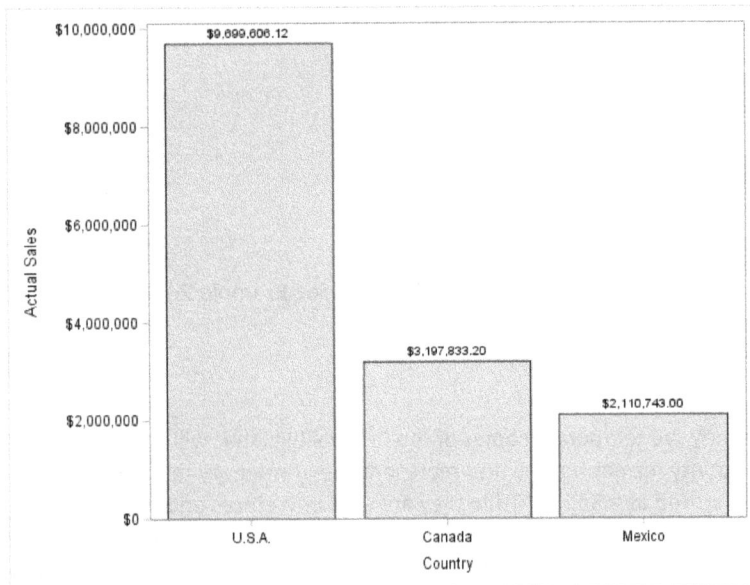

7.2.4 Not Just FEDSQL on CAS: DATA Step and Others

This subsection about running programs in CAS summarizes notes about using FEDSQL, the DATA step, and others in CAS. Use MSGLEVEL=I to observe what's going in detail.

SQL:
- Will not run in CAS.

FEDSQL:

- *Can* execute in CAS.
- **Requirements:** To run in CAS, the input needs to be a CAS table, and the FEDSQL code needs a SESSREF= option.
- **Caveat:** Certain FEDSQL language elements cause the CAS table to be downloaded to SPRE for processing.

DATA step:

- *Can* execute in CAS.
- **Requirements:** If the tables in SET (input) and DATA (output) are both CAS tables, the data is processed in CAS.
- **Caveat: Language statements not allowed in CAS:** Such language elements might return the CAS table to SPRE for processing, for example, using the "where=(...)" data set option after DATA for output tables; for

CAS, use a subsetting IF instead. A "where=(...)" for input tables is allowed. Also not allowed are MODIFY, REMOVE, and REPLACE statements; process them in SPRE instead. Some formats are not supported such as WORDS and WORDSDF; the processing takes place in SPRE. Other formats might be replaced by NLS equivalent formats for CAS.

- **Caveat: WORK:** Certain language elements can prevent CAS processing and return the CAS table to SPRE for processing. If the table in DATA is a WORK table, then the DATA step is run in SPRE. When a table is processed in SPRE, its sorting order is different to CAS; within CAS the sorting order changes depending on which thread finishes first.
- **Caveat: BY statement (BY processing):** In CAS, sorting is not required. DATA steps with a BY statement processed on CAS will group data according to the levels of the BY variable(s) and distribute them on threads accordingly. Previous PROC SORT steps are no longer necessary. However, the sorting order is different in CAS and might require some small refactoring of the processing logic.
- **Caveat: Threading:** You might need to consider whether data processing is single- or multiple-threaded. A DATA step with SINGLE=YES still runs in CAS if it uses CAS-enabled programming statements. Both input and output tables are CAS tables. In the case of the sum statement, just changing the library from SAS to CAS is not enough. You might get undesired results if a formerly single-threaded program sequentially processed on SAS is not properly adjusted to run in multiple threads in CAS. Depending on the data volume, you might need to refactor the processing logic in case the DATA step option SINGLE= YES is not enough.

Table 7.2-3: CAS-enabled SAS Procedures

Name	Description
APPEND	Adds rows from a CAS table to the end of a SAS data set and adds rows from a SAS data set to the end of a CAS table.
CONTENTS	Shows the contents of a CAS table and prints the directory of the caslib.
COPY	Copies tables to and from libraries.
DATASETS	Manages also CAS tables.
DELETE	Deletes SAS data sets and CAS tables.
DS2	Runs DS2 code.
EXPORT	Writes a SAS data set to delimited or JMP files.
FEDSQL	Submits a limited set of FedSQL code.
FORMAT	Creates user-defined formats.
HTTP	Processes data from the web.
IMPORT	Reads external data into a SAS data set.
JAVAINFO	Shows information about the version of Java on your system.
OPTIONS	Lists the current settings of SAS options in the SAS log.
PRINT	Prints SAS data sets and CAS tables.
PRINTTO	Redirects output.
PRODUCT_STATUS	Lists the SAS products that are installed on your system.
PWENCODE	Encodes passwords.
SGPANEL	Creates a panel of graph cells for the values of one or more classification variables.
SGPLOT	Creates statistical graphics such as histograms and regression plots, in addition to simple graphics such as scatter plots and line plots.
SGSCATTER	Creates a paneled graph of scatter plots for multiple combinations of variables.
SORT	Sorts SAS data sets in the local SAS session.
TRANSPOSE	Creates a CAS table by restructuring the values in the CAS table and transposing selected columns into rows.

Other SAS procedures:

- **PROC FEDSQL** and **PROC DS2:** Submit their language code to CAS for processing. Besides these two, there are other CAS-enabled procedures.
- **CAS-enabled procedures:** Access and process CAS tables and table metadata in the local SAS session.
- **Caveat:** CAS-enabled does not mean equivalent. For example, some options might not run in CAS but might run in SPRE.
- Some not CAS-enabled SAS procedures will still process in CAS if they have a CAS input table. WHERE clauses might direct a subset of the table to SPRE for processing.

7.2.5 CAS Procedures: Statistics, Data Mining, and Machine Learning

These statistical, data mining, and machine learning procedures process data in memory in CAS.

Table 7.2-4: New CAS Procedures: Statistics

Name	Description
ASSESS	Assesses and compares supervised learning models.
BINNING	Performs binning, which is a common data preparation step in the model-building process.
CARDINALITY	Determines a variable's cardinality or limited cardinality.
GENSELECT	Provides model fitting and model building for generalized linear models.
KCLUS	Performs clustering, which is a common step in data exploration.
LOGSELECT	Fits binary and binomial response models.
MDSUMMARY	Basic descriptive statistics in parallel for CAS tables.
NLMOD	Uses nonlinear least squares or maximum likelihood to fit nonlinear regression models.
PARTITION	Simple random sampling, stratified sampling, and oversampling to produce a table that contains a subset of observations or partitioned observations.
PCA	A multivariate technique for examining relationships among several quantitative variables.
PLSMOD	Fits models by using any one of a number of linear predictive methods, including partial least squares (PLS).
QTRSELECT	Chooses and fits quantile regression models.
REGSELECT	Fits and performs model selection for ordinary linear least squares models.
TREESPLIT	Builds tree-based statistical models for classification and regression.
VARIMPUTE	Performs numeric variable imputation.
VARREDUCE	Performs both supervised and unsupervised variable selection.

Table 7.2-5: New CAS Procedures: Data Mining and Machine Learning

Name	Description
ASTORE	Creates score code that can use code from an analytic store as well as DS2 code.
BOOLRULE	Extracts Boolean rules from large-scale transactional data.
FACTMAC	Implements the factorization machine model.
FOREST	Creates a predictive model called a forest, which consists of several decision trees.
GRADBOOST	Creates a predictive model called a gradient boosting model.
NETWORK	Provides a number of network analysis algorithms that take an abstract graph or network as input, helps explain network structure, and computes important network measures.
SVMACHINE	Implements the support vector machines (SVM) algorithm, which computes support vector machine learning classifiers for the binary pattern recognition problem.
TEXTMINE	Integrates natural language processing and statistical analysis to analyze large-scale textual data.
TMSCORE	Scores textual data. In text mining, scoring is the process of applying parsing and singular value decomposition projections to new textual data.

7.3 FEDSQL in DS2 Programs

A chapter on FEDSQL definitely needs to be supplemented with a section on the DS2 language. Why? And what is DS2 anyway?

DS2 is a SAS proprietary language for advanced data manipulation. DS2 has more functionalities than SQL, FEDSQL, and even the DATA step. Some of the more technical benefits might be that it is threaded for faster processing, which means if you use the DS2 THREAD statement, you could process data in parallel simultaneously, not just sequentially. DS2 runs natively in parallel across all available nodes. DS2 offers the precision of ANSI SQL data types with up to 52 digits of precision. New ANSI SQL numeric data types are DECIMAL (up to 52 significant digits), BIGINT (up to 19), DOUBLE (16), FLOAT (16), INTEGER (10), SMALLINT (5), or TINYINT (3). For other data types please refer to the DS2 documentation.

DS2 also runs in engines of hosting environments where SQL or DATA step do not, for example, in SAS Decision Services, SAS Micro Analytic Service, or SAS Event Stream Processing. But why do I mention DS2 in the context of PROC FEDSQL and PROC SQL? DS2 offers the charming option to write SQL code directly into so-called packages. For example, DS2 can accept the result set of an SQL query as input to the SET statement. The next subsection will provide a brief introduction to the structure of a DS2 program to facilitate the reading of the examples presented later.

Recent benchmark tests confirmed that DS2 multi-threading performs faster than a functionally equivalent, single-threaded DATA step in certain, very specific computationally intensive tasks, especially those in which the computation-to-observation ratio or computational complexity is high. However, in other instances, the DS2 procedure performs slower than an equivalent DATA step while consuming more system resources (Hughes, 2019). SAS users can't help but weigh up whether the potential performance gain justifies the training and rewriting of selected SAS programs in DS2 language.

The following subsections present:

- **Building blocks of DS2** (Subsection 7.3.1).
- **Applications of DS2** (Subsection 7.3.2) using higher-precision data types, threaded processing, PROC DSTODS2, and many more.

7.3.1 Building Blocks of DS2

A DS2 program is usually made of at least one of three program blocks. I call them building blocks. You need these building blocks depending on what you try to accomplish.

The building block PACKAGE is for programs that create variables and methods stored in SAS libraries. Packages are much like libraries for user-written methods. DS2 also provides pre-defined packages. The PACKAGE building block starts with PACKAGE and ends with ENDPACKAGE.

The building block THREAD manipulates input data sets to produce output result sets that are returned to a data program. THREAD processes data rows in parallel and assumes there are several cores available DS2 could use for parallel processing. The THREAD building block starts with THREAD and ends with ENDTHREAD.

The building block DATA manipulates input data sets to produce output result sets. They can accept input from tables, thread program result sets, or SQL query result sets. The DATA building block starts with DATA and ends with ENDDATA.

Program 7.3-1: Building Blocks of DS2

```
proc ds2;

/* Program Block PACKAGE */
   package package_name;
<DS2 programming statements to create the package here>
endpackage;
run;
```

```
/* Program Block THREAD */
thread thread_name;
<DS2 programming statements to create the thread here>
endthread;
run;
/* Program Block DATA */
data output_dataset_name;
<DS2 programming statements to process data here>
enddata;
     run;

quit ;
```

One of the most common statements used in the DS2 building blocks is the METHOD statement. The METHOD statement defines a block of code that can be called and executed multiple times. A method is similar to a SAS function or CALL routine in that it is a user-written method to perform a computation and return a value. Every DS2 program contains the INIT, RUN, and TERM methods. If you do not specify METHOD INIT or METHOD TERM statements, DS2 automatically provides them.

There are several similarities between the DATA step, DS2, PROC SQL, and PROC FEDSQL. Unfortunately, with similarities come also dissimilarities. For example, Jordan (2018) demonstrates the functional differences between the DATA step MERGE BY, the DS2 join, and the SQL join. (See Example 2.) Also good to know is that the DS2 languages usually uses single quotation marks; you only need double quotation marks to delimit an identifier.

7.3.2 Applications of DS2 Programs

This subsection presents several DS2 applications. Each of these examples highlights another important feature of DS2 programming, especially when it comes to FEDSQL:

- *Example 1:* SQL in DS2: Simple combining of two data sets (SET)
- *Example 2:* Differences in combining data sets (merge by, join)
- *Example 3:* Defining a method in DS2 and applying it in FEDSQL as a function
- *Example 4:* Writing an SQL query in a DS2 program (SET statement)
- *Example 5:* Writing a DS2 connection string and processing several DATA steps in one DS2 program
- *Example 6:* Submitting FEDSQL syntax using DS2 (SQLSTMT statement)
- *Example 7:* Get DS2 to access SAS dictionaries
- *Example 8:* DS2 and higher-precision data types
- *Example 9:* Threaded processing using DS2
- *Example 10:* Let SAS write DS2 programs (PROC DSTODS2)

Example 1: SQL in DS2: Simple Combining of Two Data Sets (SET)

The next two DATA steps create data sets LEFT and RIGHT for the programs under Example 1 and Example 2.

Program 7.3-2: Demo Data LEFT	**Program 7.3-3: Demo Data RIGHT**
``` data LEFT ; input ID A B C ; datalines; 01 1 2 3 02 1 2 3 03 99 99 99 04 1 2 3 05 1 2 3 ; run ; ```	``` data RIGHT ; input ID E F G ; datalines; 01 4 4 4 02 4 4 4 03 99 99 99 05 4 4 4 06 4 4 4 ; run ; ```

Please note the different entries under ID. They will result in differences in Example 2 about merging.

Program 7.3-4: Example 1: Simple Combining of Two Data Sets   **Figure 7.3-1: Output: Example 1**

```
proc DS2 ;
   data LEFT_RIGHT4 (overwrite=yes) ;
      method run();
      set { select LEFT.ID, LEFT.A, RIGHT.F
      from work.LEFT, work.RIGHT
      where LEFT.ID = RIGHT.ID } ;
      output ;
      end ;
   enddata ;
run ;
quit ;
```

ID	A	F
1	1	4
2	1	4
3	99	99
5	1	4

The DATA building block creates the output data set LEFT_RIGHT4, based on the two input data sets LEFT and RIGHT. From the bold lines of code in this DS2 example, you can easily see that the expression of the METHOD statement contains ANSI SQL in the braces ("{...}") after SET.

Isn't it nice to see how easily SQL code can be used in other SAS languages? The following example presents a DS2 version of a DATA step. The program structures do not look very different; the next example demonstrates that the functionalities in the background unfortunately do.

Why did I use "SQL" in the text although I state in the chapter title that it's actually FEDSQL? The book is about SQL (be it SAS or ANSI). I figure that it might make a transfer easier if you can start from something known. Also, the syntax outline (see Section 7.1.3) might have indicated that the PROC FEDSQL syntax is, roughly speaking, a subset of PROC SQL language elements.

Example 2: Differences in Combining Data Sets (Merge by, Join)

DATA step, DS2, SQL, and FEDSQL unfortunately differ in how they operate in merging data sets. (See Jordan, 2018; Chapter 9.4 for differences between DATA Step and SQL in updating.)

The following three programs merge LEFT and RIGHT using basic SQL, FEDSQL, and DS2 programming steps. For a change, the DS2 program contains DATA step code.

Program 7.3-5: Example 2: SQL Variant   **Figure 7.3-2: Output: Example 2: SQL**

```
proc sql ;
   create table LEFT_RIGHT  as
   select LEFT.ID, LEFT.A,RIGHT.F
   from LEFT full join RIGHT
      on LEFT.ID=RIGHT.ID ;
quit ;
```

ID	A	F
1	1	4
2	1	4
3	99	99
4	1	.
5	1	4
.	.	4

Program 7.3-6: Example 2: FEDSQL Variant   **Figure 7.3-3: Output: Example 2: FEDSQL**

```
proc fedsql ;
   drop table  LEFT_RIGHT2 force ;
   create table LEFT_RIGHT2  as
   select LEFT.ID, LEFT.A,RIGHT.F
   from LEFT full join RIGHT
      on LEFT.ID=RIGHT.ID ;
quit ;
```

ID	A	F
1	1	4
2	1	4
3	99	99
4	1	.
5	1	4
.	.	4

Program 7.3-7: Example 2: DS2 variant	Figure 7.3-4: Output: Example 2: DS2

```
proc ds2;
   data LEFT_RIGHT3
   (keep=(ID A F))/overwrite=yes ;
      method run() ;
      merge LEFT RIGHT ;
      by ID;
      end;
   enddata ;
   run ;
quit ;
```

ID	A	F
1	1	4
2	1	4
3	99	99
4	1	.
5	1	4
6	.	4

The results for SQL and FEDSQL match. The rudimentary DATA step variant in DS2 achieves a different result. However, not just the DS2 code generating LEFT_RIGHT3 needs to be checked, but also the requirements: which result corresponds to the specifications? Is it the results LEFT_RIGHT1 to LEFT_RIGHT3, or even LEFT_RIGHT4 of Example 1? Or another variety?

Example 3: Defining a Method in DS2 and Applying it in FEDSQL as a Function

Example 3 consists of three separate elements.

The DS2 program contains two of them: a PACKAGE and a DATA building block. The PACKAGE building block creates the package MY_DS2_PCKG. The PACKAGE building block also contains the instruction ("method") MY_DS2_SUM to add up the fields A, B, C, and to multiply C by 2 before. The DATA building block creates the temporary SAS data set MY_SASDATA. This data set contains the three fields A, B, and C (not yet multiplied by 2) and ten rows. Sample outputs before and after passing the FEDSQL program are shown after Program 7.3-8.

The FEDSQL program calls the method MY_DS2_SUM of the MY_DS2_PCKG package and uses it like a user-written function on A, B, and C, and creates SUM_ABC.

Program 7.3-8: Example 3: Defining a Method in DS2 and Applying It in FEDSQL

```
proc ds2;
   package MY_DS2_PCKG / overwrite =yes;
      method MY_DS2_SUM( double A, double B , double C) returns double;
         return A + B + C*2;
      end;
   endpackage;
   data MY_SASDATA / overwrite = yes;
      dcl double A B C;
      method init();
      dcl int i;
      do i = 1 to 10;
        A = i; B = i; C = i;
        output;
      end;
   end;
   enddata;
run;
quit;

proc fedsql;
   create table work.MY_DS2_RESULT as
   select A, B, C,
         work.MY_DS2_PCKG.MY_DS2_SUM( A, B, C ) as SUM_ABC
   from work.MY_SASDATA;
quit;
```

These sample outputs present the data set MY_SASDATA created by the DATA package (before entering the FEDSQL) on the left. On the right, you see the created table MY_DS2_RESULT after the method MY_DS2_SUM was applied to A, B, C of MY_SASDATA.

Figure 7.3-5: Output: Example 3: MY_SASDATA

A	B	C
1	1	1
2	2	2
3	3	3
4	4	4
5	5	5
6	6	6
7	7	7
8	8	8
9	9	9
10	10	10

Figure 7.3-6: Output: Example 3: MY_DS2_SUM Package: SUM_ABC=A+B+C*2

A	B	C	SUM_ABC
1	1	1	4
2	2	2	8
3	3	3	12
4	4	4	16
5	5	5	20
6	6	6	24
7	7	7	28
8	8	8	32
9	9	9	36
10	10	10	40

Example 4: Writing an SQL Query in a DS2 Program (SET Statement)

The following DS2 program impressively highlights the advantages of SQL. You can use SQL language after the SET statement as if you would write a query using the SQL or FEDSQL procedures. This DS2 program accesses the temporary SAS data set MY_DS2_RESULT created in Example 3 and simply keeps only rows where A is greater than 5. The result of the first DS2 variant is printed directly to the SAS output. The second DS2 variant creates a temporary SAS data set MY_DS2_RESULT2. If you want to simplify the second DS2 variant, you could even drop the DCL ("declare") line. The third DS2 variant below accesses SASHELP.CLASS; you will find details after that example. Make sure you use braces {...} after SET, otherwise DS2 will encounter a compilation error.

Program 7.3-9: Example 4: Writing an SQL Query: Printing to Output

```
proc ds2;
data;
   method run();
      set {select * from work.MY_DS2_RESULT where A > 5};
   end;
enddata;
run;
quit;
```

Figure 7.3-7: Output: Example 4: SAS Output and Log

A	B	C	SUM_ABC
6	6	6	24
7	7	7	28
8	8	8	32
9	9	9	36
10	10	10	40

Program 7.3-10: Example 4: Writing an SQL Query: Creating a Data Set

```
proc ds2;
data MY_DS2_RESULT2 / overwrite = yes;
dcl double A B C ABC_SUM ;
   method run();
      set {select * from work.MY_DS2_RESULT where A > 5};
   end;
enddata;
run;
quit;
```

The DS2 program below accesses the example data set CLASS in SAS 9.4, which is usually accessed by referring to SASHELP.CLASS. Because DS2 drivers cannot access preset LIBREFs like SASHELP or DICTIONARY, you could set your own LIBNAME (see "mybase) pointing to the storage location of the CLASS table. Example 5 illustrates another approach using a DS2 connection string. Example 7 demonstrates how to access SAS dictionaries.

Program 7.3-11: Example 4: Writing an SQL Query for Accessing a Preset LIBREF

```
libname mybase v9 'C:\Program Files\SASHome\SASFoundation\9.4\core\sashelp\';
proc ds2;
data MY_DS2_QUERY1 / overwrite = yes;
dcl double AGE HEIGHT WEIGHT ;
```

```
dcl char(1) SEX ;
dcl char(8) NAME ;
   method run();
       set {select AGE, HEIGHT, WEIGHT, SEX, NAME
                  from mybase.class
                  where AGE > 13 and
                        HEIGHT > 65.0};
   end;
enddata;
run;
quit;
```

The DS2 program above accesses MYBASE.CLASS (formerly known as SASHELP.CLASS), keeps the selected fields, assigns required formats, filters rows using two conditions, and creates the temporary table MY_DS2_QUERY1.

Example 5: Writing a DS2 Connection String and Processing Several DATA Steps in One DS2 Program

Another option to access tables residing in preset libraries like SASHELP or DICTIONARY is to copy them to user-defined libraries and access them with DS2 connection strings. Example 5 contains two programs. Using a single simple connection, Program 7.3-12 illustrates how to write a DS2 connection string. Program 7.3-13 will then demonstrate how to use a more complex DS2 connection string containing several "LIBREFs" and how to process several DATA steps in one DS2 program. In its simplest form, a DS2 connection statement looks like Program 7.3.12. In order to "translate" a LIBNAME statement into DS2 parameters, we need to specify it and submit it just one time, together with the MSGLEVEL system option, which prints the current active librefs into the SAS log. We copy the parameters from the log to specify our connection string.

Program 7.3-12: Example 5: Writing a DS2 Connection String
```
libname MY_DATA v9 "D:\SASTest";
options msglevel=i; proc ds2; quit;

proc ds2
      NOLIBS
      CONN="DRIVER=DS2;
          CONOPTS= ((DRIVER=BASE;
                      CATALOG=MY_DATA;
                      SCHEMA=(NAME=MY_DATA;
                      PRIMARYPATH={D:\SASTest}))))" ;
```

Let us look now at some of the details used in the DS2 connection string. NOLIBS turns off default data source connections like SASHELP or SASUSER. CONN= specifies the data source connection string. CONN= is supported only for SAS libraries. After CONOPTS= you specify your connection options. It's actually here where you paste the connection string from the SAS log; just watch out for the brackets and the quotation marks. DRIVER= specifies the BASE driver for Base SAS data. CATALOG= is an arbitrary identifier (name) for an SQL catalog ("MY_DATA") that groups logically related schemas. It does not have the same name as the folder (for example, "SASTest"). The connection contains a SCHEMA= specification named MY_DATA. NAME= is an alias for the physical location of the SAS library, similar to the SAS libref. MY_DATA was specified in the LIBNAME statement. PRIMARYPATH= specifies the path where we physically store our data.

Program 7.3-13 consists of two major parts: a DS2 connection string and a series of DATA steps. The example assumes that the PRDSALE table from SASHELP was copied to a folder that is referenced by "MY_DATA" (see the LIBNAME statement above), which will be used to connect to this table. This DS2 connection string might appear a bit more complex than the one above but this is just because of the six LIBREFS that it contains. However, their structure is uniform and easy to read. Please note that there is no DS2 link to SASHELP. At the beginning, we find a connection to our user-defined library of *permanent tables* (MY_DATA); toward the end we find a connection to a *temporary* library (WORK) of *temporary* tables. Both are bold in the DATA steps below. Please note that the prefix of the current SAS Temporary File folder is different on every computer and, more importantly, subject to change (see bold "**_TD6716**"). Depending on your requirements, you might want to change to a more dynamic, less hardcoded approach. After you specified your DS2 connection string, you do not need the above LIBNAME statement anymore; you can remove it from the SAS code.

Program 7.3-13: Example 5: Processing Several DATA Steps in One DS2 Program

```
proc ds2
        NOLIBS
        CONN="DRIVER=DS2;
              CONOPTS= (DRIVER=FEDSQL;
CONOPTS=
  ((DRIVER=BASE;CATALOG=MY_DATA;SCHEMA=(NAME=MY_DATA;PRIMARYPATH={D:\SASTest}));
   (DRIVER=BASE;CATALOG=MAPS;SCHEMA=(NAME=MAPS;PRIMARYPATH=
            {C:\Program Files\SASHome\SASFoundation\9.4\maps}));
   (DRIVER=BASE;CATALOG=MAPSSAS;SCHEMA=(NAME=MAPSSAS;PRIMARYPATH=
            {C:\Program Files\SASHome\SASFoundation\9.4\maps}));
   (DRIVER=BASE;CATALOG=MAPSGFK;SCHEMA=(NAME=MAPSGFK;PRIMARYPATH=
            {C:\Program Files\SASHome\SASFoundation\9.4\mapsgfk}));
   (DRIVER=BASE;CATALOG=SASUSER;SCHEMA=(NAME=SASUSER;PRIMARYPATH=
            {C:\Users\cfgsc\Documents\My SAS Files\9.4}));
   (DRIVER=BASE;CATALOG=WORK;SCHEMA=(NAME=WORK;PRIMARYPATH=
            {C:\Users\cfgsc\AppData\Local\Temp\SAS Temporary Files\_TD6716_MC-T410-1_
})))" ;

        data WORK.RAW_DATA (keep=(YEAR MONTH COUNTRY ACTUAL PREDICT ))
             WORK.AGG_DATA (keep=(COUNTRY N SUM_ACTUAL SUM_PREDICT ACTPRED_RATIO ))
                                                      /overwrite=yes;
             dcl float YEAR MONTH ACTUAL PREDICT  ;
             dcl char(10) COUNTRY ;
             dcl float N SUM_ACTUAL SUM_PREDICT ACTPRED_RATIO  ;
             method run();
                     set {select YEAR, MONTH, ACTUAL, PREDICT, COUNTRY
                                 from MY_DATA.prdsale} ;
                     by COUNTRY;
     /* (1) Output raw data RAW_DATA  */
                         if YEAR=1994 and
                            COUNTRY in ('U.S.A.')
                                then output WORK.RAW_DATA ;
     /* (2) Output aggregated data AGG_DATA  */
                         if first.COUNTRY then
                           do;
                           N=0; SUM_ACTUAL =0; SUM_PREDICT=0 ; ACTPRED_RATIO=0;
                           end;
                         N+1 ; SUM_ACTUAL+ACTUAL; SUM_PREDICT+PREDICT ;
                             ACTPRED_RATIO=(SUM_ACTUAL-SUM_PREDICT)/SUM_ACTUAL;
                         if last.COUNTRY
                             then output WORK.AGG_DATA ;
                 end;
             enddata;
        run;
quit;
```

When we submit Program 7.3-13, the first part, the DS2 connection string, establishes connections to libraries that the DATA steps need to access to read and write. The DATA steps require two libraries, WORK and MY_DATA. The FEDSQL code after SET accesses PRDSALE using MY_DATA. The second part, a series of DATA steps, first declares several columns to be queried and generated. METHOD RUN() calls a method where DS2 code runs in an implicit loop like in a DATA step. Typically, the RUN method contains the main DS2 program code, DATA step, or SQL. After SET follows some FEDSQL code. BY sorts the data by COUNTRY. The first IF-THEN DO clause creates a subset of raw data (RAW_DATA). The second IF-THEN DO clause creates an aggregated data set (AGG_DATA) grouped by COUNTRY using the sum statement and FIRST/ LAST. Here are the contents of the aggregated data set AGG_DATA.

Figure 7.3-8: Output: Example 5: Contents of AGG_DATA

COUNTRY	N	SUM_ ACTUAL	SUM_ PREDICT	ACTPRED_ RATIO
CANADA	480	246990	233019	0.056565
GERMANY	480	245998	231554	0.058716
U.S.A.	480	237349	241722	-0.018424

Example 6: Submitting FEDSQL Syntax Using DS2 (SQLSTMT Statement)

Only FedSQL statements can be used with the SQLSTMT package. DBMS-specific SQL cannot be used. The SQLSTMT package provides methods for passing dynamic FedSQL statements to a database for execution. The SQLSTMT package follows the "bring the CODE to the DATA" scenario, not "bring the DATA to the CODE," and transfers, stores, or executes the FEDSQL statement in the database. For an appropriate LIBNAME, please see Examples 4 and 5.

Program 7.3-14: Example 6: Submitting FEDSQL Syntax Using DS2 (SQLSTMT Statement)

```
proc ds2 ;
data _null_ ;
dcl package sqlstmt
    MY_DS2_QUERY('select AGE, HEIGHT, WEIGHT, SEX, NAME
                 from mybase.class
                 where AGE > 13 and
                     HEIGHT > 65.0' );
dcl double AGE HEIGHT WEIGHT ;
dcl char(1) SEX ;
dcl char(8) NAME ;
method init();
      MY_DS2_QUERY.execute();
      MY_DS2_QUERY.bindresults([AGE HEIGHT WEIGHT SEX NAME ]);
do while (MY_DS2_QUERY.fetch() = 0);
   put AGE= HEIGHT= WEIGHT= SEX= NAME= ;
end;
end;
enddata;
run;
quit;
```

In the DCL PACKAGE statement, the "variable" MY_DS2_QUERY is specified after SQLSTMT. The content of the MY_DS2_QUERY variable is a string containing FEDSQL statements. The next DCL specifies the columns to be queried from MYBASE.CLASS. After INIT(), the EXECUTE method executes the MY_DS2_QUERY query. The BINDRESULTS method binds a list of variables to the columns of the result set of the FEDSQL statements. In the DO WHILE loop, the FETCH method fetches the next data row from the result set of the FEDSQL statement and puts the data rows that meet the query conditions into the SAS log.

SAS Log 7.3-1: Output Example 6: Submitting FEDSQL Syntax Using DS2 (SQLSTMT Statement)

```
AGE=14  HEIGHT=69    WEIGHT=112.5 SEX=M  NAME=Alfred
AGE=15  HEIGHT=66.5 WEIGHT=112   SEX=F  NAME=Mary
AGE=16  HEIGHT=72    WEIGHT=150   SEX=M  NAME=Philip
AGE=15  HEIGHT=67    WEIGHT=133   SEX=M  NAME=Ronald
AGE=15  HEIGHT=66.5 WEIGHT=112   SEX=M  NAME=William
```

Example 7: Get DS2 to Access SAS Dictionaries

FEDSQL cannot access preset LIBREFs like DICTIONARY. In the following example, we use a DS2 approach to access the DICTIONARY folder. There is nothing new in this DS2 code. We use the RUN() method, some code to be run, a WHERE condition, and we also know that the code in braces is in fact FEDSQL, not SQL.

Program 7.3-15: Example 7: Get DS2 to Access SAS Dictionaries

```
proc ds2;
   data MY_DS2_DIC_TBLS (overwrite=yes) ;
     method run();
     set {select *
             from dictionary.tables
          where TABLE_NAME like 'GEO%'} ;
     end;
   enddata ;
run;
quit;

proc print data= MY_DS2_DIC_TBLS ;
run ;
```

What might be new, however, is something in the background: the FEDSQL code in PROC DS2 does not access the DICTIONARY tables we already know from working with PROC SQL. (See Section 9.2.) They are FEDSQL DICTIONARY tables and have different structures and contents. The contents of the FEDSQL table DICTIONARY.TABLES is shown in Figure 7.3-9.

Figure 7.3-9: Output: Example 7: Contents of the FEDSQL Table DICTIONARY.TABLES

CAT	TABLE_ SCHEM	TABLE_ TABLE_NAME	TABLE_ TYPE	REMARKS	NATIVE_ CAT
MY_DS2	MY_DS2	GEOEXM	TABLE		MY_DS2
MY_DS2	MY_DS2	GEOEXP	TABLE		MY_DS2
MY_DS2	MY_DS2	GEOEXS	TABLE		MY_DS2
MAPS	MAPS	GEORGIA	TABLE		MAPS
MAPS	MAPS	GEORGIA2	TABLE		MAPS
MAPSSAS	MAPSSAS	GEORGIA	TABLE		MAPSSAS
MAPSSAS	MAPSSAS	GEORGIA2	TABLE		MAPSSAS
MAPSGFK	MAPSGFK	GEORGIA	TABLE		MAPSGFK
MAPSGFK	MAPSGFK	GEORGIA_ATTR	TABLE		MAPSGFK

To achieve a comparable SQL result, you would need to rewrite the DS2 into the PROC SQL program shown in Program 7.3-16. Make sure to use MEMNAME in SQL instead of TABLE_NAME in FEDSQL. The SAS output would contain 9 rows, but 41 columns compared to the 6 of FEDSQL.

Program 7.3-16: Example 7: PROC SQL Variant of the FEDSQL Dictionary Query

```
proc sql inobs=10;
    select *
            from DICTIONARY.TABLES
            where MEMNAME like 'GEO%'  ;
quit;
```

Example 8: DS2 and Higher-Precision Data Types

In SAS, the default data type for all numeric values in DS2 is DOUBLE. You can use the following DS2 program as a utility to check whether SAS reads and writes digit sequences accurately. (See Subsection 7.1.1 about precision.) The following digit sequences were taken from Mohammed et al. (2015).

Program 7.3-17: Example 8: DS2 and Higher-Precision Data Types

```
proc ds2 ;
   data _null_ ;
         method init() ;
      dcl real A ;
      dcl double B ;
      dcl decimal(30,5) C ;
    A=4611686018427387904n ;
    B=2222222222222222222n ;
    C=2312346890765434687n ;
put A=; put B= ; put C= ;
   end ;
enddata ;
run;
quit;
```

SAS Log 7.3-2: Log for REAL, DOUBLE, and DECIMAL

```
A=4611686018427387904.00000
B=2222222222222222336
C=2312346895369895936
NOTE: Execution succeeded. No rows affected.
```

SAS Log 7.3-3: Log for BIGINT

```
A=4611686018427387904
B=2222222222222222222
C=2312346890765434687
NOTE: Execution succeeded. No rows affected.
```

The n suffix declares a numeric constant using the higher-precision NUMERIC data type. The first SAS log shows the effect of the data types REAL, DOUBLE, and DECIMAL. The second SAS log shows the result of using only the BIGINT data type. Values with a smaller number of digits (16 or 17) do not require the n suffix.

Example 9: Threaded Processing Using DS2

DS2 is multi-threaded by nature, which means you can submit DS2 code as a program but also as a thread. In threaded processing, data are processed in parallel, which should give you massive performance gains. If you want to process data in multi-threading, you need to do two steps: First, you need to define a thread. (See "Step 1".) A thread is defined by the THREAD statement and ended by ENDTHREAD statement. Second, you need to apply that thread by using the DCL THREAD statement and execute it using the SET FROM statement. (See "Step 2".)

The example below defines the thread MY_DS2_THREAD in the first DS2 program. The example uses the SASHELP data set BWEIGHT after having connected via a DS2 driver before (see above) and creates some flags. The created thread MY_DS2_THREAD is stored in the data set WORK.MY_DS2_THREAD. The second DS2 program accesses WORK.MY_DS2_THREAD and applies the thread MY_DS2_THREAD (SET FROM). After THREAD=*n*, you can set the maximum number of threads DS2 should try to use. Vary this parameter as needed.

This two-step program has been intentionally kept simple to emphasize the steps required. The performance gains are most apparent with data volumes of several million data rows.

Program 7.3-18: Example 9: Threaded Processing Using DS2

```
/* Step 1: Define the thread MY_DS2_THREAD */
proc ds2 ;
thread MY_DS2_THREAD /overwrite=yes ;
   method run() ;
   DCL integer CHECK_ME i ;
   set MY_DS2.BWEIGHT ;
     CHECK_ME = 0;
     if MOMSMOKE = 1 and WEIGHT < 3000 then CHECK_ME = 1 ;
     else if CigsPerDay > 20 and WEIGHT < 3000 then CHECK_ME = 1 ;
     if CHECK_ME then output ;
   end ;
endthread;
run ;
run ;quit ;

/* Step 2: Apply the thread */
proc ds2 ;
   data DS_RESULTS /overwrite=yes ;
```

```
dcl thread WORK.MY_DS2_THREAD MY_DS2_THREAD ;
   method run() ;
set from MY_DS2_THREAD threads=4 ;
end ;
enddata ;
run ;
quit ;
```

Example 10: Let SAS Write DS2 Programs (PROC DSTODS2)

For beginners in the DS2 language, SAS offers the option to "translate" programs from the DATA step to DS2. After the IN=argument, PROC DSTODS2 just requires an input file (and a path) containing the source code to be translated, for example, FEDSQL2DS2.sas. After the OUT= argument, specify the location (path) and name of an output where the DS2 code is to be written. This output file doesn't need to exist; PROC DSTODS2 will create it.

The following PROC DSTODS2 lines will be submitted two times. The first time, the translation of the code fails. The second time, the translation of the code is successful. The difference between the two times is the contents of the input file FEDSQL2DS2.sas. The first example contains a few lines of PROC SQL code. (See Program 7.3-19.) The successful second example in Program 7.3-20 contains some DATA step lines.

```
proc dstods2 in="D:\FEDSQL2DS2.sas" out="D:\PROCSQLinDS2.ds2";
run ;
```

In the first example in Program 7.3-19, the translation of the SQL code into DS2 fails. The explanation is simple: PROC DSTODS2 does not work for other procedures like PROC SQL. Unfortunately, PROC DSTODS2 also does not support several statements including ATTRIB, DATALINES, DELETE, MODIFY, UPDATE, or WHERE.

Program 7.3-19: Example 10: Contents of Unsuccessful Example	**SAS Log 7.3-4: Output: Example 10**

Program 7.3-19: Example 10: Contents of Unsuccessful Example

```
proc sql ;
   create table MYDATA as
   select REGION, SUBSIDIARY,
         PRODUCT, SALES,
(SALES/sum(SALES)*100) as PROZENT
   from SASHELP.SHOES
      where REGION="Africa" and
            SUBSIDIARY="Addis Ababa"
      order by PROZENT desc ;
quit ;
```

SAS Log 7.3-4: Output: Example 10

```
2     proc sql ;
      ----
      180
3       create table MYDATA as select
        ------
        180
ERROR: Could not compile source code.
9     quit;
      ----
      180
ERROR 180-322: Statement is not valid or it
is used out of proper order.

NOTE: The SAS System stopped processing this
step because of errors.
NOTE: PROCEDURE DSTODS2 used (Total process
time) :
           real time         7:15.92
           cpu time          12.60 seconds
```

In Program 7.3-20, the translation of the DATA step code into DS2 is successful. On the left, you find the original DATA step code. On the right, the generated code. You just need to adjust the code a little bit and then it is ready to run in DS2. In Figures 7.3-10 and 7.3-11, you find the adjusted DS2 code and the output generated, the table BOYS.

Program 7.3-20: Example 10: Contents of Successful Example: IN (FEDSQL2DS2.sas, DATA Step)

```
data BOYS GIRLS ;
 set mybase.class;
     select (SEX) ;
       when ('M')
           output BOYS ;
       when ('F')
           output GIRLS ;
     otherwise;
       end;
run;
```

Figure 7.3-10: Output: Example 10: Contents of Successful Example: OUT (PROCSQLinDS2.ds2, DS2 Step)

```
data BOYS GIRLS;
method run();
  set MYBASE.CLASS;
     select (SEX);
       when ('M')
           output BOYS ;
       when ('F')
           output GIRLS ;
     otherwise ;
       end ;
         ;
_return: ;   ...
```

Program 7.3-21: Example 10: Enhanced DS2 Step Version

```
proc ds2;
   data BOYS GIRLS ;
   dcl double AGE ;
    dcl double HEIGHT ;
    dcl double WEIGHT ;
    dcl char(8) NAME ;
    dcl char(1) SEX ;
    method run();
   set MYBASE.CLASS;
      select (SEX);
        when ('M') output BOYS;
        when ('F') output GIRLS;
   otherwise ;
     end;
     end;
   enddata;
run;
quit;
```

Figure 7.3-11: Output: Example 10: Contents of Enhanced DS2 Step Version

Name	Sex	Age	Height	Weight
Alfred	M	14	69.0	112.5
Henry	M	14	63.5	102.5
James	M	12	57.3	83.0
Jeffrey	M	13	62.5	84.0
John	M	12	59.0	99.5
Philip	M	16	72.0	150.0
Robert	M	12	64.8	128.0
Ronald	M	15	67.0	133.0
Thomas	M	11	57.5	85.0
William	M	15	66.5	112.0

Usually, the resulting DS2 program might not be syntactically complete or working at all. The code in Program 7.3-21 required only minor changes to the original code. Now a "PROC DS2" precedes the code. The data is followed by a series of DCLs. (You could also omit them.) I also removed the superfluous semicolon and the following _return:, and added a RUN; at the end. If there were comments, PROC DSTODS2 would remove them. PROC DSTODS2 usually compiles and executes without error, especially if you use new DATA step syntax and as few comments and commented-out program sections as possible. The above example was intentionally made somewhat simple. PROC DSTODS2 has a few limits that you need to be aware of:

- PROC DSTODS2 does not support all DATA step statements, especially older statements.
- PROC DSTODS2 does not support other procedures such as PROC SQL.
- PROC DSTODS2 might even fail to compile and to execute for certain functions and formats.

7.4 FEDSQL Syntax (Details)

Section 7.4 gives you an overview of the PROC FEDSQL syntax of SAS 9.4.

- **Subsections 7.4.1 and 7.4.2** present an outline of FEDSQL syntax and a brief overview of FEDSQL statements. As you will see, many SQL statements and language elements can also be used in FEDSQL expressions for SAS and CAS.

- **Subsections 7.4.3 and 7.4.4** provide descriptions of connection and processing options.
- **Subsections 7.4.5 and 7.4.6** list details of the CONN= and CNTL options that might be relevant for connecting to data sources like Amazon Redshift, Google BigQuery, Greenplum, Microsoft SQL Server, Netezza, PostgreSQL, Salesforce, SAP HANA, SAP IQ, Snowflake, Yellowbrick, and so on.

7.4.1 Outline of FEDSQL Syntax

This subsection gives a simplified overview of the FEDSQL syntax. The numbers in brackets refer to sections where you can find more detailed information.

PROC FEDSQL Outline

PROC FEDSQL

    <Connection options> [see 7.4.3 and 7.4.5]
    <Processing options> [see 7.4.4 and 7.4.6] ;

  [FEDSQL statements, see 7.4.2] ;

QUIT ;

7.4.2 Short Overview of FEDSQL Statements

This subsection gives a simplified overview of the FEDSQL syntax of version SAS 9.4 and SAS Viya 3.5. As you will see, many SQL statements and language elements can also be used in FEDSQL expressions for SAS and CAS. The following FEDSQL statements are supported in SAS; the shorter list below is supported in CAS.

FEDSQL Statements Supported in SAS 9.4

  ALTER TABLE

  BEGIN

  COMMIT

  CREATE INDEX | TABLE| VIEW

  DELETE

  DESCRIBE TABLE | VIEW

  DROP INDEX | TABLE | VIEW

  EXECUTE

  INSERT

  ROLLBACK

  SELECT

  UPDATE

The FEDSQL statements available for CAS server are a subset of those available for SAS.

> **CREATE TABLE** with the **AS** query expression
>
> **SELECT**
>
>> **UNION** (SAS Viya 3.5 and later)
>
> **DROP TABLE**

At the moment, the following SELECT statement features are **not** supported in CAS:

> **EXCEPT**
> **INTERSECT**
> correlated subqueries (however, see Example 6 in Subsection 7.2.3)
> dictionary queries
> views
> DS2 user-defined functions (package expressions).

7.4.3 Short Description of FEDSQL Connection Options

CONN= "connection-string"
> Specifies a string for a data source connection. CONN= is supported only for SAS libraries. If both CONN= and SESSREF= (or SESSUUID=) are specified in the procedure statement, CONN= is ignored. If you want to limit your request to a specified data source, use LIBS= instead of CONN=. If both CONN= and LIBS= are specified, the last option on the procedure statement is applied.

LIBS= libref | (libref1libref2 ...librefn)
> Restricts the default data source connection to the specified libref(s). All other librefs are ignored. LIBS= is useful when multiple LIBNAME statements are defined in a SAS session, to limit the scope of the FEDSQL request to only the LIBNAME statement(s) to which the request applies. It can also avoid duplicate catalog errors when connecting to data sources that support native catalogs, such as Netezza. LIBS= is supported only for SAS libraries.

NOLIBS
> Turns off the default data source connection. NOLIBS is intended for use with CONN=. Using NOLIBS with CONN= overrides the default data source connection with the specified connection string. NOLIBS is supported only for SAS libraries.

SESSREF=
> Identifies the CAS session by its session name. A CAS server must be configured for your system. This option is supported in SAS Viya 3.1 and later, and in SAS 9.4M5 and later.

SESSUUID= "session-uuid"
> Identifies the CAS session by its universally unique identifier (UUID). A CAS server must be configured for your system. This option is supported in SAS Viya 3.1 and later, and in SAS 9.4M5 and later. If both SESSREF= or SESSUUID= and **LIBS=** (or CONN=) are specified on the procedure statement, SESSREF= or SESSUUID= is applied, and LIBS= is ignored.

7.4.4 Short Description of FEDSQL Processing Options

_METHOD

Generates a brief text description of the nodes and stages in the query plan. The information is written to the SAS log. _METHOD is supported on the CAS server only.

_POSTOPTPLAN

Prints an XML tree that illustrates each stage of the FEDSQL query plan and the results from each execution stage. _POSTOPTPLAN is supported on the CAS server only.

ANSIMODE

Specifies that nonexistent values in CHAR and DOUBLE columns are processed as ANSI SQL null values. By default, PROC FEDSQL processes nonexistent values in CHAR and DOUBLE columns as missing values. ANSIMODE is supported for SAS libraries only.

AUTOCOMMIT | NOAUTOCOMMIT

Specifies whether updates are automatically committed (that is, saved to a table) after a default number of rows are updated and whether rollback is available. NOAUTOCOMMIT is supported for SAS libraries only.

CNTL= (*parameter***)**

Specifies optional control parameters for FEDSQL query planning and query execution in CAS. Multiple parameters are allowed inside of the parentheses. See supported parameters and details in Subsection 7.4.6.

ERRORSTOP | NOERRORSTOP

Specifies whether the procedure stops executing if it encounters an error. In a batch or noninteractive session, **ERRORSTOP** (batch or noninteractive session) instructs the procedure to stop executing the statements but to continue checking the syntax after it has encountered an error. **NOERRORSTOP** (interactive session) instructs the procedure to execute the statements and to continue checking the syntax after an error occurs. NOERRORSTOP is useful if you want a batch job to continue executing SQL procedure statements after an error is encountered.

EXEC | NOEXEC

Specifies whether a statement should be executed after its syntax is checked for accuracy. EXEC specifies to execute the statement. NOEXEC specifies not to execute the statement.

LABEL | NOLABEL

Specifies whether to use the column label or the column name as the column heading.

MEMSIZE= *n* | *n*M | *n*G

Specifies a limit for the amount of memory that is used for an underlying query (such as a SELECT statement), so that allocated memory is available to support other PROC FEDSQL operations. Generally, specifying a memory limit is not necessary unless FedSQL reports a memory problem error.

NOPRINT

Suppresses the normal display of results. NOPRINT affects the value of the SQLOBS automatic macro variable, which contains the number of rows that are executed by a statement.

NUMBER

Specifies to include a column named Row, which is the row (observation) number of the data as the rows are retrieved.

STIMER
> Specifies to write a subset of system performance statistics, such as time-elapsed statistics, to the SAS log. Default: No performance statistics are written to the SAS log.

XCODE= ERROR (default)| WARNING | IGNORE
> Controls the behavior of the SAS session when an NLS transcoding failure occurs.

7.4.5 Attributes for the **CONN= Option** (Connection Option)

The data source connection string is added to the default data source connection, unless you specify NOLIBS with CONN=. The data source connection string can have the following attributes. The CONN= attributes are data source dependent.

API_TRACE= YES | NO (default)
> Specifies to trace API usage by your FEDSQL program. Salesforce limits the number of API calls that are allowed per organization. When you specify API_TRACE=YES, additional messages are written to the log that report the number of API calls used from the allocated call limit. Notes are also written that include the exact SOQL that was sent to Salesforce.

AUTHENDPOINT= "string"
> Specifies a URL that represents the authorization end point that is used to authenticate a user's Salesforce account.

CATALOG= catalog-identifier;
> Specifies an arbitrary identifier for an SQL catalog, which groups logically related schemas. A catalog name can be up to 32 characters long. Google BigQuery connects with catalog=* by default.

CLASSPATH= "path-to-JDBC-driver-class";
> Specifies the path to the JAR files that are used by the interface. Use the conventions of the host environment. Multiple locations should be separated with a colon (:) for UNIX or a semicolon (;) for Windows. Enclose the string in single or double quotation marks.

CLIENT_ID= {*ID*}
> Specifies the ID for the OAuth client for which you are obtaining a refresh token.

CLIENT_SECRET= value
> Specifies the secret value for the OAuth client identified in CLIENT_ID=.

CONOPTS= (database-connection-string);
> Specifies an ODBC-compliant database connection string using ODBC-style syntax. These options must specify a complete connection string to the data source. Data sources: Amazon Redshift, Greenplum, Microsoft SQL Server, Netezza, ODBC, PostgreSQL, SAP HANA, SAP IQ, Snowflake, Yellowbrick.

CRED_PATH= "path and filename"
> Specifies the path to a credentials file that enables authentication to Google Cloud Platform. Data sources: Google BigQuery.

DATABASE= database-name ;
> Specifies the database to which you want to connect. Data sources: Amazon Redshift, Greenplum, Microsoft SQL Server, Netezza, ODBC, PostgreSQL, SAP HANA, SAP IQ, Snowflake, Yellowbrick.

DBMAX_TEXT= integer
> Specifies the maximum length to be used for STRING columns. The default DBMAX_TEXT= in Hive is 32,768 characters. The default DBMAX_TEXT= in Spark is 1,024 characters. Data sources: Hive, Spark.

DBQ= 'domain'

Specifies the name of an existing SPD Server domain to which the connecting user has access. Data source: SPD Server.

DRIVER= driver name ;

Specifies the data source that you want to connect to. Data source: All.

DRIVERCLASS= driver class ;

Specifies the driver class name. DRIVERCLASS enables you to specify a driver other than the default driver. SAS/ACCESS software uses open source drivers where possible. Data sources: Hive, JDBC.

DSN= data source identifier ;

Specifies the data source name to which you want to connect. Data sources: Greenplum, Netezza, PostgreSQL, Snowflake, Yellowbrick.

HD_CONFIG= 'path' ;

Specifies the path to a Hadoop configuration file. When the SAS_HADOOP_CONFIG_PATH environment variable is set, there is no need to set this option.

HOST= 'host name' ;

Specifies the host name or IP address of the computer hosting the SPD Server name server. Data source: SPD Server.

LOCALE= 'locale name' ;

Specifies a name that invokes a set of attributes that reflect the language, local conventions, and culture for a geographical region for the SPD Server session. A LOCALE= specification is required. Data source: SPD Server.

MAX_BINARY_LEN= value (default: 2.000)

Specifies the maximum length in bytes to allocate for binary columns. Binary data in Google BigQuery is stored in BYTES columns, which do not have a fixed length. Data source: Google BigQuery.

MAX_CHAR_LEN= value (default: 16.384)

Specifies the maximum number of characters to allocate for character-based columns. Specify a length that is long enough to avoid truncation. Data source: Google BigQuery. Use NUMBYTESPERCHAR= with MAX_CHAR_LEN= when the client encoding is UTF-8.

NUMBYTESPERCHAR= maximum client bytes

Specifies the maximum number of bytes per single character in the database client encoding, which matches the SAS encoding. This multiplying factor is used to adjust column lengths for STRING columns. In Google BigQuery, when a SAS UTF-8 encoding is used, the length of STRING variables is 4 times the length of the default character length, or 4 times the length that is specified in the MAX_CHAR_LEN= connection attribute. That is, when MAX_CHAR_LEN=8 is specified, a length of 32 bytes of storage is used. Data source: Google BigQuery.

ODBC_DSN= ODBC dsn name ;

Specifies a valid ODBC-compliant database DSN that contains connection information for connecting to the ODBC-compliant database. Data source: ODBC.

PASSWORD= password ;

Specifies an optional password that is associated with the specified database or server user ID. The password is case sensitive. Data sources: SPD Server, Amazon Redshift, Hive, JDBC, MongoDB, Microsoft SQL Server, Oracle, PostgreSQL, Salesforce, SAP HANA, Snowflake, Spark, Vertica, Yellowbrick.

PATH= database specification ;
> Specifies the Oracle connect identifier. A connect identifier can be a net service name, a database service name, or a net service alias. Data source: Oracle.

PORT= port number ;
> Specifies the listen port of the server where the database resides. Data sources: SPD Server, Amazon Redshift, Hive, MongoDB, Netezza, PostgreSQL, SAP HANA, Snowflake, Spark, Vertica, Yellowbrick.

PROJECT= Google Cloud Platform project ID'
> Specifies the project ID for a Google Cloud Platform project. Data source: Google BigQuery.

PROPERTIES=
> Specifies Hive configuration variables for this session. Data sources: Hive, Spark.

PROXY= "URL" ;
> Specifies the URL of a proxy server that is used to connect to Google BigQuery. Data source: Google BigQuery.

PROXY_HOST= "host name" ;
> Specifies the host name or IP address of a proxy server that is used to connect to Salesforce. Data source: Salesforce.

PROXY_PASS= "password" ;
> Specifies the password that is associated with PROXY_USER=. Data source: Salesforce.

PROXY_PORT= port number (default: 8080)
> Specifies the port number of the proxy server. Data source: Salesforce.

PROXY_USER= "user id"
> Specifies the user ID that is used to access the proxy server (optional). Data source: Salesforce.

SAPHANA_DSN
> Specifies a configured SAP HANA ODBC data source to which you want to connect. You must have existing SAP HANA ODBC data sources configured on your client. Data source: SAP HANA.

SCHEMA= value
> Specifies a SCHEMA= in which to create or read data. SCHEMA= is specified when connecting to Base SAS, SPD Server, Google BigQuery, Hive, JDBC, Snowflake, Spark, and Vertica. When connecting to Base SAS data, value must be in the form: (*NAME*=value; *PRIMARYPATH*=value). **NAME=** specifies an arbitrary identifier for an SQL schema. **PRIMARYPATH=** specifies the physical location for the SAS library. In most operating environments, this is a directory path. **HOST=** Specifies the host name or IP address of the computer hosting the SPD Server name server (optional). **SERV=** Specifies the listen port for the SPD Server (optional). **NAME=** specifies an arbitrary identifier for an SQL schema. A schema name must be a valid SAS name and can be up to 8 characters long. **DBQ=** specifies the name of an existing SPD Server domain to which the connecting user has access. For data source specific details, please see the SAS 9.4 and SAS Viya 3.5 documentation.

SCHEMA_COLLECTION= 'collection name'
> Specifies the MongoDB collection where the schema file is stored. There can be only one schema per collection. As a best practice, provide a descriptive name for your schema. Schema names are case-sensitive. Multiple users can access the same stored schema by specifying the same schema name. Data source: MongoDB.

SCHEMA_DB= 'database name'

> Specifies the MongoDB database where the schema file is stored, if it is different than the MongoDB database where the data resides. Data source: MongoDB.

SCHEMA_PWD= 'password'

> Specifies the password that is associated with SCHEMA_UID=. The password is case sensitive. Data source: MongoDB.

SCHEMA_PORT= port number

> Specifies the listen port of the MongoDB server where the schema file is stored, if it is different than the listen port of the MongoDB server where the data resides. Data source: MongoDB.

SCHEMA_SERVER= 'server name'

> Specifies the name of the MongoDB server where the schema file is stored, if it is different than the MongoDB server where the data resides. Data source: MongoDB.

SCHEMA_UID= 'user id'

> Specifies the MongoDB database or server user ID that is used to access the stored schema file, if it is different from the user ID that was used to access the MongoDB database or server where the data resides. If the user ID contains blanks or national characters, enclose them in quotation marks. Data source: MongoDB.

SERV= 'port number' ;

> Specifies the initial listen port for the SPD Server. Data source: SPD Server.

SERVER= server-name ;

> Specifies the name of the server where the database resides. Data source: Amazon Redshift, Greenplum, Hive, MongoDB, Netezza, PostgreSQL, SAP HANA, Snowflake, Spark, Teradata, Vertica, Yellowbrick.

SHOW_METADATA= YES | NO (default)

> Specifies to display metadata tables and columns. By default, metadata tables are not displayed by SAS. Data source: Salesforce.

SHOW_RECYCLED= YES | NO (default)

> Specifies to include rows that are marked for deletion. By default, rows that are marked for deletion are not displayed by SAS. Data source: Salesforce.

TOKEN= {value}

> Specifies a refresh token for connecting to Google BigQuery. Data source: Google BigQuery.

TPT_DATA_ENCRYPTION= NO (default) | YES

> Sets the TPT_DATA_ENCRYPTION session attribute for connections that the Teradata TPT (Teradata Parallel Transporter) creates for the LOAD, UPDATE, and STREAM operations. This attribute encrypts all SQL requests, responses, and data. Can be overridden with the TD_DATA_ENCRYPTION= table option. Data source: Teradata.

TPT_UNICODE_PASSTHRU= NO (default) | YES

> Sets the session attribute TPT_UNICODE_PASSTHRU= session attribute for connections that the Teradata TPT (Teradata Parallel Transporter) creates for the LOAD, UPDATE, and STREAM operations. Data source: Teradata.

URI= "JDBC connection string"

> Specifies the JDBC connection string as a URL. URI= enables you to explicitly provide the connection string for complex environments where automatic generation of a URL is problematic. The format of the URL is specific to the database. Data source: Hive, JDBC, Spark.

USE_INFORMATION_SCHEMA= YES (default) | NO

By default, SAS/ACCESS to Google BigQuery uses this argument to get metadata for the list of schemas and tables or views, if the names contain a pattern. Disable INFORMATION_SCHEMA to improve performance. Data source: Google BigQuery.

USE_NATIVE_NAMES= YES | NO (default)

In Salesforce, custom table and column names have the suffix "__c". By default, SAS strips this suffix and the suffix must be omitted in queries, unless the suffix removal causes a duplicate name. To display (and use in queries) the full, suffixed names, set USE_NATIVE_NAMES=YES. Data source: Salesforce.

USER= user id ;

Optionally specifies a database or server user ID. If the user ID contains blanks or national characters, enclose it in quotation marks. Data sources: SPD Server, Amazon Redshift, Hive, JDBC, MongoDB, Microsoft SQL Server, Oracle, PostgreSQL, Salesforce, Snowflake, Spark, Vertica, Yellowbrick.

7.4.6 Parameters for the **CNTL= Option** (Processing Option)

Multiple parameters are allowed inside of the CNTL= parentheses. Each parameter is separated with a space.

DISABLEPASSTHROUGH

Disables implicit FedSQL pass-through in CAS. This option can save processing time for FedSQL requests that contain functions that are specific to SAS, or whose tables have already been loaded into the CAS session.

DYNAMICCARDINALITY

Instructs the FEDSQL query planner to perform cardinality estimations before selecting a query plan. Cardinality estimation can improve the accuracy of selectivity estimates for join conditions and WHERE clause predicates, which in turn might lead to better join-order decisions and faster query execution times for some queries.

OPTIMIZEVARBINARYPRECISION

Optimizes VARBINARY precision by using a precision that is appropriate to the actual data, instead of the precision declared for the VARBINARY columns. The greatest benefits from this option are achieved when the declared precision is far larger than the precision of the actual data.

OPTIMIZEVARCHARPRECISION

Optimizes VARCHAR precision by using a precision that is appropriate to the actual data, instead of the precision declared for the VARCHAR columns. The greatest benefits from this option are achieved when the declared precision is far larger than the precision of the actual data.

PRESERVEJOINORDER

Joins tables in the specified order instead of an order chosen by the FEDSQL query optimizer.

REQUIREFULLPASSTHROUGH

Stops processing the FedSQL request when implicit pass-through of the full query cannot be achieved.

SHOWSTAGES

Writes query execution details to the SAS log. This includes the information returned by the _METHOD option. In addition, showStages prints the stage query, the number of SQL threads used by each stage, and other execution details such as time for each intermediate stage. Do not specify NOEXEC with showStages.

7.5 Summary: When to Use FedSQL and When to Use SQL

Talking about FedSQL also includes the options to use it as a procedure (PROC FEDSQL), as an action (fedsql. execdirect), and also in PROC DS2. But depending on your requirements, PROC SQL might be perfect because of its "SAS extras." No matter where you come from, FEDSQL or SQL, the other ANSI variant has so much to offer.

Table 7.5-1: When to Use PROC FEDSQL and When to Use PROC SQL

Questions to Ask Yourself	SQL	FEDSQL
Want to learn that programming language?	●	●
Work with data with more than 16 digits?		●
Work with big data?		●
Work with ANSI-1999 compliant data sources like Google, Amazon, or Salesforce?		●
Want to do federated queries (connect to multiple sources in one query)?	●	●
Want to use new ANSI data types?		●
Want to work in a cloud (CAS) environment, prepare to, or already do?		●
Want to take advantage of in-memory processing?		●
Want to process data by fully supported multi-threading (SMP, MPP)?		●
Want to use pass-throughs (implicit, explicit)?	●	●
Work with nonrelational data sources like MongoDB?		●
Work with a vendor-neutral than a SAS-specialized SQL variant?		●
Need to write complex programs?		●
Are you fine with PROC FEDSQL "limitations" compared to SQL (see next question)?		●
Want to expand from PROC FEDSQL to PROC CAS and DS2?		●
Want to use SAS specifics of ANSI-1992 like remerging, INTO, or CALCULATED?	●	
Want to benefit from the integrated interface to SAS Macro Facility (see Chapter 4)?	●	
Are you fine with PROC SQL "limitations" compared to FEDSQL such as single-thread?	●	
Precision of data is validated and OK for SQL?	●	
Work environment will remain SAS 9.4 (server, local) for the time being?	●	
No need to rewrite SQL code due to process (complexity, volume, speed)?	●	

Chapter 8: Performance and Efficiency

8.1 Introduction to Performance and Efficiency

Performance refers to the performance of users, systems, or environments. *Efficiency* might describe performance but also considers the costs for programming, user and system processing, acquisition and maintenance costs for hardware and software, compliance and security requirements, sustainability, documentation, ecological energy efficiency ("Green IT," see Schendera, 2020/2007, Chapter 13), and the like. It is fair to say that performance and efficiency are not easy to quantify. There is no single definition that applies to every context. It is often not even clear how to measure performance or even a performance gain.

This chapter presents performance optimizing measures that strive to minimize CPU time, real time, memory, disk space, or input/output (I/O) processes. The suggested measures are based on the premise that there is no single best way to optimize a SAS program (SAS Institute, 1990).

The focus of this chapter is PROC SQL, although it does touch on topics such as alternative programming languages like FedSQL, DS2, and hash objects; programming environments like client/server, grid, or cloud; and special techniques like in-memory or distributed processing. Many of these measures are so fundamental that they apply no matter the programming language, programming environment, or special processing technique.

Optimizing performance depends on factors like processing complexity, data density, data volume, and system type. However, decreasing real time might increase CPU time, decreasing disk space might increase CPU time, and so on. These trade-offs make it difficult to rank measures according to their benefits. Therefore, the following sections are ordered by fundamental measures that might help you to optimize these quantities.

Section 8.1 emphasizes the importance of approaching measures for performance and efficiency as a strategy planned with a cool head. From the outset, the medium- and long-term quantities in the context of efficiency versus performance must be identified and optimized such as, short-term efficiency in writing and testing SAS programs versus long-term performance in processing very large amounts of data.

Sections 8.2 to 8.6 present numerous suggestions for accelerated data processing with SAS. These recommendations range from having SAS generate code to special SQL-specific tuning tips. Due to the scope, complexity, and developmental dynamics of SAS, this chapter includes only general recommendations.

- **Section 8.2** recommends **reducing** the number of columns, rows, tables, structures, or characters.
- **Section 8.3** recommends **shortening** and **compressing** using LENGTH and COMPRESS.
- **Section 8.4** recommends avoiding unnecessary **sorting** and presents techniques like optimizing PROC SQL processes, indexing, BY processing, in-memory processing, or multi-threading.
- **Section 8.5** presents numerous **tuning** tips like choosing the location of processing, SQL-specific tips (e.g., when merging tables), or techniques like stored compiled DATA step programs.
- **Section 8.6** discusses the benefits of the implicit and explicit **pass-through** when it comes to deciding whether to process in SAS or in a DBMS.

Section 8.7 groups the measures presented according to whether they help to optimize I/O, CPU, or memory.

Section 8.8 presents alternative programming environments like grid- and cloud-based processing environments, SAS Grid Computing, and SAS Cloud Analytic Services (CAS).

8.1.1 The Price of Performance (Efficiency)

Commonly, the wish for performance is (1) optimal quality of results (2) in the shortest possible time (3) at minimum cost. (See Schendera, 2007, 22-23.) This is a wish that is as old as mankind. But if in IT or BI, the goal is to achieve optimal result quality, time and costs can be differentiated into two factors on the user and system side:

- The **shortest time** in terms of the system processing speed or the programming, validation, IT tuning, and so on.
- The **minimum costs** in terms of the required manpower or of the acquisition of hard- and software systems, maintenance, and so on.

Performance has its price. If costs also play a role, performance turns into efficiency. Efficiency is therefore a measure of performance, taking into account the necessary costs.

Therefore, *performant* is when the required amount of data is optimally processed in a minimum of time. *Efficient* is when the required amount of data is optimally processed using reasonable effort, time, and money. Efficiency is performance measured by costs. If we talk about efficiency, it is always as a complex interplay of the variables time, costs, and performance.

Example: Writing a program that makes optimal use of a system costs time and therefore money on the user side, but saves computing time and budgets on the system side. An appropriate ratio of the time and costs for performance-supporting, efficient programs ensures the efficiency of the interaction as a whole. A system is only as good and efficient as the entirety of its users.

The following considerations focus on the user, how he or she can take into account "efficiency" for himself or herself and the system from the very beginning when planning and programming. It goes without saying that the following considerations must be of a general nature. Not everything that is programmed quickly is also efficient in the sense that the system is able to process the program with high performance. Conversely, developing performant programs can be so costly that the cost of manpower is disproportionate to the performance gained on the system side. The performance of a PC can be tuned in completely different scales compared to a mainframe computer or a multinational distributed system landscape. The same applies to the various applications or system environments themselves. Not everything that works for one will automatically work for another on the same scale.

8.1.2 The Need for Speed: Considerations and Tips for Better Performance

When dealing with data of any size with SQL, at some point the question of performance arises. Experience shows that the question of performance arises after programs are written, and the data tables and system environment already exist. Performance is too often a process that happens after the fact. In principle, this question should be asked from the beginning. The knowledge and know-how gained from this opportunity can be transformed into a concept or a strategy, which maximizes performance from the beginning. If, for example, when testing performance, certain phases of data processing are extremely slow or memory intensive, then these phases can be specifically optimized and their performance significantly increased. The earlier users can check and successfully optimize performance, the better. Time is also money.

The following are a few notes on achieving better performance:

- Performance is **not free**. Performance is *always* at the expense of costs. Determining efficiency is ultimately measured by the most important resources: system performance, manpower, and in larger systems, hardware.
- Performance **can be optimized**. The following sections present a selection of the most common techniques.
- Most techniques for optimizing performance have **advantages and disadvantages**. Performance usually depends on many specific factors such as data density, table structure, and the system environment. *Benchmark testing* on your own data in your own system is inevitable.

- **Performance** might be considered as relative and having many facets. However, it does strive for and represent an optimum: the quality of a synchronized system, of software, hardware, data, jobs and processes, and the invested know-how of its decision-makers, operators, and users.

The most important resources on the user side are know-how, motivation, and time. In fact, at this very moment, you are already motivated to invest time to challenge and possibly expand your know-how.

The author would like to publish successful performance enhancements in future editions and would therefore be very pleased to receive feedback, especially if your successes could be achieved with the help of one of the listed techniques in this chapter, individual tips, or other approaches.

8.1.3 A Strategy as a SAS Program

A SAS program is nothing less than the implementation of a strategy. The stringency of a strategy is necessary for the transparency of a SAS program. All steps have to be clear, complete, redundancy-free, and correct.

It is generally not advisable to start programming without a plan. **Avoid ad hoc approaches.** Ad hoc approaches are proceedings that have the following characteristics:

- The approach is without any concept. It is not based on a plan; prioritization of table, performance variables and measures like programming and testing effort, disk space, CPU, I/O, and so on; nor on a concrete goal. Suboptimal working conditions and deadline constraints will cause suboptimal performance even after the installation of completely new systems, for example.
- An arbitrary selection of performance areas and quantities are tested. Thus, fewer criteria are checked and optimized than would be necessary and effective.
- The measures used to optimize the selected criteria are banal, for example, repeatedly submitting a program "by hand" instead of a much more efficient macro approach.
- The set thresholds for the performance variables are too tolerant. Instead of a 50% feasible performance increase, the performance is only improved by 20% without further justification.

Instead, it is recommended that you first get an overview of your working environment:

- Check software version including patches and hot fixes, disk space, memory, processors, and so on. Are the SAS options optimally set or disabled?
- Then draw a rough sketch of your specific approach. Consider not only the steps required (reading, subsetting, sorting, deleting, aggregating, calculating, formatting, and so on), but also their appropriate sequence to ensure logic and efficiency.
- If necessary, prioritize relevant or current tables, the required performance quantities, effort, disk space, CPU, I/O, and the measures to be examined for this purpose such as splitting tables or new hardware in a first step.
- Think long-term and sustainable. If you are not sure whether certain tables, performance quantities, or measures could become relevant at some point, it is more economical to check or optimize them at the same time.

A clear plan makes it easier to see whether a process has been efficiently programmed:

- without unnecessarily "dragged along" temporary rows, columns, or even entire tables (see Section 8.2)
- without unnecessarily wasted memory or disk space (see saving and compressing, Section 8.3)
- doing without unnecessary sorting processes (PROC SORTs) or reading operations (see Section 8.4)
- by accelerating one or other sub-processes, such as specifically introduced "abbreviations" using the SQL Option ALL (see Section 8.5)
- have data processing performed in SAS or DBMS (see Section 8.6)
- trying out different processing languages or approaches such as multi-threading using DS2 (see Chapter 7) or even migrating to different processing environments like grid or cloud (see Section 8.8)

A clear plan not only makes it easier to program, but also saves time and resources. The effort that you put into a plan as an investment, you save yourself, as experience has shown, many times over in programming afterward.

- Helpful techniques **before** include having the SAS code generated using the SQL Query Windows, SAS Studio, Enterprise Guide, SAS Data Integration Studio, or SAS Enterprise Miner. To write SAS code yourself, it is worthwhile to use the SAS procedures PROC CONTENTS and PROC DATASETS.
- Helpful techniques **during** the programming process include programming standards including rules for headers, comments, indentions, company-specific standards, protocols, documentation, security checks, integrity constraints, audit reviews, validation according to the four-eye principle, and so on. On the other hand, it is advisable to check various programming approaches on your own data for processing speed, processing location and, if necessary, further acceleration options. In the benchmark tests, each programming variant should be submitted several times, each in separate SAS sessions, with the same data. Next, the data should be varied with more rows or more numerical or string variables. Benchmark tests might well depend on the contents of the processing tables. (See also the COMPRESS function.) Configurations that achieve particularly striking slow or fast processing times must be carefully examined for possible causes. Finally, the average processing time is determined for each programming variant; the programming variant with the shortest average processing time should ultimately be applied to the data by default.
- Helpful techniques **after** programming are the regular use of the two SQL options FEEDBACK and VALIDATE. FEEDBACK is particularly useful when it is necessary to understand the functioning of SQL queries written by third parties that work in principle but might be incomprehensible. VALIDATE is particularly useful when it is necessary to check self-written SQL queries for correct functioning and to obtain recommendations from SAS.

When writing a SAS program, keep in mind that the clarity of the program is a pillar of efficiency in many ways. If a program is clearly written in terms of headings, spacing, indents, colors, and explanations, it helps you to understand and use a program more quickly. If a program is clearly written in the sense of a document or form, this helps you understand whether the logic and content of this program is correct, to correct it, and to apply it successfully. A program that is not clear in terms of structure cannot be understood in terms of logic. A program that cannot be understood cannot be applied correctly or at all. Also remember to assign labels for variables in the program and to explain any necessary transformations in comments. The investment in understanding programs that are not maintained at all is many times higher than the time needed for a clean documentation.

Performance when working with large data sets can already be increased by simple measures. Most of the measures presented in the following sections focus on the optimization of I/O processes. The optimization of I/O processes usually also optimizes the use of memory and CPU performance at the same time. Section 8.7 lists techniques for the specific optimization of the working memory or CPU performance in keywords.

Experience shows that performance and efficiency are influenced differently by specifics of data, tables, as well as system environments and hardware and therefore benefit differently from performance measures. Therefore, measures should be applied in a targeted manner, applied in a differentiated manner, and always be subject to their own benchmark tests.

8.2 Less is More: Narrowing Large Tables Down to the Essential

Efficient programming is characterized by the fact that you only keep those tables, columns, or rows of data that are really needed. Delete tables, columns, and so on that are no longer needed as soon as possible and preferably at the beginning of an SQL program or DATA step. This applies both to tables in SAS format and external data. Only the required subsets of SAS tables are transferred to the Program Data Vector (PDV), thus reducing I/O time. In this way, the storage space for the resulting output table is also reduced. SAS also processes external data faster once it is created as SAS tables. Data from external files should therefore be created as SAS tables as soon as possible, at least for the duration of processing and analysis. Even if the result of the data processing is created as an external file, the performance gain is generally considered to be significant.

The following simple measures can significantly increase storage space and processing speed:

- **Reduce the number of columns:** Use the SQL/DATA step statements KEEP= or DROP=. The functionality of keeping or excluding using the SELECT statement of PROC SQL is always implied. Place all KEEP/DROP statements at the beginning of the SQL program or the DATA Step.
- **Reduce the number of rows:** WHERE and IF should also be applied as early and as often as possible; the SQL options INOBS= and OUTOBS= are another alternative. (See data set options OBS= and FIRSTOBS=.)
- **Reduce tables:** Delete temporary auxiliary files that are no longer required by using the DELETE statement in PROC SQL or PROC DATASETS as soon as possible.
- **Reduce structures (restructuring tables, transposing):** The advantage of transposing is, depending on the "direction," two-fold: (1) This controlled conversion of the table structure into fewer columns or rows can simplify its structures. (2) In addition, the simplified structure might make it easier to reduce the volume of the table by filtering of rows using WHERE/IF or columns using KEEP/DROP.
- **Reduce characters:** Get rid of unnecessary characters when defining data tables. Tailor-made tag sets make it possible to do without quotation marks and commas, among other things.
- **Reduce by aggregating rows or columns:** Early rowwise or columnwise data aggregation using PROC SQL or PROC SUMMARY also reduces the number of columns or rows in one or more tables.

The following sections provide examples of these measures.

8.2.1 Reduce Columns: KEEP, DROP, and SELECT Statements

When reading a SAS table using the DATA step, the KEEP= statement has three advantages. KEEP= can significantly reduce the amount of storage required, reduce the time needed to process a table provided that the data can be deleted, and can also be easily used as a data set option. KEEP= should be used as data set option in preference to KEEP or DROP statements as DATA step options (without the equal sign). With KEEP=, SAS only needs to create space in the PDV for the required data; this is much more efficient than first reading in all the data and only then excluding the unneeded data by means of the DATA step. In Program 8.2-1 on the left, for example, HEIGHT is excluded from processing from the outset with DROP=. In Program 8.2-2 on the right, HEIGHT with DROP is included in the processing, stored in MYDATA, and only then excluded from MYDATA. The same applies to the WHERE statement or the subsetting IF.

Program 8.2-1 KEEP/DROP (Data Set Option)	**Program 8.2-2 KEEP/DROP= (DATA Step Option)**
| ```
data MYDATA ;
set SASHELP.CLASS(drop=HEIGHT) ;
run ;
``` | ```
data MYDATA2 ;
  set SASHELP.CLASS ;
  drop HEIGHT ;
  run ;
``` |

Depending on how you program in PROC SQL with the SELECT statement, the functionality can correspond to a KEEP, but also a DROP:

- If the variables to be kept from the table in the FROM clause are referenced explicitly or using *, the functionality of the SELECT statement corresponds to a **KEEP**.
- If the variables from the table in the FROM clause are not referenced at all, that is, neither explicitly using a column name nor using *, the functionality of the SELECT statement is equivalent to a **DROP**.

These two functions of the SELECT statement are always included in the discussion of retaining or excluding variables or columns using DROP or KEEP.

As another option, you can use the data set options in PROC SQL when reading or writing data in combination with *.

Program 8.2-3: Single KEEP/DROP (SQL)	If only KEEP= is used, four columns are kept. If only DROP= is used, four columns are kept.

```
proc sql ;
create data MYDATA3 as
    select *
from SASHELP.CLASS(drop=HEIGHT) ;
quit ;
```

Program 8.2-4: Combined KEEP/DROP (SQL)	KEEP= and DROP= can be used simultaneously. In this case, only one column is kept finally, AGE.

```
proc sql ;
create data MYDATA4 (keep=AGE)
    as select *
from SASHELP.CLASS(drop=HEIGHT) ;
quit ;
```

8.2.2 Reduce Rows: WHERE Statement, Subsetting IF, and Data Set Options

While KEEP or DROP specify the number of columns, the WHERE statement or subsetting IF control the number of rows (the DELETE DATA step option is explained in the section on the DELETE PROC SQL option). WHERE or IF can also significantly reduce the amount of storage required and the time required to process a table. Like KEEP or DROP, WHERE or IF should be applied as early and as often as possible.

A WHERE expression is generally considered to be more powerful and efficient than an IF expression. A WHERE expression selects data by conditions before it is read into the PDV, which can be advantageous for a very long character variable, whereas IF selects cases by conditions only after they have been read into the PDV. More operators can be specified in a WHERE expression than in an IF, for example, BETWEEN/AND, CONTAINS, IS MISSING/NULL, LIKE, and SAME/AND. A WHERE expression can be used as a data set or DATA step option, in SAS procedures, and also enables you to process indexed tables. For further differences between WHERE and IF including BY processing, please refer to the SAS documentation (SAS Institute, 2019).

Program 8.2-5: WHERE as Data Set Option	**Program 8.2-6: WHERE as DATA Step Option**

```
data MYDATA
   (where=(REGION="Africa"));
set
   SASHELP.SHOES(drop=SALES);
run ;
```

```
data MYDATA ;
set SASHELP.SHOES(drop=SALES) ;
    where REGION="Africa" ;
run ;
```

Program 8.2-7: WHERE (SQL, Reading)	**Program 8.2-8: WHERE (SQL, Reading and Writing)**

```
proc sql ;
create table MYDATA as
select * from
SASHELP.SHOES
   (where=(REGION="Africa"));
quit ;
```

```
proc sql ;
create table MYDATA
   (where=(PRODUCT="Boot")) as
    select * from
SASHELP.SHOES(drop=SALES
    where=(REGION="Africa"));
quit ;
```

The structures of the tables to be read or processed decide whether KEEP, DROP, WHERE, or IF are more effective. If the data to be excluded is primarily in many columns but few rows, KEEP or DROP might be more effective than WHERE and IF. If the data to be excluded is in many rows but few columns, WHERE or IF might be more effective than KEEP or DROP. Ideally, KEEP, DROP, WHERE, and IF should be applied simultaneously if possible to "tailor" the number of rows and columns to be read by selecting columns and rows. Subsetting can be done using the WHERE clause or the IF statement.

What about performance differences between the WHERE clause and the IF statement? Langston (2005) found no significant differences between these two approaches in his benchmark tests. What recommendations can be made?

- Use WHERE or IF as early and as often as possible to limit the volume of tables to be processed.
- Prefer a WHERE to an IF if a DBMS is optimized for WHERE processing (Langston, 2005).
- If processing in a DBMS is planned, the programming of SQL syntax must be adapted to the processing location and processing method of missing values. (See Section 8.6.)
- Optimize the writing of WHERE conditions. (See the recommendations listed below.) Inefficient and efficient WHERE clauses achieve the same result, logically. However, the processing times of efficient WHERE clauses are significantly lower for large data volumes. Even small differences in programming can result in larger differences in performance.

Avoid the NOT operator, provided that a logically equivalent form is possible.
Inefficient: where VALUE not > 5 ;. Efficient: where VALUE <= 5 ;.
Avoid arithmetic expressions in a predicate.
Inefficient: where VALUE > 8*4 ;. Efficient: where VALUE > 32 ;.
Avoid LIKE predicates beginning with % or _, if possible.
Inefficient: where STRING like '%ARIA' ;. Efficient: where STRING like 'M%RIA'.
Use the BETWEEN predicate instead of the >= and <= operators.
Inefficient: where VALUE >= 1 and VALUE <= 5. Efficient: where VALUE between 1 and 5.

- If necessary, SAS can split a large table during a single Read operation using DO IF into several smaller subsets for further processing within a DATA step. (See also the various approaches for splitting a table using macro or hash programming in Sections 4.5 or 6.2.3.) In this way, the I/O time can also be drastically reduced. PROC FORMAT is an extremely efficient alternative to recoding values using IF THEN ELSE. The FORMAT statement can be used to efficiently convert string and numeric values. Multi-threading can be seen as an automatic table splitting method in contrast to these manual approaches.
- The SQL options INOBS= and OUTOBS= and the data set options OBS= and FIRSTOBS= are another possibility to control the number of required rows of a SAS table before further processing.

Langston (2005) also compared the performance of SELECT/WHEN statements with IF THEN ELSE statements in a DATA step. SELECT/WHEN statements (not to be confused with CASE/WHEN in SQL) perform slightly better than IF THEN ELSE statements. However, the major differences lie in their basic functionality. SELECT/WHEN clauses are limited in comparison to IF THEN ELSE statements in that they can only process scalar values. However, if the ELSE clause is omitted in IF THEN ELSE statements, their performance deteriorates significantly.

8.2.3 Reduce Tables by Deleting (DELETE)

What KEEP and DROP mean for columns and WHERE and IF mean for rows, DELETE means for deleting temporary tables that are no longer needed. Delete temporary tables that are no longer needed as soon as possible using the DELETE statement in PROC SQL and PROC DATASETS, the KILL option ("delete all without exception"), or the SAVE statement ("delete all except …") from PROC DATASETS. This measure also increases storage space and processing speed.

DELETE as a DATA step option enables you to delete rows analogous to WHERE or IF. DELETE is more convenient for excluding. WHERE/IF, on the other hand, is more convenient for keeping rows.

8.2.4 Reduce Structures: Transposing Tables

Transposing a table from the structure "stack" ("long") into the form "unstack" ("wide") is an obvious but often overlooked measure. When transposing, the rows and columns of a file are simply "rotated." The contents of the tables are not changed. The structure of the tables is differentiated between "stack" and "unstack." In the case of "stack," related values of a case are arranged one below the other in a column. With "unstack," related values of a case are arranged next to each other in a row. The performance gain in connection with transposing should be estimated in advance on site using partial data. As investments in performance, two quantities must be considered: a) writing and testing a program for transposing the tables (see the example in Chapter 4), and b) the transposing process itself can be quite computationally intensive, especially with large tables.

Because in a horizontal "unstack" structure, related values of a case are arranged next to each other, significantly fewer rows have to be processed. An additional KEEP/DROP DATA step option for the many columns can be very effective in this variant. In a vertical "stack" structure, however, related values of a case are arranged in a column, which means that considerably fewer columns have to be processed. In this variant, an additional WHERE clause for the many lines could be very effective. Whether it is more complex to program queries or analyses for tables in a stack or unstack structure depends on the specific task to be programmed. Sometimes an analysis syntax requires certain table structures. For example, PROC GLM in SAS/STAT requires "stack" for a variance analysis, but a variance analysis with repeated measurement requires "unstack." In any case, transposing a table should be considered to reduce manpower when programming analyses or queries. The advantage of transposing is, depending on the "direction," two-fold. The conversion of the table structure into fewer columns or rows simplifies its structures in a controlled manner. The size of tables can be reduced to fewer rows using WHERE/IF or fewer columns using KEEP/DROP at the same time.

8.2.5 Eliminate Unnecessary Characters by Means of Tag Sets

With very large data volumes, it is important to process text files with as little "ballast" as possible. The default tag set template TAGSETS.CSV defines table output with comma-separated data by default. String variables are also set in quotation marks by default. If the tag set file is customized and no quotation marks or commas are used when exporting to CSV or comparable applications, the performance when exporting millions or even more data rows can be noticeably increased, especially when string variables are exported without quotation marks, among other things. An example for the application of customized tag sets can be found in Section 9.3.

8.2.6 Aggregate

Early rowwise or columnwise aggregation of data using PROC SQL or PROC SUMMARY also allows for a reduction of the number of columns or rows in a table or several tables without losing the relevant information from its contents. PROC SQL allows for the application of various aggregation functions directly to data. With the DATA step, aggregating functions can be programmed with a little effort. However, the SAS procedures SUMMARY or MEANS are unrivaled.

Program 8.2-9: Aggregating (SQL)	**Program 8.2-10: Aggregating (SUMMARY)**
```proc sql ; create table MYDATA as select REGION, SUBSIDIARY, sum(SALES) as SALE_SUM from SASHELP.SHOES group by REGION, SUBSIDIARY ; quit ;```	```proc summary data=SASHELP.SHOES ; var SALES ; class REGION SUBSIDIARY; output out=MYDATA sum=SALE_SUM ; types REGION*SUBSIDIARY ; run ;```

The two PROC SQL and PROC SUMMARY examples in Programs 8.2-9 and 8.2-10 compress the information content from the original 395 rows with sales figures per product line to only 53 rows for the totals SALE_SUM per subsidiary per country.

In contrast to the other measures proposed previously, the rows or columns are not reduced by WHERE/IF or KEEP/DROP, but primarily using aggregation functions. The procedures SQL, MEANS, and SUMMARY can also use WHERE/IF or KEEP/DROP to reduce rows or columns, however. When creating or calculating new variables, defining the variable length using LENGTH also helps to save storage space and increase processing speed. (See the next section.)

8.3 Squeeze Even More Air Out of Data: Shortening and Compression

If you use KEEP/DROP or WHERE/IF to reduce the structure of a table to the minimum number of columns and rows required, the LENGTH statement and the COMPRESS function improve performance even further.

- **Reducing variable length (LENGTH statement):** If the approximate number of digits of the values of a variable is known, you can gain considerable storage space and CPU time by specifically assigning shorter variable lengths.
- **Reducing storage space (COMPRESS function):** This SAS function compresses each table created and can also help to save storage space and computing time.

8.3.1 Shortening the Variable Length (LENGTH Statement)

A numeric variable is stored in a SAS table with the default length of 8 bytes. SAS can store a number with 15 to 16 digits in these 8 bytes. In other words, by default, SAS reserves as much space during storage as if the values were always in the order of 8,888,888,888,888 or even higher. (See UNIX.) Given the actual size of the values stored, this generosity often switches to a waste of overly scarce resources. For practical work, on the other hand, this means that you do not have to store in 8 bytes if the same purpose is achieved by using 4 or even 3 bytes. If the approximate number of digits of the values of a variable is known, storage can also be saved by assigning shorter lengths in terms of bytes. The processing speed increases due to less CPU time, as well as less time for arithmetic operations (often caused by the truncated decimal places with the price of less precision). In the following small example, the default length in four numeric variables of the SASHELP file SHOES is reduced from 8 bytes to 4 or even 3 bytes. This small adjustment reduces the storage volume of SASHELP.SHOES by almost 25% from 41 KB to 33 KB. Now imagine a scenario with much more data in the terabyte range.

Program 8.3-1: Adjusting Variable Length (LENGTH)
```
data MYDATA ;
   set SASHELP.SHOES ;
   length Stores 3 ;
   length Sales Inventory Returns 4 ;
run;
```

The following table illustrates the relationship between length in bytes, significant digits retained, and the largest integer in a numeric SAS variable (example: UNIX; source: SAS Institute).

Table 8.3-1: Relationship between Length, Significant Digits, and Largest Integer (UNIX)

Length (bytes)	Significant digits retained	Largest integer represented exactly
3	3	8,192
4	6	2,097,152
5	8	536,870,912
6	11	137,438,953,472
7	13	35,184,372,088,832
8	15	9,007,199,254,740,992

The maximum size of a number, its length in bytes, as well as the number of digits can vary depending on the operating system. The documentation of the respective operating system provides information about a length in bytes and which numbers it can exactly represent or store. There are, however, basic recommendations for using the LENGTH statement, some of which will be compiled here.

- **Integer numeric values:** Assigning shorter lengths is recommended as long as they are not large integer values. For example, if it is known that the integer values of the numeric variable are between 0 and 999,

3 bytes can be specified as the length to store the values of this variable in a space-saving way. Numerical dummy variables, which are variables that only take on the values 1/0 or 1/2, can also generally be stored in a variable with length 3.

- **Vital or monetary data (meaning of precision):** Vital data like laboratory data or monetary data like currency rates require the highest possible precision for analysis, filtering, and further processing. The more decimal places are truncated in floating point numbers, the greater the loss of precision. The highest possible precision is recommended for vital or monetary data. The required degree of precision should be specified in consultation with the responsible specialists. A loss of precision might also be caused by suboptimal hardware or differences between operating and database management systems.
- **Date values:** For date values or variables, the LENGTH statement is a must. SAS date values up to the year 5000 can be stored with a length of 4 bytes with a clear conscience. This reduces the space required for date variables by half.
- **Attention with decimal values:** The assignment of shorter lengths is only recommended if the greatest care is taken. The length reduction by assigning shorter lengths actually cuts off the stored decimal place. Assigning shorter lengths to decimal values should only be done if the decimal places and the precision in analyses and calculations influenced by them are actually irrelevant.
- **Converting to character:** Converting codes or constants to levels of string variables is recommended because text characters require 2-3 times less storage space than values of numeric variables. One-off conversions cost programming or system time on the one hand; on the other hand, levels of string variables usually cannot be included as numerical values in analyses anymore.
- **Compressing:** The advantage of reducing the length of numeric integer values by compression is accompanied by the cost of additional CPU time required to decompress compressed numeric values back to their original length as soon as they are read into a DATA step or SAS procedure.
- **Aggregating using summary functions:** Storing numeric values with a length less than 8 is one thing, aggregating them another. For example, summary functions applied to numeric variables (for example, floating point numbers) with a length less than 8 may return inaccurate values. If you have to perform calculations and truncations, apply the truncations after you have done your calculations. SAS Tech Support recommends leaving numeric variables with the default length of 8 if at all possible.

8.3.2 To Compress or Not Compress? COMPRESS Function

The COMPRESS function compresses each table created. In general, the advantage of this function is to easily gain storage space and computing time. Converting numeric codes or constants into string values is often suggested because text characters require two to three times less storage space than numeric values. The disadvantage of compression is that compression itself requires CPU time and storage space, as well as CPU time and storage space for decompression for further processing with a DATA step or SAS procedure. In SAS, the COMPRESS function is offered as a data set option and also as a system option. The concrete savings in storage space and computing time depend not only on the expenses required for the dual process of compressing, but also on characteristics such as the density and character set of the table to be compressed. In terms of CPU time, it is recommended to check what requires less CPU, that is, what is faster? Is direct access to uncompressed data and its further processing faster? Or is working with compressed data faster? So, are the two processes of compressing for saving and decompressing for further processing together still faster than direct access and further processing? As already mentioned, features such as "density" and character set of the table to be compressed also play a role:

- **"Density":** The fewer repetitive characters a table contains, the "denser" it is. The lowest level of density is when a table contains only one repeating character. The maximum density is when a table contains no repeated characters. The denser a table is, the more CPU time is required to compress or decompress it for further processing. Also, the amount of storage space gained might not be substantial. The less dense a table is, the more storage space can be gained by compressing it, and the less CPU is required for the dual operation of compressing and decompressing.
- **Character set:** Character sets distinguish between tables that contain long strings with many blanks and tables that contain long observations. For tables with long strings with many blanks, the option COMPRESS=CHAR|YES can be used. For tables that contain long observations (> 1000 bytes), COMPRESS=BINARY can be used. Compression using BINARY requires much more CPU time for decompression than CHAR.

Please see the SAS documentation for illustrative examples of compressing numerical and character data:

- Compressing a 235 MB of uncompressed, predominantly long character values with many trailing blanks results in a 54 MB table (savings: 181 MB, approx. 77% of storage space) at the cost of needing 23 seconds more CPU. The storage savings come at a cost of approx. 0.1 seconds per MB (23.19/181) and 0.3 seconds per % (23.19/77).
- Compressing 52 MB uncompressed, predominantly numerical data results in a 39 MB table (savings: 13 MB, approx. 25% of storage space) at the cost of approx. 13 seconds more CPU. The storage savings costs approx. 1.0 seconds per MB and 0.5 seconds per %.

Benchmark tests are recommended for the COMPRESS function. In addition to the number of rows and columns of the tables, their results depend on the contents of the tables, namely the number of different characters and the extent to which they are repeated.

8.4 Sorting? The Fewer the Better

If a table is reduced to minimum storage space or CPU via KEEP/DROP or WHERE/IF and the LENGTH and COMPRESS functions, other techniques still allow further performance gains. One of the most basic techniques is to dispense with sorting as far as possible or even completely, even within SQL. The widespread view that PROC SQL does not require sorting must be clarified because PROC SQL does perform internal sorting when using ORDER BY. These operations occur explicitly when they cannot be completed successfully, for example, when there is insufficient disk space for the required paging files.

SAS Log 8.4-1: Error Note When Not Sufficient Disk Space Available

```
File is full and may be damaged.
ERROR: Write to file WORK._tf0014.UTILITY failed. File is full and may be damaged.
NOTE: Error was encountered during utility-file processing. You may be able to execute the
SQL statement successfully if you allocate more space to the WORK library.
ERROR: There is not enough WORK disk space to store the results of an internal sorting
phase.
ERROR: An error has occurred.
```

Sorting (including *implicit* sorting) is extremely computationally intensive, and it requires sufficient disk space and appropriate processor capacity. Sorting should be avoided as far as possible. Instead, techniques and tricks should be used, such as

- **No sorting:** Dispensing with sorting as much or even completely as possible is a basic technique. However, this technique also implies checking PROC SQL programs for unnecessary sorting. If necessary, alternative programming approaches could be used like hash objects or FedSQL. (See Chapters 6 and 7.)
- **Minimal sorting:** Indexing is very effective and is considered the industry standard for achieving performance gains when handling even very large amounts of data.
- **Minimal reading** (BY-processing): BY-processing can reduce the number of times a SAS procedure has to be called because one Read operation is sufficient to subsequently execute several operations, analyses, or reports for each group per BY-level.
- **Sorting supported by hardware** (multi-threading): If the amount of data is so large that memory or CPU is insufficient for sorting even within PROC SQL or PROC SORT, the processing or sorting of SAS tables can be done in parallel on different drives.

8.4.1 No Sorting: Do Without ORDER BY

One of the most basic techniques is to do without sorting as far as possible or even completely. This technique also implies checking PROC SQL programs to see whether they contain unnecessary sorting. Even in PROC SQL you should avoid the keyword ORDER BY when creating tables and views, if possible. For ORDER BY, the data must be re-sorted each time, which can be very computationally intensive with large data sets or complicated nesting of views or tables.

Another possibility is in-memory processing using hash objects or FedSQL. (See Chapters 6 and 7.) The processing of data using hash objects takes place in memory and is therefore considered one of the most powerful processing techniques, especially for very large amounts of data. For example, no sorting or indexing is required when loading data into the hash object (Dorfman et al., 2010, 2009). Subsection 7.2.3 is dedicated to in-memory processing on CAS. Subsections 6.2.2 and 6.2.4 present examples of sorting using hash programming.

Whether a PROC SQL, DATA step, or in-memory approach is used for programming a join of tables might depend on the data volume and system environment, technology, memory, and so on, but also on characteristics of the tables to be processed and will be discussed further in Section 8.5.

8.4.2 Minimal Sorting Using CREATE INDEX: Indexing Instead of Sorting

Indexing is used when large tables undergo WHERE or BY processing. Since the involved sequential search (WHERE) or sort (BY) is extremely computationally intensive, the indexing technique aims to save as much effort as possible for these processes. The disadvantage of indexing is on the one hand that, like any other technique, it requires CPU, I/O, RAM, or even disk space. On the other hand, indexing generally compensates by the performance gains achieved. Indexing is very effective. Sadof (2000, 1999) reported a quadrupling of the processing speed by means of a WHERE condition at 500,000 cases. A merge with SQL on an indexed table (250,000 rows) is more than three times as fast despite the additional indexing process that is always required.

What Is an Index?

An index created during indexing is part of an internal search mechanism and is a type of "table of contents" in a separate file depending on the operating system, which is created for the respective SAS table in order to access certain observations directly. The index in the index file creates values for finding observations in the associated SAS table. Without an index, SAS would have to read every single observation in the SAS table one after the other sequentially until a specific observation is found. With an index, SAS goes through the index file using a binary search, finds the corresponding index value much faster, and then accesses the corresponding observation in the SAS table directly without having to read the entries in the SAS table sequentially. From this perspective, an index is a "shortcut" for a search-and-find operation.

Recommendations

SAS makes the following recommendations for creating an index:

- Working with **large amounts of data:** You should create an index if only a small subset of large amounts of data is to be retrieved, for example, < 15%.
- Working with **small amounts of data:** You should *not* create an index if a table is less than three pages long. In this case, the usual sequential access would be faster. The length of a table in pages can be displayed by PROC DATASETS or PROC CONTENTS. The number and names of any existing indexes in a SAS table can also be displayed.
- **Variables for an index:** For a single index with only one variable, variables should be used that equal an ID (primary key). For a composite index with several variables, variables should be used that are used in a query, especially in WHERE or BY processing. When creating an index according to one or more variables, the data should be sorted according to these variables.
- Working with **frequently changing data:** With a data table that changes frequently, you should take into account that the respective update of the index might compensate for the advantages in processing speed.
- **Indexes for views:** Indexes cannot be created directly in views. However, views can be provided with indexes using a workaround. The trick is to first assign an index to a SAS table and then create a view from the SAS table.

By the way, PROC SORT does not use indexes; SAS tables with indexes can be sorted using PROC SORT. However, SAS tables created with OUT= do not automatically have an index.

Besides PROC SQL and its CREATE INDEX, indexes can be created with PROC DATASETS in the DATA step or other SAS procedures or applications. In the following PROC SQL examples in Programs 8.4-1 and 8.4-2, a case in a table can be uniquely identified by sorting the columns REGION, SUBSIDIARY, and PRODUCT; the index INDEX created "instead" does the same. Similar to a primary key, an index enables you to identify a case (row) by a single value. Sorting a table using a single index in a WHERE or BY processing is obviously always faster than sorting the same table using two or even more variables. A composite index can be used instead of several variables that are always used together. In this way, an index greatly increases performance for large data sets, especially in WHERE and BY processing.

Program 8.4-1: Creating an Index (INDEX)

```
proc sql;
create index INDEX on
 sashelp.shoes (REGION,SUBSIDIARY, PRODUCT);
quit ;
```

Note: CREATE INDEX creates indexes for columns in tables. CREATE INDEX cannot be used if the table is accessed with an engine that does not support UPDATE processing. CREATE INDEX is used to create a compound index (INDEX) for the columns REGION, SUBSIDIARY, and PRODUCT in the data record SASHELP.SHOES (see Program 8.4-1). A created index value corresponds to concatenated values from REGION, SUBSIDIARY, and PRODUCT. The composite index INDEX can thus be used instead of the three sorting variables REGION, SUBSIDIARY, and PRODUCT used together. Another advantage of an index is that it makes BY and WHERE processing more effective and is easy to program.

Once an index is created, it is treated as part of the data record. If cases are then assigned to the data set or deleted from it, the index is automatically updated, which in turn involves updating effort. In addition, if a UNIQUE is placed between CREATE and INDEX as shown in the following example, SAS is prompted to reject any change that would cause more than one row to have the same index value.

Program 8.4-2: Creating an Index (UNIQUE)

```
proc sql;
create unique index INDEX on
        sashelp.shoes(REGION,SUBSIDIARY, PRODUCT);
quit ;
```

SAS is prompted to reject any change that would cause more than one row to have the same index value. To handle missing values, the NMISS option in the PROC DATASETS statement INDEX can be used.

All these advantages are matched by one requirement, which cannot even be called a disadvantage of the index but must be emphasized as a fundamental requirement of the variables to be "replaced." On the basis of the selected variables (REGION, SUBSIDIARY and PRODUCT), the index must actually be able to uniquely identify a case (row) comparable to a primary key. If the previously assigned variables do not do so because only REGION and SUBSIDIARY were assigned as sorting variables, the created index "only" continues an error made by the user at an earlier point.

To create an index, the structure, nesting, and sorting of the table to be indexed and the number and sequence of the sorting variables must be known exactly. When working with indexes, the data set option DBINDEX= is of interest. (See Subsection 8.5.2.) If used incorrectly, however, it can not only impair performance, but under certain circumstances (for example, duplicate entries) can also affect the result of a query.

SAS offers further options for tuning efficient sorting, depending on the operating system. For z/OS, see the SAS system options SORTPGM=, MEMSIZE=/MEMLEAVE=, or if necessary also HIPERSPACE. For UNIX, see SORTSIZE= or REALMEMSIZE, DBMS. For ORACLE, see the SQL*Loader Index. For other installations, for example, DFSORT from IBM, see SORTBLKMODE, among others as OPTIONS statement. SAS reports a reduction of CPU by up to 25% when using the SORTBLKMODE option. For further possibilities and information, please refer to the SAS documentation.

In SAS Cloud Analytic Services (CAS), indexes also do improve the performance of a WHERE clause. However, indexes should not be used for anything else such as join optimization, BY-group processing, CAS action, and SAS Visual Analytics.

8.4.3 Minimal Reading (BY Processing)

BY-processing can reduce the need to call a SAS procedure several times because one reading process is sufficient to subsequently execute several operations, analyses, or reports for each group per BY-level. Ideally, a SAS procedure is called only once.

In the context of multi-threading (see next section) the data set option DBSLICE might be of interest. DBSCLICE allows the data load to be distributed to separate threads on the basis of explicitly specified levels of a BY variable. In contrast to BY processing, which reads large amounts of data once, the effect of DBSLICE can be described as reading large data volumes according to the BY values in parallel ("parallel processing"). In BY processing, the data from SASHELP.SHOES would be read once and then processed one after the other for each level of REGION (for example, "Europe"). In parallel processing using DBSLICE=, the data in the table SASHELP.SHOES is divided according to the completely specified levels of REGION, a separate thread is set up for each level, and the data is made available in parallel.

Program 8.4-3: Data Set Option (DBSLICE=)

```
proc print data=sashelp.shoes
    (dbslice=("REGION='Africa'" "REGION='Asia'"
              "REGION='Canada'" ));
    where SALES > 3020 ;
run ;
```

Applying DBSLICE= requires that the BY-values of the data to be read are known and that these are specified completely in the parenthesis expression. The above example is simplified and does not include, for example, the REGION values "Central America/Caribbean", "Eastern Europe", and so on. A further requirement is that the data volume in all REGION instances must be approximately the same. This requirement is necessary to ensure that the processing load is distributed more or less equally among the threads. If a variant contains a disproportionately high amount of the data to be processed, efficient parallel processing according to BY levels cannot be guaranteed. Distributed processing using parallel threads means shorter processing overall due to faster processing because of fewer processes per CPU. If there is no even distribution of data, users can create it themselves using PROC SQL in almost any granularity.

8.4.4 Sorting Supported by Hardware (Multi-Threading)

If sorting is unavoidable, but memory, storage space, or CPU are not sufficient for sorting even within PROC SQL or PROC SORT, the processing of SAS tables can be carried out on several drives in parallel using multi-threading. SMP therefore requires more than one drive, at least two drives per CPU should be expected. The difference between multi-threading and the techniques already presented is that multi-threading benefits less from a relatively time-consuming optimization of data or programming, but mainly from an investment in thread-enabled hardware, especially several CPUs, drives, and thread-enabled software. In addition to multi-threading, SAS offers further technologies for high-performance computing like in-memory analytics, in-database computing, and grid computing. Multi-threading distributes the processing load over several CPUs. Depending on the threading approach, the distribution can take place according to data packets ("boss-worker") or program steps ("pipeline").

In an ideally optimized, scalable SMP environment, the number of n CPUs might well correspond to the corresponding proportional n-fold acceleration of processes. In contrast, there are also threading approaches that already work with only one CPU but require several drives. Other SAS procedures that support the use of multiple processors include MEANS, TABULATE, and REPORT. The current SAS documentation shows which other thread-enabled hardware, applications, and procedures SAS Institute provides. The SAS SPD Engine optimized for multi-threading or the SAS SPD Server should be mentioned here. SPD stands for "Scalable Performance Data." Both engine and server are optimized on several levels so that they provide applications with even very large amounts of data quickly and reliably. Even the use of the global SAS option THREADS can lead to substantial performance gains in real-time. In order to document the specific performance gain, the user subjects I/O, CPU, or even RAM of his system to various comparative benchmark tests. In view of increasing data volumes and usually simultaneously decreasing hardware prices, multi-threading should be seriously considered as a long-term technological strategy. The advantage of multi-threading is not only to speed up sorting, but in the long run to speed up all conceivable processes of data processing and analysis. Multi-threading is an ideal transition to the next topic, the acceleration of selected, especially SQL processes by a not so complex, additionally optimizing programming.

8.5 Accelerate: Special Tricks for Special Occasions (SQL and More)

If a table is reduced to minimum storage or CPU via KEEP/DROP or WHERE/IF and the LENGTH and COMPRESS function, other techniques still allow further performance gains. However, all techniques require that you are aware of the characteristics of the process and the tables to be processed in it. For example, you might want to consider whether to use PROC SQL, the DATA step, or programming alternatives like FedSQL or hash objects when merging tables. The following techniques can be used if certain processes, transformations, or applications are to be accelerated. Subsection 8.5.1 introduces more **general techniques**; Subsection 8.5.2 moves on to more **SQL-specific techniques**. Subsection 8.5.3 recommends **other techniques**, such as Stored Compiled DATA Step Programs and working with macro variables and programs.

8.5.1 Fundamental Techniques

This subsection introduces several basic techniques for accelerating the processing of data, from in-memory and threaded processing using hash objects, FedSQL, or DS2; adjusting processing location and time; fine-tuning of SAS; to storing data as SAS tables.

In-memory and Threaded Processing

A basic technique is to apply a technically completely different processing approach such as in-memory processing using hash objects (Chapter 6) or FedSQL programming variants (Chapter 7). The processing of data takes place in memory and not on the disk and is therefore considered one of the most powerful processing techniques, especially for very large amounts of data. SAS offers CAS and enables you to program hash objects explicitly. In-memory processing is so versatile, flexible, and high-performant that this volume dedicates two separate chapters to these options. (See Chapters 6 and 7.) If you use multi-threaded instead of single-threaded processing, you will improve performance even more. In multi-threaded processing, data are not processed one after the other but in parallel, which should result in a massive performance gain. (See Subsection 7.3.2 for multi-threaded processing using DS2.)

Explicit or Implicit Pass-Through: Accessing a DBMS

If you want to access a DBMS, operations can be passed to the DBMS using the implicit or the explicit pass-through. Both facilities differ in programming, functionality, and performance. (See Section 10.3.) The implicit pass-through automatically *translates* statements that have been sent into the DBMS-specific SQL. In contrast, the explicit pass-through passes ANSI SQL-compatible aggregation functions directly to the DBMS. The advantages and disadvantages of both approaches, the syntax, and programming examples are explained in Section 10.3. Suggestions for *optimizing* the processing are made in Section 8.6.

Knowing When and Where (PROC UPLOAD, PROC DOWNLOAD)

If users are working in a SAS server environment with a host and a local area, it is recommended to first compare the average processing speeds of both potential processing environments. Host environments are usually but not exclusively more performant at processing very large amounts of data than a local PC. In this case, it is recommended that you use PROC UPLOAD and SUBMIT to transfer the data and processes from the local work station to the host and process them there. On the other hand, host environments are usually shared with other SAS users, with the result that processing on the host can take on quite unpleasant performance losses, especially at certain times of the day and month (for example, during lunchtime or at the end of the month or quarter).

In this case, transfer the data and processes from the host to the local work station via PROC DOWNLOAD and RSUBMIT and have them processed there, independent of the external risks of performance impairment. In addition to the actual processing time, the additional times for uploading and downloading data must also be taken into account.

After this rather general recommendation to narrow down a place and time of data processing, the further recommendations will again focus on SQL and SAS and tuning the options of SAS.

Fine-tuning of SAS Options

The number of accesses during data processing can also be reduced by fine-tuning SAS. SAS offers general and SQL-specific global SAS options to optimize SQL performance.

SQL-specific System Options

Since SAS 9.2, SAS offers the following SQL-specific global SAS options to optimize SQL performance. Similarly named functions are also available as PROC SQL functions, as SAS options SQLREDUCEPUT and NOSQLREMERGE, and as PROC SQL options NOREMERGE and REDUCEPUT.

- **DBIDIRECTEXEC** controls the SQL optimization for SAS/ACCESS engines.
- **SQLCONSTANTDATETIME** specifies whether PROC SQL replaces references to the DATE, TIME, DATETIME, and TODAY functions in a query with equivalent constant values before it is executed.
- **SQLMAPPUTTO** specifies whether the PUT functions (for example, SQLREDUCEPUT and so on) in PROC SQL are processed by SAS or the SAS_PUT() function from Teradata (see the later section on query acceleration).
- **SQLREDUCEPUT** specifies the engine type for a query that is to be optimized by replacing the PUT functions with a logically equivalent expression.
- For PROC SQL with SQLREDUCEPUT=NONE, **SQLREDUCEPUTOBS** specifies the minimum number of rows in a table to establish whether it is necessary to optimize the PUT functions in a query.
- Also, for PROC SQL with SQLREDUCEPUT=NONE, **SQLREDUCEPUTVALUES** specifies the maximum number of SAS format values that can occur in an expression with PUT functions to establish whether optimizing the PUT functions in a query is necessary.
- **SQLREMERGE** specifies that PROC SQL should process queries that might require remerging. Remerging can occur when an aggregation function is used in a SELECT or HAVING clause. Remerging is I/O intensive because two data passes are required. NOSQLREMERGE specifies that PROC SQL does not perform remerging.
- **SQLUNDOPOLICY** specifies whether PROC SQL keeps or discards updated data if errors occur during the update process.

Global System Options

In addition to SQL specific options, SAS also offers general global SAS options to optimize performance. SAS offers various options for this purpose, including BUFNO=, BUFSIZE=, and SASFILE.

- Setting **BUFNO** values higher enables you to process more data in fewer passes and can therefore speed up the processing of a SAS table.
- The default **BUFSIZE** value is already optimized for sequential access. Elevating BUFSIZE values for direct access causes a larger page to be created for a SAS table. For small tables, the BUFSIZE value can also be lowered. The page size is the amount of data that can be transferred into a buffer for an I/O operation.
- The **SASFILE** statement enables faster processing by reading the SAS table in question into memory, thereby reducing the number of reading steps.

For further options such as IMPLMAC together with MAUTOSOURCE and SPOOL/NOSPOOL, please refer to the SAS documentation.

Create SAS Tables to Accelerate Processes

The creation of SAS tables is possible in three scenarios:

- **For external data:** Experience has shown that external data can be processed more quickly once it is created as SAS tables.
- **With old SAS data:** SAS data in old SAS versions could possibly be processed faster in more recent SAS environments if the old SAS data is first created or converted into a SAS table in the current SAS format.
- **For repeated, longer processes:** Instead of running very long processes from start to finish, it might be an option to create "interim files." Later runs no longer start at the very beginning, but at the last saved "interim file" (storage point).
- Already specifying the engine when specifying the storage location of data using LIBNAME, SAS is spared various process steps to check for the appropriate engine.

Program 8.5-1: LIBNAME Without and With Engine

LIBNAME without engine	LIBNAME with engine
libname PFAD "C:\";	libname PFAD v9 "C:\";

Note: Whenever possible, views should be created instead of tables, which would save additional I/O and disk space. Under certain circumstances, however, temporary SAS tables can have better performance than views. Benchmark tests help in case of doubt.

8.5.2 SQL-specific Techniques

This section introduces SQL-specific techniques for accelerating processing steps with SQL. In addition to more basic approaches such as joins, subqueries, or queries or when disk space is more important than speed (views versus tables), specific techniques for accelerating joins and queries are presented. Among other things, the following will be discussed:

- Joins, subqueries, or queries?
- Is disk space more important than speed? Views instead of tables.
- General considerations on joining tables.
- Merging tables: SQL instead of the DATA step? It depends...
- Accelerated merging of tables (SQL).
- Accelerating queries (SQL).

Joins, Subqueries, or Queries?

Many queries can be formulated as joins or subqueries. Joins are generally considered to be easier to process. If data from several tables or views is to be accessed, joins must be executed. If a desired result requires several queries, subqueries should be used. Subqueries should also be used when you query for category levels. A subquery must be used for the NOT EXISTS condition. Subqueries, on the other hand, are generally not considered to be particularly high-performant; you should check whether they can be rewritten into simpler or more performant queries.

Is Disk Space More Important than Speed? Views Instead of Tables

Views are only views on data, and unlike tables do not require physical storage space. The advantage of disk space generally comes at the expense of processing speed. A disadvantage of many views might be that including views in DATA steps might still be slower than DATA steps that only work with tables. Langston (2005, 4-5) reports that creating and importing a DATA step view is slower overall than a DATA step approach that only works with tables. Own benchmark tests are also indispensable here.

General Considerations on Joining Tables

Joining two or more tables **is** creating SAS tables. This section is followed by a summary of possible considerations for deciding whether PROC SQL or the DATA step should be used for merging tables. The reasons to use the DATA step are given in the next section. If you have decided to use PROC SQL, further possibilities to speed up SQL processes are presented in the following list.

- **Accelerated "appending"** of tables: SQL keyword ALL: Specify the keyword ALL for set operations, for example, if no multiple data rows occur in the tables, or if the occurrence of multiple data rows in the output table is irrelevant.
- SQL queries should preferably be formulated as **joins**, or subqueries if necessary. Joins are generally considered easier to process.
- Joins can be additionally **optimized** using the DBINDEX= and DBKEY= data set options.
- If **disk space** is more important than speed when working with PROC SQL, views can be used instead of tables.

Merging Tables: SQL Instead of DATA Step? It Depends…

Programmed properly, SQL is more efficient than a DATA step merge. It saves sorting (processor load), interim steps (additional tasks), and auxiliary files (disk space). A disadvantage of PROC SQL compared to MERGE might be the number of tables that can be merged in one step at once, depending on the method of merging. With inner joins, a maximum of 256 tables can be joined simultaneously in one step (from SAS Version 9.2); all other join variants are always sequences of two-table joins (apart from subqueries). In comparison, a DATA step can process an unlimited number of tables simultaneously. From the syntax to be programmed, the DATA step appears more attractive first.

However, the merging of only two tables without carefully checking the work steps is inadvisable – all the more so when several tables are involved. If one considers the effort required to check "difficult" tables, especially if they are non-proprietary but from third-party providers, then the advantage of a "user-friendly" syntax and the gain in programming time no longer makes a difference.

The question of whether to use SQL, DATA step, or alternatives like hash objects in a DS2 or DATA step for joins is a question that users can only answer for themselves after checking their tables, including their volume, structures, and data quality. Table 8.5-1 is an orientation to table characteristics.

Table 8.5-1: When to Use SQL, the DATA Step, or Alternatives

Approach	Data volume	Many tables	DQ not existent	Structural inconsistency	Sorting required	CPU time relevant
PROC SQL	●	●			●	
DATA Step	●	●	●	●		
Hash Objects	●		●	●	●	●

All approaches, whether PROC SQL, DATA step, or hash objects, can handle large data volumes. Operations for many tables can be programmed easily with PROC SQL or the DATA step. If, however, additional operations are required to check and guarantee data quality, you might have to switch to the flexibility of the DATA step or hash objects. The same applies if the structural consistency is suboptimal for joining and would have to be established first. If, on the other hand, data needs sorting, hash objects and PROC SQL tend to be preferable to the DATA step. If the CPU time plays a decisive role, especially with a few, very large SAS tables, the remaining option is alternatives like in-memory or threaded processing using hash objects, DS2, or FedSQL.

Whether to use PROC SQL when joining tables instead of the DATA step depends on characteristics of the tables to be processed, including volume, number, consistent structure and sorting, as well as the desired advantages of the approaches like uncomplicated programming, flexibility and/or efficiency in processing. It depends…

Accelerated Merging of Tables (SQL)

When merging tables, you can accelerate both the joining and the appending of tables:

- **Joining of tables: ON/WHERE clauses (Inner Joins)**
  The ON or WHERE clauses in inner joins (types 1 and 2) should be formulated as *where t1.ID = t2.ID ;* and not as *where t1.ID - t2.ID = 0 ;* . SAS processes the first variant more efficiently than the second.
  In addition, joins can be further optimized using the DBINDEX= and DBKEY= data set options. DBINDEX identifies and verifies that indexes exist in a DBMS table. DBKEY names a key variable for optimizing a DBMS query. However, if used incorrectly, the DBINDEX= and DBKEY= options can not only worsen performance, but under certain circumstances (for example, duplicate entries) can also affect the result of a query. For more information about these options and the performance gain achieved, see the SAS technical documentation.

- **Appending of tables: SQL keyword ALL**
  The keyword ALL should always be used in set operations if the tables to be appended to one another do not contain multiple data rows or if the occurrence of multiple data rows in the output table is irrelevant. By default, set operators usually automatically exclude multiple rows from the created tables (output tables). The optional keyword ALL keeps multiple rows from the created tables, shortens the execution by the step of exclusion (also, if there are no multiple rows at all), and can significantly accelerate the performance of the query expression. The interesting thing about this measure is that, in contrast to the techniques described earlier using KEEP/DROP, WHERE/IF or DELETE, multiple lines are kept in order to accelerate a process by shortening an execution by one step (the exclusion).

- **Indexing (INDEX)**
  As of SAS 9.2, indexing can also accelerate the joining of tables. As a prerequisite, you must check whether you are using indexing and whether the columns involved in the join are represented by the indexes.

Accelerating Queries (SQL)

For the acceleration of queries, SAS offers various options that can be used individually or in combination:

- **Indexing (INDEX):** The indexing approach can also accelerate queries. As a prerequisite, you need to check whether you are using indexing and whether the columns involved in the query are represented by the indexes. Since SAS 9.2, SAS accelerates the processing of SELECT DISTINCT statements with the help of index files.
- **Optimizing the PUT function (REDUCEPUT=, SQLREDUCEPUT=):** Queries on formatted data can be accelerated by optimizing the PUT function. Using the REDUCEPUT= SQL option or SQLREDUCEPUT= system option, SAS optimizes the PUT function before the query is executed and replaces it by a logically equivalent expression. SQL queries on formatted data, especially with PUT functions in WHERE or HAVING clauses such as the following example, would be optimized by these options.

Program 8.5-2: SQL Query Benefitting from Optimizing a PUT Function
```
select A, B from &my_lib.
    where (PUT(A, yesno.) in ('yes', 'no'));
```

Reducing the PUT functions would also make it possible to pass a larger part of the query to the DBMS for a more performant processing, which would also speed up the processing of a WHERE clause if the query contained one.

- **Replacing reference to the SAS functions DATE, TIME, DATETIME, and TODAY (CONSTDATETIME):** If queries contain time or date values, you can optimize query performance by reducing the reference to the SAS functions DATE, TIME, DATETIME, and TODAY, for example. With the appropriate setting (SQL: CONSTDATETIME, System Option: SQLCONSTDATETIME), PROC SQL executes the query only once and uses the obtained time or date values as equivalent constants throughout the query. This procedure ensures consistent results even if the time and date functions are queried several times during a query or if the query executes the time and date functions close to a time or date threshold. In addition, this approach also enables you to transfer a larger part of the query to the DBMS for a more performant processing.

Since SAS 9.2, the %INDTD_PUBLISH_FORMATS macro for Teradata also enables you to publish the implemented PUT function as a named SAS_PUT() function.

- **Avoiding remerging (NOSQLREMERGE):** I/O intensive remerging (reunion) might occur when aggregation functions are used in a SELECT or HAVING clause. Avoiding remerging avoids I/O.

8.5.3 Other Techniques

This section introduces techniques for accelerating processing steps. These are presented as further possibilities to accelerate processes: self-written functions versus SAS features, compiling and storing DATA steps (stored compiled DATA step programs), as well as macro variables and programs for automating recurring processes.

Do It Yourself or Let SAS Do It? Self-written Functions versus SAS Features

Many SAS users have probably asked themselves what is more efficient: writing a function yourself using PROC FCMP, PROC PROTO, PROC IML, operators, or arrays? Or using a function offered by SAS? Even with simple, self-written functions like summation, the performance gain is modest (Langston, 2005) provided that the self-written function is programmed correctly, which experience shows is not self-evident. When programming with Base SAS, many paths lead the user to the same goal, but often at different speeds. When programming more complex functions, the user runs the risk of not having implemented the most efficient approach right from the start, not to mention subtle programming errors. For more complex functions, it is recommended to use already existing SAS functions instead of self-written programs whenever possible. (See an overview in Chapter 10.) With the advantage of higher performance and functionality, users save not only the effort of self-programming but also testing and documentation. Legal aspects such as warranty or even liability issues should not even be mentioned.

Compiling and Storing DATA Steps

By compiling and storing repeatedly used DATA steps as stored compiled DATA step programs, CPU performance can also be optimized. With stored compiled DATA step programs, the process of repeatedly calling DATA steps is accelerated by using the PGM= statement because DATA steps are only compiled and stored once in storage and call. SAS programs stored and compiled in this way are nothing more than "normal" DATA steps, which however relieve the CPU of efforts involved in their repeated compilation.

Phase I: Creating a Stored Compiled DATA Step Program (Principle)	**Phase II: Calling a Stored Compiled DATA Step Program (Principle)**
Program 8.5-3: Phase I: Creating	**Program 8.5-4: Phase II: Including**

Phase I:
```
libname STORED 'D:\';

data MYDATA
 / pgm=stored.sample;
   set SASHELP.SHOES ;
   if REGION="Africa" then
      do;
         ... ;
      end;
   else
   if REGION="Asia" then
      do;
         ... ;
      end;
run ;
proc print data=MYDATA;
run ;
```

Phase II:
```
libname STORED 'D:\';

data pgm=stored.sample;
   redirect
input SASHELP.SHOES=MYDATA ;
run;

proc print data=MYDATA;
run ;
```

Note: The program sequence on the left accomplishes the same as the program sequence on the right.

Once created as a stored compiled DATA step program **(Phase I)**, the included DATA steps do not need to be compiled again when called **(Phase II)**. Further advantages of stored compiled DATA step programs are that for users, the "actual" SAS program becomes much more manageable through the "outsourcing" of DATA steps, which can be seen in the two different lengths of Programs 8.5-3 and 8.5-4. Other programs and SAS applications can also access the "outsourced" DATA steps. This way of outsourcing is currently only possible for DATA step programs. For other applications, it is recommended to use SAS macro variables and programs. SAS macros, in turn, can be integrated and executed via %INCLUDE, the Stored Compiled Macro Facility, and the Autocall Facility (for the required system options, see Section 4.2).

Macro Variables and Programs for Repetitive Processes

Macros are the automation of recurring processes within the same application. Applied to SAS, macros are the automated processing of the same application for many different variables, the efficient processing of different applications for one and the same variable, or often a combination of both. If, for example, the same SAS program is applied to other variables or data sets with minimal or no adjustments, this is an ideal case for rewriting the original SAS program into a SAS macro. With the help of simple but effective examples, the chapter on macro programming describes how both the programming and the execution of programs can be accelerated many times over. (See Subsection 4.2.3.) The main advantages of SAS macro programming are:

- **Scope of performance:** In principle, macros cover all the features of conventional programming using SAS, including arrays, loops, and so on, including validity, automatability, and reusability.
- **Increase of efficiency:** SAS macros exponentiate the efficiency of syntax programming. Macros save valuable time during programming and are less prone to errors later during execution.
- **Speed:** Macros usually run faster than repeatedly executed normal statements and are often programmed many times faster.

Using SAS macros can extend the performance scope of PROC SQL in such a way that this book dedicates a separate, extensive chapter to macro programming. (See Chapter 4.)

The time saved during multiple execution, especially for SAS macros, and during programming, especially for stored compiled DATA step programs, usually outweighs the time required for single compilation and execution.

8.6 Data Processing in SAS or in the DBMS: Tuning of SQL to DBMS

The flexibility of SAS is characterized by the fact that SAS either processes the data itself or passes it to the DBMS for processing, depending on which processing location ensures higher performance. SAS does not always process the data itself but tries to take advantage of the processing benefits of the DBMS by always transferring certain SQL operations to the DBMS. DBMS can be: DB2 UNIX/PC, Google BigQuery, Greenplum, Hadoop, Informix, ODBC, OLE DB, Microsoft SQL Server, Oracle, PostgreSQL, Redshift, SAP HANA, Snowflake, and Teradata.

8.6.1 Transfer of SQL to the DBMS

As soon as data can be processed in a DBMS, this reduces the effort to load the data and increases performance, especially when PROC SQL passes statements to the DBMS to limit the number of rows for very large amounts of data. Using SAS/ACCESS, operations can be passed to the DBMS for processing in at least two ways. (See Section 10.3.) Weighing the respective advantages and disadvantages of these accesses is a prerequisite for choosing the appropriate efficient approach for your own requirements.

- **Implicit pass-through** (LIBNAME statement): With help from special routines that are provided by the particular SAS/ACCESS engine, SQL translates the submitted statements into the DBMS-specific SQL and then passes this SQL to the DBMS for further processing. Therefore, the performance of the implicit pass-through can be achieved without having to write DBMS-specific SQL, especially when directly passing

ANSI SQL functions. Further options allow more control of the DBMS, including table locks (READ_LOCK_ TYPE=, UPDATE_LOCK_TYPE=) or the optimization of joins or WHERE clauses. The option DIRECT_EXE also enables you to pass an SQL statement (for example, DELETE) directly to a DBMS.

- **Explicit pass-through** (CONNECT TO component): If the explicit pass-through is used, the advantage is that ANSI SQL can be passed directly to the DBMS for execution. In addition, the explicit pass-through also accepts non-ANSI SQL and can pass DBMS-specific SQL directly to the DBMS. In this respect, the explicit pass-through is considered to be extremely high-performant. Further options allow the optimization of joins or queries. In interaction with SAS/AF applications, the explicit pass-through also enables control over COMMIT and ROLLBACK. A disadvantage of this approach can be, depending on the requirements, that DBMS-specific SQL might have to be written by the user.

8.6.2 Tracing of SQL Passed to the DBMS (SASTRACE)

It is often helpful to check which SQL code was passed to the DBMS and whether an operation was processed in SAS or in the DBMS. The SAS tracing options (initialized with SASTRACE=) are very helpful in determining which SQL code (and thus which operation: query, join, and so on) was actually passed to the DBMS, and whether an operation is processed in SAS or was passed to the DBMS for processing.

- For example, PROC SQL is not always able to **transfer** a query completely or at all to the DBMS. If PROC SQL is not able to pass a query completely to the DBMS, PROC SQL sometimes tries to pass a subquery to the DBMS afterward.
- If you want to prevent certain **operations** from being processed in the DBMS, you can use the LIBNAME option DIRECT_SQL.
- **Certain circumstances** can prevent operations from being passed to the DBMS at all, but to be executed in SAS itself. These include INTO clauses, various settings for joins (see below), remerges (reunions), union joins, and truncated comparisons using EQT, GTT, LTT, and so on.

The next sections describe PROC SQL operations that SAS can pass to the DBMS for processing, including functions, joins, operators, and WHERE clauses.

Finally, the DIRECT_EXE option enables you to pass an SQL statement in the LIBNAME statement directly to a DBMS. The performance is significantly increased by using DIRECT_EXE.

8.6.3 PROC SQL Operations that SAS can Pass to the DBMS

The following sections describe various PROC SQL operations that SAS can pass to the DBMS for processing, including

- **Accesses and functions** (ANSI SQL compatible versus DBMS-specific)
- **Joins** (variants, operators, restrictions)
- **WHERE clauses** (missings and SAS functions)

Accesses and Functions

This section describes how SAS can use SAS/ACCESS to pass aggregating functions in PROC SQL, ANSI SQL compatible, or DBMS specific to the DBMS for processing.

- **Implicit pass-through:** SAS/ACCESS translates other SAS functions directly into DBMS-specific functions, which are then passed on to the DBMS if the DBMS supports these functions. Because of this translation process, the directly passing explicit pass-through is often considered to be more performant. However, you can use the LIBNAME option DIRECT_EXE to pass an SQL statement directly to a DBMS. If SAS/ACCESS cannot translate a SAS function in a WHERE clause into a DBMS function, SAS retrieves all rows from the DBMS and applies the WHERE clause to them itself. The use of the implicit pass-through also affects the use of the join operators DISTINCT and UNION or the table locks. (See below.)

- **Explicit pass-through:** If the explicit pass-through is used, SAS/ACCESS can pass the aggregating functions MIN, MAX, AVG, MEAN, FREQ, N, SUM, and COUNT directly to the DBMS because these are ANSI SQL compatible aggregating functions. The explicit pass-through can also pass DBMS-specific SQL directly to the DBMS. A disadvantage of the explicit pass-through is that, unlike the implicit pass-through, many SAS functions and routines cannot be passed directly to the DBMS; it might therefore be necessary to write DBMS-specific SQL yourself.

The type and number of SAS functions that are translated or directly passed can differ from DBMS to DBMS. For more information, please refer to the SAS documentation.

Joins

SQL join statements are among the SQL statements that SAS/ACCESS tries to pass directly to the DBMS for processing by default. The processing of joins in the DBMS significantly increases performance, especially with large amounts of data. If a join of two or more tables is to be made, SAS/ACCESS can in many cases pass the join to the DBMS for processing. The following join variants are possible for a pass to the DBMS:

- **for all DBMS:** Inner joins between two or more tables.
- **for DBMSs that support (standard) ANSI outer join (standard) syntax:** Outer joins between two or more DBMS tables.
- **for ODBC and Microsoft SQL Server:** Outer joins between two or more tables. Outer joins might not be combined with inner joins within a query.
- **DBMSs that support nonstandard outer join syntax (Oracle, SAP ASE, Netezza, and Informix):** Oracle supports nonstandard outer-join syntax with the following restrictions: Full outer joins are not supported, and only a comparison operator is allowed in an ON clause. SAP ASE also supports nonstandard outer-join syntax with the following restrictions. Full outer joins are not supported, and only the "=" comparison operator is allowed in an ON clause. Netezza supports a cross-database JOIN operation. For Informix, the maximum number of tables that you can specify to perform a join is limited to 22. Additional DBMS-specific restrictions might apply. For example, SAP ASE evaluates multi-joins that have WHERE clauses differently than SAS (here, use the DIRECT_SQL=LIBNAME option to process the joins internally).

Before the join is executed, PROC SQL uses certain criteria to check whether the DBMS is capable of doing this at all. If possible, PROC SQL passes the join to the DBMS that executes the join and returns only the results to SAS. If the DBMS is not able to execute the join, PROC SQL processes the join. For DBMS-specific criteria, see the SAS technical documentation.

If a join cannot be passed to the DBMS for direct processing, PROC SQL attempts to join the tables in SAS itself. The options DBINDEX=, DBKEY=, and MULTI_DATASRC_OPT= can help to optimize performance. However, the options DBKEY= and DBINDEX= can, under certain circumstances, also worsen performance and even affect the result of a query, for example, in case of duplicate entries. For more information on these options, please refer to the current SAS technical documentation.

With a LIBNAME statement, DISTINCT and UNION are processed in the DBMS and not in SAS. If PROC SQL detects a DISTINCT operator, it passes it directly to the DBMS. After checking for multiple rows, the DBMS returns only the unique rows to SAS.

Restrictions When Passing Joins to the DBMS

A successful join depends on the way the SQL is written, and whether and to what extent the DBMS accepts the generated syntax. The following are the most common reasons why join statements in PROC SQL cannot be transferred directly to a DBMS for further processing:

- The generated SQL syntax is not accepted by the DBMS for various (mainly **DBMS-specific**) reasons. In this case, PROC SQL tries to join the tables in SAS.

- The SQL query requires **several librefs**, but they do not have common connection characteristics such as different servers, user IDs, and so on. In this case, PROC SQL does not attempt to pass the SQL statements to the DBMS for direct processing.
- Using **data set options** in a table referenced by the query automatically results in not passing the statement to the DBMS for direct processing.
- Using certain **LIBNAME options** also automatically results in not passing the statement to the DBMS for direct processing. These LIBNAME options include table locks (READ_LOCK_TYPE= LIBNAME, UPDATE_LOCK_TYPE= LIBNAME).
- Certain **DIRECT_SQL= settings** also cause statements being partially or not at all passed to the DBMS. With YES (default), PROC tries to pass SQL joins directly to the DBMS. If the DBMS is not able to execute the join, PROC SQL processes the join in SAS. With NO, no SQL joins are passed to the DBMS but other SQL statements are. With NOWHERE, no WHERE clauses are passed to the DBMS. With NOFUNCTIONS, no functions are passed to the DBMS. With NOGENSQL, PROC SQL does not attempt to pass the generated syntax for joins directly to the DBMS. With NOMULTOUTJOINS, multiple outer joins are not passed to the DBMS. With NONE, no statements are passed to the DBMS at all.
- **SAS functions** in the SELECT clause can also cause joins not to be passed to the DBMS, for example, if the DBMS does not support these functions.

WHERE Clauses

SAS tries to take advantage of the processing capabilities of DBMS by translating WHERE clauses into generated SQL code. The performance gain, especially with very large amounts of data, can be considerable. We have already referred to various recommendations for writing high-performing WHERE clauses. In interaction with DBMS, further recommendations are:

- **WHERE** is preferable to IF if a DBMS is optimized for WHERE processing (Langston, 2005).
- **Tuning** the programming of SQL syntax to the location and method of processing missing values (NULL values, NULL). Since most DBMSs exclude missing values in a WHERE clause in a query (whereas SAS does not), inconsistent handling of missing values can lead to undesired results. You must therefore decide between adding the condition "... and VALUE is not missing" for processing in SAS and possibly omitting this condition for further processing of data in non-SAS DBMS. (See Chapter 2.)
- If implicit pass-through's **LIBNAME** statement is used, SAS functions in WHERE clauses can often be translated into DBMS-specific functions so that they can be passed to the DBMS. If SAS cannot translate a SAS function in a WHERE clause into a DBMS function, SAS retrieves all rows from the DBMS and applies the WHERE clause to them itself.
- In order **not** to pass WHERE clauses, SAS functions, or generated SQL code to the DBMS, the DIRECT_SQL= settings NOGENSQL, NOWHERE, or NOFUNCTIONS can be used. If appropriate and used correctly, the options DBINDEX=, DBKEY=, and MULTI_DATASRC_OPT= might improve performance.
- Depending on the circumstances, creating and using an **index** is also recommended if the SAS functions SUBSTR and TRIM are used in the WHERE clause.

8.6.4 Passing an SQL Statement Directly to a DBMS (DIRECT_EXE)

The DIRECT_EXE option enables you to pass an SQL statement directly to a DBMS. The performance can be significantly increased by using DIRECT_EXE. For example, the DELETE statement can be passed directly to the DBMS for execution using the LIBNAME statement and DIRECT_EXE=DELETE (provided it does not contain a WHERE clause). SAS no longer has to read the entire table and delete row by row, but PROC SQL passes the DELETE statement directly to the DBMS, which then deletes all rows (but not the table).

8.6.5 More Tips and Tricks for Performance: VIO Method

The principle of the VIO method (VIO="virtual I/O") is based on the fact that temporary SAS directories like WORK are addressed as virtual I/O files. The VIO method is intended to reduce I/O operations and save disk space. To initialize the VIO method, you only need to specify UNIT=VIO as an engine option in a LIBNAME statement or function. The VIO method is currently only available for the z/OS operating environment.

8.7 Step by Step to Greater Performance: Performance as a Strategy

The measures presented in this chapter can be classified according to whether they help to optimize I/O, CPU, or memory. In combination, they are able to optimize SAS toward even higher performance.

8.7.1 Optimization of I/O

Input/output is one of the most important factors for performance. I/O are repeated processes when reading or searching for data in the management or analysis of data. To optimize I/O, SAS offers the following options and measures:

- **Reduce the data load** to the required tables, data, or bytes by using DELETE, KEEP/DROP, WHERE/IF, INOBS/OUTOBS, and OBS/FIRSTOBS, or LENGTH.
- **Write programs** in such a way that fewer accesses and thus fewer read or search processes are required by indexing, "abbreviations," views, or specifying engines.
- **Set the system** to read or search more data as soon as I/O processes are required, for example, by BUFNO/BUFSIZE, CATCACHE, or COMPRESS, and SASFILE.

8.7.2 Optimization of Memory Usage

Memory usage can also be optimized. Available memory can be increased or limited for other operations. Most other measures are at the expense of I/O or CPU. SAS offers the following options and measures to optimize the use of the main memory:

- **Increase** the available memory by MEMSIZE/MEMLEAVE; however, this reduces the time required for reading data into the memory.
- **System options** such as SORTSIZE= and SUMSIZE= limit the memory that is made available for other operations such as sorting or aggregating.
- Available memory can also be optimized at the expense of I/O or CPU, provided that the use of the main memory actually has **priority** over I/O or CPU.

8.7.3 Optimization of CPU Performance

CPU performance can be optimized by optimizing I/O processes by reducing their number or increasing memory. Thus, when I/O processes or memory are optimized, CPU time is generally also optimized. Depending on the optimization of the memory, other processes might have unintentionally less memory available. To optimize CPU, SAS offers the following possibilities and measures:

- **Use In-memory Processing.** When using approaches like hash objects or DS2, data is processed in memory and not on the disk. (See Chapters 6 and 7.)
- Work with so-called **Stored Compiled DATA Step Programs** instead of repeatedly calling DATA steps. DATA steps can be written once, compiled, and then called several times.
- Use the **LENGTH statement**. The number of bytes should correspond to the available memory space and the required precision of variables. However, the higher the byte number of variables, the slower the associated read and save processes.
- Optimization of the **search times** (for example, the PATH= system option). The most frequently used directories should be placed at the beginning of the list, the least frequently used directories at the end.
- **Parallel processing.** Data is processed simultaneously, not sequentially. (See Chapter 7.)

Taken together, these measures can support the optimization of I/O, CPU, and memory. However, a last unknown quantity is the testing of the efficiency of the measures and program examples on the tables and data of the user's system. This side of performance depends on many system-specific factors like the number of tables, data density, table structure, as well as software and hardware. You cannot avoid your own benchmark testing, on your own data, in your own system on site.

8.7.4 Plan the Program, Program the Plan, Test the Program

Performance is more than narrowing down large tables by filtering, compressing, or reformatting. These techniques would only tackle the tables themselves. In principle, performance starts with the user first and then with the system. Performance is the result, the goal of a strategy. At the beginning of a strategy there is the need to get an overview and to secure resources.

Get an overview of the checks and measures to be carried out. Create priorities. Each application area can have different priorities. In general, relevant tables have priority over interesting data. In data warehouses, data with a high ROI has priority over data with a low ROI, and so on. In research, primary variables have priority over secondary variables. If you are not a decision maker, try to agree on a generally accepted priority list with all parties.

This is not always easy in practice because different requirements can often conflict, or the same requirement is often assigned different priorities. A related problem is that although criteria for performance can be objective and context-independent, the requirements made on these criteria by those involved are often subjective and context-dependent.

The definition of performance criteria is one of the most demanding requirements, and experience has shown that this can often be met by several difficulties at once:

- Decision-makers and sometimes also users cannot or do not always want to assess the **status quo** prior to a measure for more performance. The reasons are often lack of time, lack of patience, or insufficient technical understanding.
- Decision-makers and sometimes also users cannot or do not always want to **communicate the required complexity** of the performance measures. Experience has shown that this situation is more demanding when these measures are to be communicated to external third parties who have little technical knowledge or a pseudo-understanding.
- Necessary **performance measures** often cannot be **prioritized** sufficiently. Some decision-makers want everything at once, and as immediately as possible. Some decision-makers might content themselves with short-sighted, low-cost criteria, but in the long run, this will be inadequate and cause even more expensive consequences.

Translate the priorities into specific individual measures. If it is not possible to agree on the maximum requirements in each case, then minimum performance requirements should be formulated. Use benchmark tests to check the success of your measures based on CPU time, disk space, time spent, hardware costs, additional data volumes, and so on. Communicate your successes, create even more acceptance, and ask for or demand even more support. A final option is…

8.7.5 Go Beyond SQL and Programming

Despite all these optimizations, the data load might be still too large or the processing too complex that it is time to consider switching to a more powerful system. At that point, it's not about the performance of individual processes, memory, or space requirements anymore but about the overall system-wide performance of all processes. For an individual user, this might mean the purchase of a more powerful desktop computer, but for employees or departments in a company, this might mean the transition to more powerful environments such as grid or cloud computing.

8.8 Beyond SQL and Programming: Environments

We cannot close a chapter about performance and efficiency without going beyond the focus on SQL and programming to talk about processing in alternative environments like cloud-based or grid-based programming. Both environments offer high-performance processing and are scalable. The fundamental conceptual difference between grid and cloud computing is that grid environments are usually owned and managed by a company, whereas cloud environments are usually leased and managed by providers. Grid environments appear more system-oriented like distributed computing, in contrast to a more service-oriented view like SaaS. SAS Cloud Analytic Services (CAS) were already introduced in Chapter 7. A short recap of CAS follows an introduction to SAS Grid Programming.

8.8.1 SAS Programming on a SAS Grid

The typical programming environment is an individual SAS license for a user's computer or one or more servers shared by several users. Many of the solutions presented above are for typical problems of these environments. An individual's computer might not have enough resources in disk, memory, and so on to run complex analytic processes. Several users or departments compete for resources on the servers, while some might be underused due to suboptimal load balancing.

SAS offers a SAS Grid as a grid computing environment, a shared file system that is made of a set of servers. The shared file system can be used to share SAS permanent data libraries, SAS programs, and macros. Users submit their programs ("jobs") to run on the grid; they do not submit their SAS programs to a specific server. After a job has been dispatched ("grid session"), the SAS Grid Manager finds the best server for the respective SAS request.

Users in a SAS Grid Computing environment benefit from massive advantages: fast processors, a lot of memory, and being able to process big data volumes at a high velocity on the same operating system at a high availability. In the case of small data, processing in the grid computing environment might take longer though, as starting the grid sessions tends to add extra time to the overall processing.

SAS Grid Manager uses either the IBM Platform LSF or Apache Hadoop YARN. Grid technology is part of several multiple solutions and products in the SAS High-Performance Analytics area and SAS Viya. Typical SAS clients are SAS Studio, SAS Enterprise Guide, SAS Enterprise Miner, SAS Data Integration Studio, or SAS Display Manager. In the case of Enterprise Guide, most of the newer grid environments use servers and settings that force Enterprise Guide to use grid sessions. A job runs on the grid automatically without any changes.

You might be tempted to think that if you work in a grid environment, you don't need to optimize your SAS code anymore. However, servers in the grid process SAS code the same as computers or servers not in a grid. This means that the optimization potential remains unchanged. Every measure such as reducing the size of tables that benefits the processing on a single computer also benefits the processing on the grid.

Most existing SAS programs run on the grid without major adjustments, with maybe the exceptions of changes in path, file, and library statements to use shared data locations. To take advantage of the grid computing environment, you might add SAS statements like GRDSVC_ENABLE(), SIGNON/SIGNOFF, and RSUBMIT/ ENDRSUBMIT (mostly SAS/CONNECT). The GRDSVC_ENABLE() function identifies and enables grid resources. Worth mentioning is PROC SCAPROC, which executes an existing, fully executable SAS program and mostly automatically creates a fully executable SAS code version that takes advantage of the SAS Grid Computing environment.

SAS Studio is a web interface and can be used in a grid environment and in a cloud environment. Like Enterprise Guide, SAS Studio offers process flows for visual programmers. A process flow is a graphical representation (node) of a process, typically a SAS program. SAS programmers can write and submit programs in noninteractive, interactive, and batch mode.

Starting with Enterprise Guide 8.2, a desktop client, you can also directly connect to SAS Viya and submit SAS code to CAS without having to install a full SAS 9 environment.

8.8.2 SAS Cloud Analytic Services (CAS)

SAS CAS is a server that provides the cloud-based run-time environment for data management and analytics with SAS. CAS is part of SAS Viya, an open, cloud-enabled platform that supports high-performance analytics. A SAS Viya license is required. Open means that this platform is designed to work with SAS 9.4 and other languages such as Java, Python, or R. SAS Viya does not replace SAS 9.4. You can submit SAS 9.4 programs from the SAS 9.4 environment, from the SAS Windowing environment, or from SAS Studio to SAS Viya for processing. Typical web applications are SAS Visual Analytics, SAS Environment Manager, or SAS Studio.

In CAS, some of the previously mentioned recommendations do apply such as partitioning and sorting optimize BY-group operations on the partition keys and Order-By operations on the SORT fields. Also, indexes do improve the performance of a WHERE clause.

However, some of the recommendations do not apply in CAS. For example, indexes should not be used for anything else like join optimization, or BY-group processing, CAS action, and SAS Visual Analytics. Also, indexes can be created only by a CAS action using the casOut option.

Chapter 9: Tips, Tricks, and More

This chapter consists of four sections on different facets of working with PROC SQL:

- **Section 9.1** introduces runtime logging of SAS programs as a key to performance. (See Chapter 8.) Four scenarios are presented with their respective advantages and disadvantages, how the runtime could be captured with the SAS option FULLSTIMER, as well as an example using ARM macros.
- **Section 9.2** introduces working with SAS dictionaries. Dictionaries contain information about columns and variables from SAS files, directories, SAS tables and views, SAS catalogs and their entries, SAS macros, and current settings of SAS. This chapter presents three applications of dictionaries.
- **Section 9.3** presents several PROC SQL applications for data handling and data structuring, including creating "exotic" column names, creating a primary key (using MONOTONIC and safer options), segmenting a SAS table (MOD function), defining a tag set for exporting SAS tables to CSV format, and protecting SAS tables with passwords.
- **Section 9.4** compares table updating using PROC SQL with the DATA step. The DATA step UPDATE statement shows a behavior with multiple entries and/or missing values in key ID that could be undesirable. However, PROC SQL is flexible enough to perform the desired updates.

9.1 Runtime as the Key to Performance

In general, performance can be assessed using three basic criteria:

- real time
- CPU time
- memory

Real time is the chronologically elapsed time for processing data. The real time is influenced by the load of the system, the CPU, and other factors. The more users access the same system, the less CPU is available to a single user. A more informative measure is therefore the CPU time.

CPU time is the actual time a CPU needs to process data, regardless of load and other factors. (For a differentiation of user and system CPU time, see the approach to the FULLSTIMER option.) For an assessment of real and CPU time, it is helpful to form a relation or a quotient from the measurements for real and CPU time. (See Scenario II.) If the real time corresponds to the CPU time, the quotient should ideally be 1:1. In systems that are at the limit of their capacity utilization, ratios such as 20:1 may be achieved. In peak times, when yearly, quarterly, and monthly financial statements coincide, even worse ratios might occur.

The following scenarios present several ways to measure the runtime of SAS processes. The scenarios are preceded by a sequence of DATA step and SQL steps whose runtime is to be determined using the different approaches. The derivation of actions is not the subject of this section.

In principle, the lower the CPU time and real time, the less time SAS needs for processing data. For less favorable ratios of real time to CPU time, processing adapted to a suboptimal system in specially set up time windows and system-side optimizations is recommended. Keywords describe advantages and disadvantages of the presented approaches.

9.1.1 Overview

- **Scenario I:** Default setting: CPU time versus real time.
- **Scenario II:** FULLSTIMER option: CPU time, real time, and memory.
- **Scenario III:** Aggregating calculation of the runtime (macro).
- **Scenario IV:** Differentiated calculation of the runtime (ARM macros).

Program 9.1-1 is an example program sequence whose runtime is to be determined. This program sequence is always referenced in the SAS code with "*/* place program block to be measured here */*".

Program 9.1-1: SAS Sample "Program Block to Be Measured Here"

```
data test1 ;
   do i = 1 to 10000000;
      x = 10 * ranuni(1234);
      y = 1 + 2 * sqrt(x) + .5 * rannor(5678);
      output ;
      end ;
 run ;

 data test2 ;
   do i = 1 to 10000000;
      x = 10 * ranuni(1234);
      y = 1 + 2 * sqrt(x) + .5 * rannor(5678);
      output ;
      end ;
 run ;

data test3 ;
   do i = 1 to 10000000;
      x = 10 * ranuni(1234);
      y = 1 + 2 * sqrt(x) + .5 * rannor(5678);
      output ;
      end ;
 run ;

proc sql ;
     create table TEST_ALL as
           select * from TEST1
                union all
           select * from TEST2
                union all
           select * from TEST3 ;
 quit ;
```

Note: Users can achieve significantly less favorable runtimes if they initiate additional CPU-intensive processes in the background in parallel to the execution of this program sequence if their SAS environments allow this.

9.1.2 Scenario I: Default Setting: CPU Time versus Real time

Advantages:

- Default setting, does not need to be specified explicitly.
- Measures the runtime of individual program sections.

Disadvantages:

- The total runtime of all individual programs is not calculated. Only measures the runtime of individual program sections.
- Other parameters (for example, memory) are not obtained.

Program 9.1-2: Scenario I: Default

```
/* place program block to be measured here  */
/* place program block to be measured here  */
```

Figure 9.1-1: Output in SAS Log: Scenario I

```
                        (Output abbreviated)
NOTE: The data set WORK.TEST1 has 10000000 observations and 3 variables.
NOTE: DATA statement used (Total process time):
      real time             2.21 seconds
      cpu time              1.64 seconds
                        (Output abbreviated)
NOTE: Table WORK.TEST_ALL created, with 30000000 rows and 3 columns.
NOTE: PROCEDURE SQL used (Total process time):
      real time             12.34 seconds
      cpu time              8.21 seconds
```

As shown In Figure 9.1-1, to create the WORK.TEST_ALL table, 12.34 seconds passed in real time. However, the CPU only needed 8.21 seconds for this task.

9.1.3 Scenario II: FULLSTIMER Option: CPU time, Real time, and Memory

Advantages:

- Only needs to be specified once as a system option, provided you have the appropriate administrator rights. Other operating systems might require different settings.
- Measures the runtime of individual program sections.
- Further parameters such as memory are obtained.

Disadvantages:

- The total runtime of all individual programs is not calculated.

Program 9.1-3: Scenario II: FULLSTIMER option

```
options fullstimer ;


/* place program block to be measured here  */
/* place program block to be measured here  */
```

Figure 9.1-2: Output in SAS Log: Scenario II: FULLSTIMER option

```
                        (Output abbreviated)
NOTE: The data set WORK.TEST1 has 10000000 observations and 3 variables.
NOTE: DATA statement used (Total process time):
      real time             1.72 seconds
      user cpu time         1.31 seconds
```

```
          system cpu time        0.29 seconds
          memory                 395.53k
          OS Memory              11508.00k
          Timestamp              01/05/2021 11:48:40 AM
          Step Count                        5   Switch Count   0

                              (Output abbreviated)
NOTE: Table WORK.TEST_ALL created, with 30000000 rows and 3 columns.
NOTE: PROCEDURE SQL used (Total process time):
          real time              12.81 seconds
          user cpu time          8.15 seconds
          system cpu time        0.78 seconds
          memory                 5887.28k
          OS Memory              16632.00k
          Timestamp              01/05/2021 11:48:59 AM
          Step Count                        8   Switch Count   0
```

To create the WORK.TEST_ALL table, 12.81 seconds passed in real time. However, the user CPU time (the time required by the CPU to execute the SAS code) required only 8.15 seconds for this task. The system CPU time (the CPU time used for other tasks related to the SAS process being executed) was 0.78 seconds for this task. Memory describes the amount of memory required to run a step. For example, to create WORK.TEST_ALL required 5887.28k. OS Memory, the maximum amount of memory that a step requested, was 16632.00k.

Note: When multi-threading and processing with multiple CPUs, the CPU time might be higher than real time. In addition, the used memory is output. Expanding the memory using the SAS software option MEMSIZE or hardware can reduce the CPU time. At the end of this chapter, I will show how to query detailed information about memory usage using PROC OPTIONS.

9.1.4 Scenario III: Aggregating Calculation of the Runtime (Macro)

Advantages:

- Must only be specified at the beginning and end of a block of programs to be executed.
- Macro measures the total runtime of the executed programs.

Disadvantages:

- Does not measure the runtime of individual program subsections. To obtain the runtime of individual program subsections, the macro would have to be placed at the desired positions in the SAS code.
- Further parameters such as memory are not determined.

Program 9.1-4: Scenario III: Aggregating Calculation of Runtime

```
%let start=%sysfunc(datetime(),);
%put -------- Begin of runtime measurement:
    %sysfunc(putn(&start,datetime40.))--------;

/* place program block to be measured here  */
/* place program block to be measured here  */

%let end=%sysfunc(datetime(),);
%put -------- End of runtime measurement:
%sysfunc(putn(&end,datetime40.))--------;

%put -------- Total duration:
    %sysfunc(putn(%sysevalf(&end - &start),time12.4) )--------;
```

Figure 9.1-3: Output in SAS Log: Scenario III: Aggregating Calculation of Runtime

```
-------- Begin of runtime measurement: 05JAN2021:12:00:37--------

           (Output abbreviated)

-------- End of runtime measurement:  05JAN2021:12:00:55--------
-------- Total duration:     0:00:17.1320--------
```

The difference between the start ("05JAN2021:12:00:37") and end ("05JAN2021:12:00:55") of the runtime measurement is automatically calculated and output to the SAS log as the total runtime of the executed programs, here as total duration of exactly 17.1320 seconds. The output is accurate to four digits after the second decimal point. If three zeros after the second decimal point are output, it is a coincidence and not caused by truncation.

9.1.5 Scenario IV: Differentiated Calculation of the Runtime (ARM Macros)

Advantages:

- In the simplest variant, the measurement of the programs only needs to be initialized via SAS options.
- Macros measure the runtime of individual program subsections and sub-processes.
- ARM macros measure the total runtime of the executed programs.
- Further parameters can be obtained.
- Parallel processes can also be measured.

Disadvantages:

- Separate ARM macros, %ARMPROC and %ARMJOIN, must be initialized to analyze the collected measurements. (See the postprocessing settings in Program 9.1-5.)
- The definition of various parameters is not clearly documented.
- The SAS programs to be measured might have to be extended by calls to the ARM macros.
- The flexibility of measuring even parallel processes is accompanied by an increasing complexity of programming and analysis.

Program 9.1-5: Scenario IV: Differentiated Calculation of Runtime (ARM)

```
*----------------------*
| Settings for Logging  |
*----------------------*;
options armloc="C:\ARM_Analysis\ARMLOG.LOG"
        armsubsys=(ARM_ALL);


* ------------------------*
| ARM monitored SAS process |
* ------------------------*;
/* place program block to be measured here  */
/* place program block to be measured here  */


* --------------------------*
| Settings for Postprocessing |
*--------------------------*;
libname ARM_data v9  "C:\ARM_Analysis";
filename ARMLOG "C:\ARM_Analysis\ARMLOG.LOG";
%armproc(lib=ARM_data, log=C:\ARM_Analysis\ARMLOG.LOG);
%armjoin(libin=ARM_data,libout=ARM_data);
run;
```

```
* -------------------*
| Logging Analysis  |
*-------------------*;
title "ARM Logging Analysis";
      proc print data=ARM_data.UPDATE ;
      run;
      proc summary data=ARM_data.UPDATE  nway missing ;
      var txusrcpu ;
      output out= ARM_STATS
            N=N_Measurements
            sum=txusrcpu_SUM ;
      run ;

      proc print data= ARM_STATS noobs ;
      var  N_Measurements txusrcpu_SUM ;
      run ;
      proc summary data=ARM_data.UPDATE nway missing   ;
      class txshdl ;
      var txusrcpu ;
      output out= ARM_STATS2
            N=N_Measurements
            sum=txusrcpu_SUM
            mean=txusrcpu_MEAN;
      run ;

proc print data= ARM_STATS2 noobs ;
var txshdl N_Measurements txusrcpu_SUM txusrcpu_MEAN ;
run ;
proc print data=ARM_data.init ;
run ;
proc options group=memory ;
run ;
```

The first PROC PRINT outputs the contents of the SAS file UPDATE.

The first PROC SUMMARY stores the number and total sum of measurements from TXUSRCPU in the fields "N_Measurements" and "txusrcpu_SUM" in the file ARM_STATS. TXUSRCPU is the user CPU time for a specific transaction. The second PROC PRINT outputs the contents of the SAS file ARM_STATS.

The second PROC SUMMARY calculates the number, sum, and average of TXUSRCPU measurements for each start handle and stores the aggregates in the fields "N_Measurements", "txusrcpu_SUM", and "txusrcpu_MEAN" of the file ARM_STATS2. The third PROC PRINT outputs the contents of the SAS file ARM_STATS2. The fourth PROC PRINT outputs the contents of the SAS file INIT.

Finally, using the GROUP=MEMORY, PROC OPTIONS queries detailed information about memory usage.

Program 9.1-6: SAS Code to Print the Contents of ARM Log ARMLOG.LOG:

```
proc document name=MY_LOG_DOC(write);

    import textfile="C:\ARM_Analysis\ARMLOG.LOG" to logfile;
run;

options nocenter nodate nonumber;
ods rtf file="C:\ARM_Analysis\ARM_LOG.rtf" ;
    replay;
run;
ods rtf close;
options center date number;
run;
quit;
```

Program 9.1-6 routes the contents of ARMLOG.LOG (text format) to ODS RTF and the SAS Output. Using the IMPORT statement, PROC DOCUMENT replays the contents of ARMLOG.LOG to the RTF ODS destination.

Figure 9.1-4: Output: Scenario IV: Content of ARM Log ARMLOG.LOG

```
I,1925469221.968000,1,0.937500,1.156250,SAS,cfgsc
G,1925469221.968000,1,1,MVA_DSIO.OPEN_CLOSE,DATA SET OPEN/
CLOSE,LIBNAME,ShortStr,MEMTYPE,ShortStr,NOBS,Count64,NVAR,Count64,NOBSREAD,
Count64,MEMNAME,LongStr
G,1925469221.968000,1,2,PROCEDURE,PROC START/STOP,PROC_NAME,ShortStr,PROC_IO,
Count64,PROC_MEM,Count64,PROC_LABEL,LongStr
S,1925469221.984000,1,2,1,0.937500,1.171875,DATASTEP,0,0,
S,1925469222.022000,1,1,2,0.937500,1.203125,WORK    ,DATA    ,0,0,0,TEST1
U,1925469222.022000,1,1,2,0.937500,1.203125,2,VAR(1,i),DEF
U,1925469222.022000,1,1,2,0.937500,1.203125,2,VAR(1,x),DEF
U,1925469222.022000,1,1,2,0.937500,1.203125,2,VAR(1,y),DEF
P,1925469223.567000,1,1,2,2.406250,1.296875,0,WORK    ,DATA    ,10000000,3,0,
TEST1
P,1925469223.567000,1,2,1,2.406250,1.296875,0,DATASTEP,246037519,10465280,
S,1925469223.577000,1,2,3,2.421875,1.296875,DATASTEP,0,0,
S,1925469223.593000,1,1,4,2.421875,1.312500,WORK    ,DATA    ,0,0,0,TEST2
U,1925469223.593000,1,1,4,2.421875,1.312500,2,VAR(1,i),DEF
(output abbreviated)
```

The contents of Figure 9.1-4 are part of the later postprocessed file ARMLOG.LOG. The ARM Logging Analysis analyzes the contents of that log. (See Figure 9.1-5.)

Logging Analysis

PROC PRINT outputs the contents of the SAS file UPDATE. UPDATE (N=21 rows) also contains the columns METRVAL1 to METRVAL6 and STRVAL1. The METRVAL columns contain values of a user-defined metric in character format. The STRVAL1 column contains a user-defined string value. As there were no user-defined metrics and string values defined before, these columns are empty and were removed from the output below for space reasons.

Figure 9.1-5: Output: Scenario IV: "Logging Analysis" (ARM_data.UPDATE)

```
                    ARM Logging Analysis

                f                              t
                m                              x
        t       t                     u        u
    a c x       2                     p        s       t
    p l s       B                     d        r       x
  O p s h       u                     t        c       c
  b i i d       f                     d        p       p
  s d d l       f                     t        u       u


  1 1 1   2 VAR(1,i),DEF   05JAN2021:12:33:42.022 0:00:00.937500 0:00:01.203125
  2 1 1   2 VAR(1,x),DEF   05JAN2021:12:33:42.022 0:00:00.937500 0:00:01.203125
  3 1 1   2 VAR(1,y),DEF   05JAN2021:12:33:42.022 0:00:00.937500 0:00:01.203125
  4 1 1   4 VAR(1,i),DEF   05JAN2021:12:33:43.593 0:00:02.421875 0:00:01.312500
  5 1 1   4 VAR(1,x),DEF   05JAN2021:12:33:43.593 0:00:02.421875 0:00:01.312500
  6 1 1   4 VAR(1,y),DEF   05JAN2021:12:33:43.593 0:00:02.421875 0:00:01.312500
  7 1 1   6 VAR(1,i),DEF   05JAN2021:12:33:45.176 0:00:03.796875 0:00:01.468750
  8 1 1   6 VAR(1,x),DEF   05JAN2021:12:33:45.176 0:00:03.796875 0:00:01.468750
```

```
 9 1 1   6 VAR(1,y),DEF  05JAN2021:12:33:45.176 0:00:03.796875 0:00:01.468750
10 1 1   8 VAR(1,i),SEL  05JAN2021:12:33:51.626 0:00:05.593750 0:00:01.625000
11 1 1   8 VAR(1,x),SEL  05JAN2021:12:33:51.626 0:00:05.593750 0:00:01.625000
12 1 1   8 VAR(1,y),SEL  05JAN2021:12:33:51.626 0:00:05.593750 0:00:01.625000
13 1 1   9 VAR(1,i),SEL  05JAN2021:12:33:51.626 0:00:05.593750 0:00:01.625000
14 1 1   9 VAR(1,x),SEL  05JAN2021:12:33:51.626 0:00:05.593750 0:00:01.625000
15 1 1   9 VAR(1,y),SEL  05JAN2021:12:33:51.626 0:00:05.593750 0:00:01.625000
16 1 1 10 VAR(1,i),SEL  05JAN2021:12:33:51.626 0:00:05.593750 0:00:01.625000
17 1 1 10 VAR(1,x),SEL  05JAN2021:12:33:51.626 0:00:05.593750 0:00:01.625000
18 1 1 10 VAR(1,y),SEL  05JAN2021:12:33:51.626 0:00:05.593750 0:00:01.625000
19 1 1 11 VAR(1,i),DEF  05JAN2021:12:33:51.642 0:00:05.593750 0:00:01.640625
20 1 1 11 VAR(1,x),DEF  05JAN2021:12:33:51.642 0:00:05.593750 0:00:01.640625
21 1 1 11 VAR(1,y),DEF  05JAN2021:12:33:51.642 0:00:05.593750 0:00:01.640625
```

Legend

appid: App ID. **clsid:** Txn Class ID.

txshdl: Start Handle **fmt2Buff:** Format 2 Data Buffer

updtdt: Txn Update Datetime. **txusrcpu:** Txn User CPU Time

txcpu: Txn System CPU Time.

PROC SUMMARY determined the number and sum of TXUSRCPU measurements. There are 21 measurements ("N_Measurements") and a sum of 1 minute and 28.6 seconds ("txusrcpu_SUM").

Figure 9.1-6: Output: Scenario IV: "Logging Analysis" (WORK.ARM_STATS)

```
     N_Measurements       txusrcpu_SUM
          21              0:01:28.593750
```

PROC SUMMARY also determined the number, sum, and average of TXUSRCPU measurements *per start handle.* The overview shows that start handles 2 and 4 are processed the fastest and require the least user CPU time overall. The PROC PRINT outputs the contents of the SAS file ARM_STATS2.

Figure 9.1-7: Output: Scenario IV: "Logging Analysis" (WORK.ARM_STATS2)

txshdl	N_Measurements	txusrcpu_SUM	txusrcpu_MEAN
2	3	0:00:02.812500	0:00:00.937500
4	3	0:00:07.265625	0:00:02.421875
6	3	0:00:11.390625	0:00:03.796875
8	3	0:00:16.781250	0:00:05.593750
9	3	0:00:16.781250	0:00:05.593750
10	3	0:00:16.781250	0:00:05.593750
11	3	0:00:16.781250	0:00:05.593750

PROC PRINT finally returns the contents of the SAS file INIT. Among other things, the name of the application ("SAS"), its user, and the date are logged.

Figure 9.1-8: Output: Scenario IV: "Logging Analysis" (ARM_data.init)

Obs	appid	appname	appuser	initdt
1	1	SAS	cfgsc	05JAN2021:12:33:41.968

The final output shows detailed information about memory usage.

Figure 9.1-9: SAS Log: Memory Usage

```
    SAS (r) Proprietary Software Release 9.4   TS1M7

Group=MEMORY
 SORTSIZE=1073741824 Specifies the amount of memory that is available to the SORT
                     procedure.
   SUMSIZE=0         Specifies a limit on the amount of memory that is available for
                     data summarization procedures when class variables are active.
 MAXMEMQUERY=0       Specifies the maximum amount of memory that is allocated for
                     procedures.
 MEMBLKSZ=16777216   Specifies the memory block size for Windows memory-based
                     libraries.
 MEMMAXSZ=2147483648 Specifies the maximum amount of memory to allocate for using
                     memory-based libraries.
 LOADMEMSIZE=0       Specifies a suggested amount of memory that is needed for
                     executable programs loaded by SAS.
 MEMSIZE=2147483648  Specifies the limit on the amount of virtual memory that can be
                     used during a SAS session.
 REALMEMSIZE=0       Specifies the amount of real memory SAS can expect to allocate.
NOTE: PROCEDURE OPTIONS used (Total process time):
       real time          0.07 seconds
       user cpu time      0.00 seconds
       system cpu time    0.00 seconds
       memory             21.18k
       OS Memory          35320.00k
       Timestamp          01/05/2021 01:20:43 PM
       Step Count                      69  Switch Count  0
```

The individual parameters for the memory are output and commented. In mixed-language systems like mine, you might encounter SAS output in English and German at the same time.

9.2 To Be in the Know: SAS Dictionaries

Albert Einstein is credited with the *bon mot*: "One does not have to know everything. But you have to know where to look it up." The same applies to the information in SAS. So where can the extensive information about directories, SAS tables and views, and SAS catalogs be looked up? SAS stores such information in dictionaries. Dictionaries are available in the form of read-only SAS tables and SAS views and contain information about

- SAS data directories,
- columns and variables in SAS files and their attributes,
- SAS tables and views,
- SAS catalogs and their entries,
- SAS macros, and
- current settings in the SAS system options, indexes, ODS styles, and so on.

The content of these dictionaries is useful as information (for example, as the result of an SQL query), but also as further input for far-reaching applications. For demonstration purposes, three applications are compiled at the end of this section:

- Example 1: The macro WHR_IS_VAR for finding the storage location of certain variables
- Example 2: The macro N_ROWS for querying the number of rows in a specific table
- Example 3: Renaming complete variable lists using the view SASHELP.VCOLUMN

Further examples for retrieving system information are compiled in the chapter on SAS macros. Section 4.7 presents various macros for retrieving information from the SAS dictionaries, including searching for the name of an option, setting, or description.

Accessing dictionaries using PROC SQL is the most efficient way to output metadata including existing directories and the type and number of files they contain. Dictionaries can also be queried with the DATA step, other SAS procedures, and also as SAS macros, but access via PROC SQL is considered to be the most efficient and fastest.

When a dictionary is queried, a process is started during which SAS temporarily compiles the current information belonging to this table. Using SQL, access to a dictionary can look like Program 9.2-1. The prefix "DICTIONARY" (query as table) must always be specified and "SASHELP" when querying as SAS view.

Program 9.2-1: Access a Dictionary Using SQL

```
proc sql;
   describe table
        DICTIONARY.MEMBERS;
```

Each time a dictionary is queried, SAS initializes a new process of updating and compiling information. If the same dictionary will be queried several times in a row, it is more efficient to create the dictionary as a temporary data set and then continue working with it. In this way, repeated updating of the information is no longer necessary.

Table 9.2-1 is an overview of the meta-information available in SAS.

Table 9.2-1: Examples of SAS Dictionaries

SAS Table version	...contains information about...	SAS View version
DICTIONARY.CATALOGS	SAS catalogs and their entries	**SASHELP.VCATALG**
DICTIONARY.CHECK_CONSTRAINTS	Known check constraints	**SASHELP.VCHKCON**
DICTIONARY.COLUMNS	Column and variables in SAS files and their attributes	**SASHELP.VCOLUMN**
DICTIONARY.DICTIONARIES	All DICTIONARY tables	**SASHELP.VDCTNRY**
DICTIONARY.EXTFILES	Filerefs and external storage locations of external data	**SASHELP.VEXTFL**
DICTIONARY.INDEXES	Indexes for certain SAS files	**SASHELP.VINDEX**
DICTIONARY.MACROS	Currently defined macro variables	**SASHELP.VMACRO**
DICTIONARY.MEMBERS	SAS objects like data sets **SASHELP.VMEMBER** contains information for all member types. The other SASHELP views are specific to particular member types.	**SASHELP.VMEMBER** for example, **SASHELP.VSACCES, SASHELP.VSCATLG, SASHELP.VSLIB, SASHELP.VSTABLE**
DICTIONARY.OPTIONS	Current settings in the SAS system options. **SASHELP.VALLOPT** also includes graphics options.	**SASHELP.VOPTION SASHELP.VALLOPT**
DICTIONARY.STYLES	ODS styles	**SASHELP.VSTYLE**
DICTIONARY.TABLE_CONSTRAINTS	Integrity constraints in all known tables	**SASHELP.VTABCON**
DICTIONARY.TABLES	SAS tables	**SASHELP.VTABLE**
DICTIONARY.VIEW_VIEW_SOURCES	Tables or other views referenced by the SQL or DATASTEP view	*Not available.*
DICTIONARY.VIEWS	SAS views	**SASHELP.VVIEW**
DICTIONARY.XATTRS	Extended attributes	**SASHELP.VXATTR**

Note: SAS provides the same content as a SAS table and a SAS view as shown in the left and right columns of Table 9.2-1; however, there are exceptions. SAS provides many more tables and views as dictionaries for the following:

- constraints (DICTIONARY.REFERENTIAL_CONSTRAINTS; SASHELP.VREFCON)
- information maps (DICTIONARY.DATAITEMS, DICTIONARY.FILTERS, DICTIONARY.INFOMAPS; SASHELP. VDATAIT, SASHELP.VFILTER, SASHELP.VINFOMP)
- titles (DICTIONARY TITLES / SASHELP.VTITLE)

Since SAS 9.2, several new features have been added, including FUNCTIONS, DESTINATIONS, and XATTRS; others have been extended in their functionality such as EXTFILES in DICTIONARY terminology.

9.2.1 Query a Dictionary as a Table and View

To view a dictionary as a table, submit a DESCRIBE TABLE command. To query a dictionary as a SAS view, submit a DESCRIBE VIEW command. In the following two examples, the catalog for SAS objects is called—once as table MEMBERS and once as view VMEMBER. The DESCRIBE TABLE/VIEW statement writes a SELECT statement for views into the SAS log. For example, the SQL syntax submitted in Programs 9.2-2 and 9.2-3 returns the output shown in Figures 9.2-1 and 9.2-2.

Program 9.2-2: Syntax for Querying a Table

```
proc sql;
   describe table
        DICTIONARY.MEMBERS;
```

Figure 9.2-1: Output in SAS Log Result of Syntax for Querying a Table

```
NOTE: SQL table DICTIONARY.MEMBERS was created like:

create table DICTIONARY.MEMBERS
  (
   libname char(8) label='Library Name',
   memname char(32) label='Member Name',
   memtype char(8) label='Member Type',
   dbms_memtype char(32) label='DBMS Member Type',
   engine char(8) label='Engine Name',
   index char(3) label='Indexes',
   path char(1024) label='Pathname'
  );
```

Program 9.2-3: Syntax for Querying a View

```
proc sql;
   describe view
        SASHELP.VMEMBER ;
```

Figure 9.2-2: Output in SAS Log Result of Syntax for Querying a View

```
NOTE: SQL view SASHELP.VMEMBER is defined as:

     select *
       from DICTIONARY.MEMBERS;
```

Note: The output SELECT statement to query a dictionary as a SAS view indicates that SASHELP views are a subset of the file dictionary MEMBERS.

9.2.2 Refining the Query of a Dictionary (WHERE)

Once the SAS internal definition of a dictionary is known, you can refine SQL queries by using more specific dictionary columns and WHERE clauses to obtain more specific information in return.

The following two SQL examples pass selected columns (LIBNAME, MEMNAME, and so on) and a WHERE clause to SAS to query a dictionary.

Program 9.2-4: Refining the Query of a Dictionary (View, WHERE, LIBNAME)

```
proc sql;
    select libname, memname, memtype, nobs
        from sashelp.VTABLE
    where libname="SASHELP";
quit;
```

The query of the view SASHELP.VTABLE returns "Library Name" ("SASHELP", specified in the WHERE clause), "Member Name" (table names), "Member Type" (data), and the number of rows in the respective tables ("Number of Physical Observations").

Figure 9.2-3: Output in SAS Output: Result of Querying a Dictionary (View, WHERE, LIBNAME)

Library Name	Member Name	Member Type	Number of Physical Observations
SASHELP	AACOMP	DATA	2020
SASHELP	AARFM	DATA	130
SASHELP	ADSMSG	DATA	426
SASHELP	AFMSG	DATA	1090
SASHELP	ASSCMGR	DATA	402
SASHELP	BASEBALL	DATA	322
SASHELP	BEI	DATA	24205
SASHELP	BIRTHWGT	DATA	100000
SASHELP	BMIMEN	DATA	3264
SASHELP	BMT	DATA	137
SASHELP	BURROWS	DATA	24591
SASHELP	BWEIGHT	(abbreviated)	

Program 9.2-5: Refining the Query of a Dictionary (Table, WHERE, LIBNAME)

```
proc sql;
    select libname, memname, memtype, nobs
        from dictionary.tables
    where libname="PATH";
quit ;
```

The query of the table DICTIONARY.TABLES returns the specified Library Name ("PATH", user-defined), Member Name (table names), Member Type (data), and the number of rows in the respective tables.

Figure 9.2-4: SAS Output: Result of Querying a Dictionary (Table, WHERE, LIBNAME)

Library Name	Member Name	Member Type	Number of Physical Observations
PATH	TEST_ALL	DATA	10
PATH	MYDATA1	DATA	500
PATH	MYDATA2	DATA	500
PATH	(abbreviated)		

9.2.3 Example 1: Finding the Storage Location of Certain Variables (Macro WHR_IS_VAR)

Access to SAS dictionaries is very useful when you are looking for the storage location of certain variables in the form of an SQL query or a SAS macro. The question, "where is the variable xy currently stored?" can be answered with an SQL query and then in the form of a small macro.

The SQL query in Program 9.2-6 accesses the DICTIONARY.COLUMNS dictionary. DICTIONARY.COLUMNS contains information about columns in tables. WHERE= passes the directory to be searched ("SASHELP") and the name of the searched column/variable ("ACTUAL") to SAS. This program variant is case-sensitive.

Program 9.2-6: Example 1: Finding the Storage Location of a Column

```
proc sql ;
    select libname, memname, name
        from DICTIONARY.COLUMNS
    where libname="SASHELP" and name="ACTUAL" ;
quit ;
```

As a result, SAS outputs the searched directory ("Library Name"), the names of the tables found ("Member Name", SASHELP.PRDSAL3) and the name of the column found therein ("Column Name", ACTUAL).

Figure 9.2-5: Output: Example 1: Finding the Storage Location of a Column

```
Library
Name        Member Name                        Column Name
-----------------------------------------------------------------------
SASHELP     PRDSAL2                            ACTUAL
SASHELP     PRDSAL3                            ACTUAL
SASHELP     PRDSALE                            ACTUAL
```

The WHR_IS_VAR macro also enables you to find the storage location of certain variables by querying the DICTIONARY. COLUMNS. Unlike the previous SQL query, the WHERE clause also requires the MEMTYPE to be of type "data".

Program 9.2-7: Example 1: Finding the Storage Location of a Column (Macro)

```
%macro WHR_IS_VAR(LIBRARY, COLUMN);
proc sql;
        select LIBNAME, MEMNAME
        from DICTIONARY.COLUMNS
        where LIBNAME="&LIBRARY." and NAME="&COLUMN."
                and MEMTYPE="DATA";
quit;
%MEND WHR_IS_VAR ;
%WHR_IS_VAR(SASHELP,Sales);
%WHR_IS_VAR(SASHELP,ACTUAL);
```

PROC SQL accesses the internal SAS table DICTIONARY.COLUMNS. COLUMNS contains information about columns in tables. After WHERE=, the directory ("LIBRARY"), the name of the column/variable ("COLUMN"), and the MEMTYPE ("DATA") are passed to SAS. This program variant is case-sensitive.

In the calls of the macro WHR_IS_VAR, the directories ("library name") and the searched variables are specified. In the output, SAS returns the directory and tables containing the searched-for variable as result.

Figure 9.2-6: Output: Example 1: Finding the Storage Location of a Column (Macro)

```
Library
Name        Member Name
-------------------------------------------

SASHELP     SHOES




Library
```

```
Name      Member Name
-------------------------------------------

SASHELP   PRDSAL2

SASHELP   PRDSAL3

SASHELP   PRDSALE
```

9.2.4 Example 2: Query the Number of Rows in Specific Tables (Macro N_ROWS)

By querying the DICTIONARY.TABLES, the macro N_ROWS helps to find out the number of rows in the specified table.

Program 9.2-8: Example 2: Query the Number of Rows in Specific Tables (Macro)

```
%macro N_ROWS(LIBRARY, TABLE);
  proc sql ;
    select LIBNAME, MEMNAME, NOBS
      from DICTIONARY.TABLES
        where upcase(LIBNAME)="&LIBRARY." AND
              upcase(MEMNAME)="&TABLE." AND
              upcase(MEMTYPE)="DATA";
  quit ;
%mend N_ROWS;
%N_ROWS(SASHELP,CLASS);
```

PROC SQL accesses the internal SAS table DICTIONARY.TABLES. TABLES contains information about SAS tables. After WHERE=, the directory (LIBRARY), the table (TABLE), and the MEMTYPE (DATA) are passed to SAS. This program variant is not case-sensitive.

After calling the macro N_ROWS, SAS returns the number of rows for the specified directory ("Library Name", SASHELP) and the specified data set ("Member Name", CLASS), and the number of rows, N=19 ("Number of Physical Observations").

Figure 9.2-7: Output: Example 2: Query the Number of Rows in Specific Tables (Macro)

```
                                      Number of
Library                                Physical
Name      Member Name               Observations
-----------------------------------------------------
SASHELP   CLASS                               19
```

9.2.5 Example 3: Renaming Complete Variable Lists Using Dictionaries

The content of dictionaries is also very useful as input for far-reaching applications. The following application aims to rename complete variable lists and is also based on accessing a dictionary, in this case the view SASHELP.VCOLUMN. The aim of the application is to convert all prefixes in the column names from "**T1**DAY" etc. to "**T2**DAY" etc.

Program 9.2-9: Example 3: Renaming Complete Variable Lists Using Dictionaries

```
data BEFORE ;
   input  T1DAY T1AGE T1PLACE T1LAKE T1FOOD T1FISH  ;
datalines;
12 31 4 2 34 5
 ;
run ;

proc sql ;
   select NAME||"="||"T2"||substr(NAME,3)
   into: RENAMEV separated by " "
```

```
    from SASHELP.VCOLUMN
    where Libname="WORK" and memname="BEFORE" and substr(NAME,1,1) ;
quit;

proc datasets library=work memtype=data nolist;
    change BEFORE=AFTER ;
    modify AFTER ;
    rename &RENAMEV ;
run;
proc print data=AFTER noobs ;
run;
```

Figure 9.2-8: Output: Example 3: Renaming Complete Variable Lists Using Dictionaries

T2DAY	T2AGE	T2PLACE	T2LAKE	T2FOOD	T2FISH
12	31	4	2	34	5

The listwise renaming of the variables is done in two steps. In the first step, PROC SQL accesses the SAS internal view SASHELP.VCOLUMN, which contains information about columns in tables. WHERE= and MEMNAME= specify the storage location and name of the table. The NAME entry specifies the column names in the BEFORE table. The SELECT statement generates a listwise writing of string "expressions" of the form "*oldname*=**T2***oldname*" (T1DAY=T2DAY) and stores them in the macro variable RENAMEV. PROC DATASETS will access RENAMEV next.

SAS Log 9.2-1: Example 3: Renaming Complete Variable Lists Using Dictionaries

```
246  proc datasets library=work memtype=data nolist;
247       change BEFORE=AFTER ;
NOTE: Changing the name WORK.BEFORE to WORK.AFTER (memtype=DATA).
248       modify AFTER ;
249       rename &RENAMEV ;
NOTE: Renaming variable T1DAY to T2DAY.
NOTE: Renaming variable T1AGE to T2AGE.
NOTE: Renaming variable T1PLACE to T2PLACE.
NOTE: Renaming variable T1LAKE to T2LAKE.
NOTE: Renaming variable T1FOOD to T2FOOD.
NOTE: Renaming variable T1FISH to T2FISH.
250  run;

NOTE: MODIFY was successful for WORK.AFTER.DATA.
```

In the second step, PROC DATASETS accesses the table BEFORE again (library=work, memtype=data), renames the table BEFORE to AFTER (CHANGE), and also specifies that the contents of table AFTER are still to be changed. Using RENAME, the original variable names (T1DAY) are replaced listwise with the new names using the renaming "expressions" stored in RENAMEV ("T1DAY=T2DAY" and so on) created in the SELECT statement above. Small changes to this program enable you to assign uniform suffixes or other names. Especially for renaming many variables, a more comfortable way is hardly imaginable. An extension of this SQL application is presented in Section 4.7. (See also Subsection 6.3.3.) Using this macro variant, complete variable lists are not only renamed, but also converted into a uniform type using a DATA step.

9.3 Data Handling and Data Structuring

This section presents several PROC SQL solutions and DATA step varieties for handling and structuring data.

Overview:
- **Topic 1:** Creating "exotic" column names
- **Topic 2:** Creating a primary key (MONOTONIC and safer options)

- **Topic 3**: Analyze and structure: Segmenting a SAS table (MOD function)
- **Topic 4:** Defining a tag set for the export of SAS tables into CSV format (PROC TEMPLATE)
- **Topic 5:** Protecting contents of SAS tables (passwords)

9.3.1 Topic 1: Creating "Exotic" Column Names

SAS usually assumes certain conventions for defining column names. The default (see keyword V7 for details) expects a maximum length of 32 characters, the first character is a letter or underscore, no special characters (with a few exceptions, depending on the context), and no blanks. The SAS option VALIDVARNAME allows a more flexible assignment of column names. The individual keywords provide the user with the following options:

- **V7:** *Default.* Maximum length 32 characters, the first character is a letter ("A", "B", "C", ..., "Z", "a", "b", "c", ..., "z") or underscore ("_"), without special characters (except "_", "$", "#" or "&") and blanks. The display is case-sensitive. For internal processing, column names are generally interpreted as uppercase. SAS interprets names of a column in variations of uppercase and lowercase letters as the same column name.
- **UPCASE:** Same as V7, except that column names are also capitalized for display.
- **ANY:** The keyword ANY enables you to assign almost any name to columns. In principle, the first character can be any character or a blank. For display purposes, the system is case-sensitive; for internal processing, column names are generally interpreted as uppercase. The maximum length is 32 characters. If "%" and "&" are to be used, they must be enclosed in double quotation marks to avoid entanglement with the SAS Macro Facility. In the following, "exotic" refers to column or variable names that do not conform to the "usual" SAS convention (see V7) and can be created using almost any characters.

Program 9.3-1: Topic 1: Creating "Exotic" Column Names

```
options validvarname=any ;

data   TEST   ;
      retain "!3"n ;
      set SASHELP.CLASS ;
          "!3"n="NonSASVarName";
              format "!3"n  $1.;
      run ;
```

Users of earlier SAS versions might find a note in the SAS log that ANY is still considered experimental and might cause errors.

SAS Log 9.3-1 VALIDVARNAME: Warning in Earlier SAS Versions

```
WARNING: Only Base procedures and SAS/STAT procedures have been tested for use with
VALIDVARNAME=ANY. Other use of this option is considered experimental and may cause
undetected errors.
```

Using VALIDVARNAME=, column names can be assigned locally or on a remote host. One of the author's experiences some time ago was that "exotic" column names could also be assigned on the host. However, it was not possible to transfer "exotic" column names to the local computer using PROC DOWNLOAD. In that case, an additional local renaming was required. Section 4.9 introduces a macro for creating continuous "exotic" column names ("2010", "2011", ...) as Application 6.

9.3.2 Topic 2: Creating a Primary Key (MONOTONIC and Safer Options)

A primary key is generally unique and thus uniquely identifies a row in a table, not its relative position in the table, which can always change by sorting. Primary keys are essential for working with one or more tables; for splitting, sorting, segmenting, or clustering a table; or joining two or more tables. All concepts in connection with relational databases therefore assume that a data row, observation, or tuple must be uniquely identifiable by means of a key.

If a SAS table does not contain a primary key, one of the first tasks is to add a primary key. SAS offers several possibilities for this. Two PROC SQL approaches and two Base SAS approaches are presented in this subsection. All four approaches add a primary variable called ID to the SASHELP.CLASS file. No output is provided.

Program 9.3-2: Topic 2: Creating a Primary Key (MONOTONIC)

```
/* PROC SQL approach 1 (MONOTONIC) */

proc sql ;
   create table DATA_with_ID1 as
      select MONOTONIC() as ID, *
   from SASHELP.CLASS ;
quit ;
```

In the first example, the SAS file SASHELP.CLASS is supplemented by the variable ID created using MONOTONIC and stored as table DATA_with_ID1. The SAS function MONOTONIC has not yet been documented and is unsupported. As attractive as the MONOTONIC function might seem, it has some limitations. For example, if the MONOTONIC function is used in an SQL procedure that aggregates data (for example, using GROUP BY) then the function might return non-sequential or missing results. If two different statements use the MONOTONIC function, then a separate number sequence might be returned for each statement. In some cases, users reported that the ID sequence did not even start with 1. The following options are considered safe compared to the MONOTONIC function. The first option is a DATA step and SQL hybrid.

Program 9.3-3: Topic 2: Creating a Primary Key (In-line View)

```
/* PROC SQL approach 2 (In-line view) */

data DATA_with_ID
   /view=DATA_with_ID ;
set SASHELP.CLASS ;
    ID+1 ;
run ;
proc sql ;
   create table DATA_with_ID2 as
   select ID, *
   from  DATA_with_ID ;
quit ;

proc print data= DATA_with_ID2 noobs ;
run;
```

This SQL approach is actually based on a "safe" DATA step. In the DATA step, a view with the created primary key ID is created and only included in the query of the subsequent PROC SQL step via an in-line view. SELECT with ID and * specifies "by the way" the arrangement of ID in the column sequence and places ID at the very beginning (left).

Program 9.3-4: Topic 2: Creating a Primary Key (_N_ approach)

```
/* SAS Base _N_ approach 1 (ID left in the table) */

data BASE_APPROACH1 ;
    retain ID ;
    set SASHELP.CLASS ;
    ID = _N_ ;
run ;
```

Program 9.3-5: Topic 2: Creating a Primary Key (Two +1 Approaches)

```
/* SAS Base increment approaches 2 and 3 (IDs right resp. left) */
```

```
data BASE_APPROACH2 ;                    data BASE_APPROACH3 ;
    set SASHELP.CLASS ;                      ID + 1 ;
    retain ID 0 ;                            set SASHELP.CLASS ;
    ID + 1 ;                             run ;
run ;
```

The main difference between the DATA step and SQL approaches is that the DATA step approaches use a safe, internal counting and also use significantly fewer programming lines. The first DATA step approach in Program 9.3-4 uses "_N_", an internal counting variable of the rows in a SAS table. The other two + 1 approaches use an explicit counting of the data rows during creation. All approaches achieve the same result with minor differences. The _N_ approach places the ID on the far left of the table; the incrementing approaches on the far right or left side of the table.

9.3.3 Topic 3: Analyze and Structure: Segmenting a SAS Table (MOD Function)

Data segmentation can be interesting for random cluster analyses (Schendera, 2009), as an approach to increment a data set at a desired granularity (Schendera, 2020/2007), or for performant parallel processing using multi-threading. (See Section 7.4.) The SAS data set option DBSLICE requires that the BY-values of the data to be read are known and that the amount of data in all BY-values is approximately the same. This last requirement is necessary to ensure that the processing load is distributed more or less equally among the threads. If there is no approximately consistent data distribution available, the following MOD function approach illustrates how this could be created using PROC SQL or a DATA step.

In Program 9.3-6, the SASHELP file PRDSAL2 is to be subdivided. PRDSAL2 contains exactly 23,040 entries. Using the MOD function, four fields CLUSTER_2 to CLUSTER_5 are created, and the SAS table PRDSAL2 is segmented equally into halves ("CLUSTER_2"), thirds ("CLUSTER_3"), quarters ("CLUSTER_4"), and fifths ("CLUSTER_5"). In this way, the MOD function creates regular number ranges from 4 to 0 for CLUSTER_5. The created subgroups 4 to 0 are each exactly the same size with N=4608. Finally, a simpler DATA step approach is also shown.

Program 9.3-6: Topic 3: Segmenting a SAS Table (MOD Function)

```
/* Step 1: Explicitly creating a view for ID */
data PRDS_with_ID
/view=PRDS_with_ID ;
set SASHELP.PRDSAL2 ;
      ID+1 ;
      run ;

/* Step 2: MOD function incl. integrating in-Line View */
proc sql ;
      create table _temp as
      select ID, * from PRDS_with_ID;
      create table PRD_CLUS_1 as
      select mod(ID, 2) as CLUSTER_2,
             mod(ID, 3) as CLUSTER_3,
             mod(ID, 4) as CLUSTER_4,
             mod(ID, 5) as CLUSTER_5,
         *
      from _temp ;
quit;

/* Step 3: Displaying the fields CLUSTER_2 to _5 */
proc freq data= PRD_CLUS_1 ;
table CLUSTER_2--CLUSTER_5 ;
run ;
```

Figure 9.3-1: Output: Sample Output (CLUSTER_4, CLUSTER_5)

CLUSTER_4	Frequency	Percent	Cumulative Frequency	Cumulative Percent
0	5760	25.00	5760	25.00
1	5760	25.00	11520	50.00
2	5760	25.00	17280	75.00
3	5760	25.00	23040	100.00

CLUSTER_5	Frequency	Percent	Cumulative Frequency	Cumulative Percent
0	4608	20.00	4608	20.00
1	4608	20.00	9216	40.00
2	4608	20.00	13824	60.00
3	4608	20.00	18432	80.00
4	4608	20.00	23040	100.00

Note: The creation and inclusion of an ID using an in-line view in steps I and II was explained in the previous section. Four fields are created in step II using the MOD function, see CLUSTER_2 to CLUSTER_5. The MOD function goes through the data set PRDSAL2 row by row and calculates the quotient of line number (created explicitly in the SQL approach using a preceding in-line view) and the specified second argument, 5. The MOD function returns the indivisible "remainder" of the division of the first argument (row number) by the second argument (5 for "CLUSTER_5") since SAS 9 uses "fuzzing" to avoid frequent floating point problems. For example, line number 1 divided by 5 results in 1, line number 2 divided by 5 at "CLUSTER_5" results in 2, 3 divided by 5 results in 3, for 4 also in 4, and 5 results in 0 because the indivisible remainder is 0. The results are repeated for all other line numbers. For line number 6 divided by 5, the result is 1. For 7, the result is 2. For 8, the result is 3. For 9, the result is 4, and for 10, the result is 0 because the indivisible remainder is again 0. The created field "CLUSTER_2" contains two levels 0 and 1, each with N=11520. The field "CLUSTER_3" contains three levels 0, 1, and 2, each with N=7680. "CLUSTER_4" contains four levels from 0 to 3, each with N=5760, and "CLUSTER_5" contains five levels from 0 to 4, each with N=4608. The created "CLUSTER_n" fields thus allow the SASHELP table PRDSAL2 to be exactly divided into halves, thirds, quarters, and fifths (see also the example output for "CLUSTER_4" and "CLUSTER_5").

The calculations of the MOD function are exact as long as both arguments are exact integers. The MOD function returns 0 if the rest is very close to 0 or close to the value of the second argument. The MOD function returns a missing if the remainder cannot be calculated with an accuracy of three digits or more. Since SAS 9, MOD makes additional adjustments by using "fuzzing." Results using MOD might therefore differ from those of earlier SAS versions. Some limitations of the MOD approach are that it can only assign values depending on a row number. The creation of a primary key if it is not available is mandatory for the SQL approach but not for the DATA step approach. In conjunction with a random function, random-based drawings could also be created. If, on the other hand, values are to be assigned depending on the data set content, a LAG approach could be used. Program 9.3-7 demonstrates the segmentation of a SAS table using a DATA step variant.

Program 9.3-7: Topic 3: Segmenting a SAS Table (MOD Function, DATA Step)

```
data PRD_CLUS_2 ;
set SASHELP.PRDSAL2 ;
        CLUSTER_2 = mod(_N_, 2) ;
        CLUSTER_3 = mod(_N_, 3) ;
        CLUSTER_4 = mod(_N_, 4) ;
        CLUSTER_5 = mod(_N_, 5) ;
run ;
```

The functionality of PROC SQL and the DATA step is identical except if no primary key is available. With the SQL approach, an ID must be created explicitly. In comparison, the DATA step approach allows access to the internal counter variable _N_. The DATA step approach therefore requires fewer programming lines.

9.3.4 Topic 4: Defining a Tag Set for the Export of SAS Tables into CSV Format (PROC TEMPLATE)

Since SAS 9, it is possible to define the export of SAS tables by means of page description languages. Page description languages include CSV, HTML4, and XML. In HTML, the tags in brackets format text elements in bold or according to a character set or color. In XML, the tags enable you to classify and structure data. In CSV, it is possible to use tags to specify whether data is separated with commas or with quotation marks.

The SAS procedure TEMPLATE enables the adaptation of any page description language like CSV using a provided tag set template, which in turn can be applied to the definition of the SAS tables to be exported. A tag set is a collection of "tags," which are meta or additional information that is attached to a file.

Some SAS users might ask themselves at this point why a CSV file should be customized. There are two reasons. First, proprietary third-party systems often require a CSV file in a very special format. PROC TEMPLATE makes it possible to adapt the definition of the SAS tables exactly to this external requirement. Second, with very large data volumes, it is important to process text files with as little "overhead" as possible. With millions or even more rows of data, the performance when processing string variables can be significantly improved by storing them without quotation marks in CSV files.

In the following example, the first step is to create a tag set for exporting SAS tables into CSV format. In the second step, this customized tag set will be applied to the export of SAS tables into CSV format using ODS.

Program 9.3-8: Topic 4: Defining a Tag Set to Export SAS Tables into CSV Format

```
/* Step 1: Defining the tag set */

proc template;
/* Defining the tag set                             */
    define tagset tagsets.csvnoq;
        parent=tagsets.csv;
/* Defining the header                              */
    define event header;
        put "," / if !cmp( COLSTART , "1" );
        put "" / if cmp( TYPE , "string" );
        put VALUE;
        put "" / if cmp( TYPE , "string" );
    end;
/* Defining the punctuation for the export using ODS */
    define event data;
        put "," / if !cmp( COLSTART , "1" );
        put "" / if cmp( TYPE , "string" );
        put VALUE;
        put "" / if cmp( TYPE , "string" );
    end;
    end;
run;
/* Step 2: Applying the tag set to CSV export */
ods listing close;
ods TAGSETS.CSVNOQ file="C/:TEXTDATA.txt" ;
proc print data= SASHELP.CLASS noobs ;
run;
ods _ALL_ close ;
ods listing ;
```

The first step in Program 9.3-8 is to create a tag set for the export of SAS tables into CSV format. The DEFINE TAGSET statement stores a new tag set definition named CSVNOQ. Using PARENT=, attributes and events are taken from the tag set template TAGSETS.CSV provided by SAS. TAGSETS.CSV defines table output with comma-separated data by default. String variables are also set in quotation marks by default. The newly created tag set "CSVNOQ" is supposed to store character variables without quotation marks by default. The "NOQ" in the tag set stands for "NOQUOTES", that is, string variables without quotation marks.

The following two DEFINE EVENT statements specify how the header and the data of the new CSV/TXT file are written using the new tag set.

The DEFINE EVENT HEADER statement specifies how the header of the new CSV file is written. Among other things, the header of a file contains the names of variables. COLSTART specifies the column where a cell starts. TYPE defines the type of data as STRING (DOUBLE, CHAR, BOOL, or INT are also possible). CMP ("compare") is used to compare a string for equality with a single variable or a list of variables. The "!" in front is a negation (also possible are "NOT" and "^"). This means that when defining the table header, a comma is placed before each column name (except the first), and that no quotation marks are used.

The DEFINE EVENT DATA statement specifies how the data of the new CSV file is written. COLSTART specifies the column where a cell starts. TYPE defines the type of data as STRING (DOUBLE, CHAR, BOOL, or INT are also possible). CMP ("compare") is used to compare a string for equality with a single variable or a list of variables. The "!" in front is a negation ("NOT"; "^" is also possible). This means that a comma is placed in front of each entry (except the first), and that no quotation marks are used for strings.

In the second step in Program 9.3-8, this customized tag set is applied to the export of SAS tables into CSV format via ODS. The ODS LISTING CLOSE prevents all rows from being printed in the local SAS output. ODS accesses the tag set TAGSETS.CSVNOQ and applies it when exporting data from SASHELP.CLASS to the file "TEXTDATA.txt". ODS CLOSE and ODS LISTING terminate the export with ODS. The other advantages of this approach are that it can handle "exotic" column names, and according to benchmark tests by the author, ODS seems to be more performant than PROC EXPORT.

9.3.5 Topic 5: Protecting Contents of SAS Tables (Passwords)

By default, the contents of SAS tables are not protected against unauthorized reading, writing, or modification, which can be extremely delicate or even non-compliant for sensitive information. However, with appropriate options, PROC SQL can protect the contents of SAS tables. For example, the options READ=, WRITE=, and ALTER= define passwords for reading, writing, and modifying SAS tables. Users who are allowed to read, write, or modify the protected tables can only do so if they know the specified password (PWD1, PWD2, and so on). If they have a password to read, this does not mean that they can change the SAS table. Users who do not have a password to read cannot access the SAS table. SAS issues a message in the SAS log if an unauthorized access attempt is made.

Program 9.3-9: Topic 5: Protecting the Contents of SAS Tables

```
*-----------------------*
| Rights at definition  |
*-----------------------*;
proc sql ;
create table PROTECTED (read=PWD1 write=PWD2 alter=PWD3 )
(A num label="Public",
 B num label="Secret" ) ;
quit ;

*---------------------*
| Rights at access    |
*---------------------*;
proc sql;
create table PUBLIC1 as
select A, B
from PROTECTED (read=PWD1) ;
create table PUBLIC2 (alter=PWD4) as
select A
```

```
from PROTECTED (read=PWD1) ;
create table PUBLIC3 (read=PWD5 alter=PWD6) as
select A, B
from PROTECTED (read=PWD1) ;
quit ;


*----------------------*
| Query without rights |
*----------------------*;
proc sql;
create table PUBLIC4 as
select *
from PUBLIC3 ;
quit;
```

Figure 9.3-2: Output: SAS Prompt

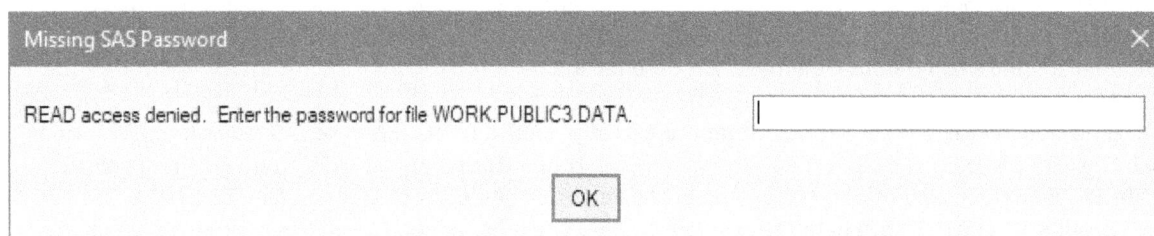

SAS Log 9.3-2: VALIDVARNAME: Warning in Earlier SAS Versions

```
108   proc sql;
109   create table PUBLIC4 as
110   select *
111   from PUBLIC3 ;
ERROR: Invalid or missing READ password on member WORK.PUBLIC3.DATA.
112   quit;
NOTE: The SAS System stopped processing this step because of errors.
NOTE: PROCEDURE SQL used (Total process time):
      real time              3:11.85
      cpu time               0.39 seconds
```

9.4 Updating Tables (SQL versus DATA Step)

PROC SQL, similar to the DATA step, can update master tables by the entries in update tables using an UPDATE statement. PROC SQL and the DATA step both require that the data sets to be merged have an identical structure and that the values in the UP_DATE data set are newer than the entries in MASTER. Note that PROC SQL might produce different results than the DATA step, but also that PROC SQL is flexible enough to enable you to program more complex update approaches. This chapter highlights special features of the DATA step UPDATE statement. PROC SQL and DATA step approaches allow users to decide which of the approaches are most appropriate for dealing with multiple IDs or missing values.

- **Scenario I:** MASTER/UPDATE without multiple IDs (system-defined missing values)
- **Scenario II:** Updating a table without multiple IDs (user-defined missing values)
- **Scenario III:** Updating a table with multiple IDs

9.4.1 Scenario I: MASTER/UPDATE without Multiple IDs: Problem: System-defined Missing Values

The values in table MASTER should be updated with the more current values from table UP_DATE. The IDs in MASTER and UP_DATE are the same. MASTER contains a nonmissing value under the ID "026", while UP_DATE contains a system-defined missing. The UPDATE data set is called UP_DATE, to make SAS code easier to read. Referring to this file as UPDATE is referring to its *function* actually, not its name.

Program 9.4-1: Scenario I: Demo Data MASTER/UPDATE

```
/* MASTER: */                      /* UPDATE: */

data MASTER;                       data UP_DATE;
input ID ACCOUNT;                  input ID ACCOUNT;
datalines;                         missing A _;
011 245                            datalines;
026 269                            011 377
028 374                            026 .
034 333                            028 374
;                                  034 A
                                   ;
```

Note: Table MASTER contains each ID only once. The UP_DATE table contains each ID from MASTER, and no duplicate or additional IDs. Look for the ID "026" in MASTER and in UP_DATE. A nonmissing value or a missing value results in different updates in DATA step and PROC SQL.

Program 9.4-2: Scenario I: Updating a Table without Multiple IDs (System-defined Missing Values)

```
/* Ia. Update using DATA Step */
data MASTER_BASE ;
update MASTER UP_DATE;
by ID;
run;
proc print data=MASTER_BASE noobs  ;
run;


/* Ib. Update using PROC SQL */
proc sql ;
update MASTER as old
     set ACCOUNT=(select ACCOUNT from UP_DATE as new
            where old.ID=new.ID)
                  where old.ID in (select ID from UP_DATE);
     select ID, ACCOUNT from MASTER ;
create table MASTER_SQL as
            select * from MASTER
            except
            select * from UP_DATE;
quit;

proc print data=MASTER_SQL noobs ;
run ;
```

Figure 9.4-1: Output: Scenario I: Updating a Table without Multiple IDs

DATA step:		SQL:	
ID	ACCOUNT	ID	ACCOUNT

11	377	11	377
26	269 ←	26	. ←
28	374	28	374
34	A	34	A

The DATA step and PROC SQL replace the nonmissing value for ID "011" with the more current value from the table UP_DATE. Both approaches also successfully update the nonmissing value for ID "034" (original: 333) with the code for a user-defined missing ("A"). However, in the DATA step, the nonmissing value for ID "026" (original: 269) was not replaced by the system-defined missing in the UP_DATE table. For comparison, in the SQL program, PROC SQL replaced the nonmissing value for ID "026" with the system-defined missing from the table UP_DATE.

Scenario I shows that nonmissing values and system-defined missing values lead to different update results in the DATA step and PROC SQL.

9.4.2 Scenario II: MASTER/UPDATE without Multiple IDs: Problem: User-defined Missing Values

The values in table MASTER should be updated with the more current values from table UP_DATE. The IDs in MASTER and UP_DATE are the same. MASTER contains a system-defined missing value under the ID "057", while UP_DATE contains a user-defined missing value.

Program 9.4-3: Scenario II: Demo Data MASTER/UPDATE

```
/* MASTER: */

data MASTER;
input ID ACCOUNT;
datalines;
011 245
026 269
028 374
034 333
057 .
;
```

```
/* UPDATE: */

data UP_DATE;
input ID ACCOUNT;
missing A _;
datalines;
011 377
026 .
028 374
034 A
057 _
;
```

Note: Table MASTER contains each ID only once. The UP_DATE table contains each ID from MASTER, and no duplicate or additional IDs. Look for the ID "057" in MASTER and in UP_DATE. User- and system-defined missing values cause different updates in the DATA step and PROC SQL.

Program 9.4-4: Scenario II: Updating a Table without Multiple IDs (User-defined Missing Values)

```
/* IIa. Update using DATA Step */
data MASTER_BASE ;
update MASTER UP_DATE;
by ID;
run;
proc print data=MASTER_BASE noobs  ;
run;


/* IIb. Update using PROC SQL */
proc sql ;
update MASTER as old
        set ACCOUNT=(select ACCOUNT from UP_DATE as new
                where old.ID=new.ID)
                        where old.ID in (select ID from UP_DATE);
        select ID, ACCOUNT from MASTER ;
create table MASTER_SQL as
                select * from MASTER
                except
                select * from UP_DATE;
quit;
proc print data=MASTER_SQL noobs ;
run ;
```

Figure 9.4-2: Output: Scenario II: Updating a Table without Multiple IDs

DATA step:			SQL:	
ID	ACCOUNT		ID	ACCOUNT

11	377		11	377
26	269 ←		26	. ←
28	374		28	374
34	A		34	A
57	. ←		57	_ ←

For the IDs "026" and "034" the already known effects occur. The DATA step and PROC SQL replace the nonmissing value for ID "011" with the more current value from the UP_DATE table. Both approaches also successfully update the nonmissing value for ID "034" (original: 333) with the coding for a user-defined missing ("A"). In contrast to the DATA step, PROC SQL replaces the nonmissing value for ID "026" (original: 269) with the missing value in the table UP_DATE. ID "057" shows a different functionality of the DATA step (UPDATE-Statement) and PROC SQL. SQL replaces the original system-defined missing value with the user-defined missing value; the DATA step does not.

Scenario II shows that nonmissing values, user-defined, and system-defined missing values lead to different update results in the DATA step and PROC SQL.

9.4.3 Scenario III: MASTER/UPDATE with Multiple IDs: Problems with Multiple IDs

The values in table MASTER should be updated with the more current values from table UP_DATE. The IDs in MASTER and UP_DATE are the same. However, MASTER and UP_DATE contain multiple IDs.

Program 9.4-5: Scenario III: Demo Data MASTER/UP_DATE

```
/* MASTER: */                      /* UPDATE: */

data MASTER;                       data UP_DATE;
     input ID ACCOUNT;                  input ID ACCOUNT;
     datalines;                         missing A _;
011 245                                 datalines;
011 245                            011 376
011 245                            011 377
026 269                            026 .
028 374                            028 374
034 333                            034 A
057  .                             057 _
057 582                            ;
;
```

Note: In the MASTER table, the ID "011" occurs three times (the values under ACCOUNT are identical in each case). In UP_DATE, the ID "011" occurs only twice (the values under ACCOUNT are different from each other and from MASTER). The ID "057" occurs twice in MASTER (once with a system-defined missing value, once with a nonmissing value). The ID "057" occurs only once in UP_DATE, with a user-defined missing value.

Look for the IDs "011" and "057" in MASTER and in UP_DATE. Multiple entries not only lead to completely different update results in DATA step and in PROC SQL, but also reveal another remarkable feature of the DATA step. Multiple entries in IDs (for example, "011") would lead to error messages in an approach like in Scenarios I or II.

SAS Log 9.4-1: Example 3: Warning in SAS Log Caused by Multiple Entries in ID

```
WARNING: The MASTER data set contains more than one observation for a BY group.
ID=11 ACCOUNT=245 FIRST.ID=0 LAST.ID=0 _ERROR_=1 _N_=3
```

If MASTER was updated using UP_DATE without the help of a COUNTER variable, the update result would be the output shown in Figure 9.4-3. Note especially ID "011".

Figure 9.4-3: Output: Scenario III: Updating a Table without COUNTER (DATA Step)

```
ID    ACCOUNT
11      377   ←
(!)
11      245   ←
11      245   ←
26      269
28      374
34       A
57       .
57      582
```

"011" (multiple): (a) If the number of multiple IDs in MASTER and UP_DATE do not match, the first value is updated. However, not by the corresponding first value from UP_DATE ("376"), as would have been expected, but by the last value from UP_DATE ("377"). (b) The other entries of the multiple IDs in MASTER are *not* updated by values from UP_DATE, even if there were values in UP_DATE.

Further special features are explained in the example with COUNTER.

Due to this effect, it appears to be helpful to increment multiple entries in a key variable and store the result in a COUNTER column (if not yet available) and include this column in the BY statement after ID.

Program 9.4-6: Scenario III: Adding Counter Columns to MASTER and UP_DATE

```
/* MASTER: */                          /* UPDATE: */

proc sort data=MASTER ;                proc sort data=UP_DATE ;
      by ID ACCOUNT ;                        by ID ACCOUNT ;
run ;                                  run ;


data MASTER_2 ;                        data UP_DATE_2 ;
set MASTER ;                           set UP_DATE ;
by ID ACCOUNT ;                        by ID ACCOUNT ;
retain COUNTER ;                       retain COUNTER ;
if first.ID then COUNTER=1;            if first.ID then COUNTER=1;
else COUNTER=COUNTER+1 ;               else COUNTER=COUNTER+1 ;
run;                                   run;
proc print data=MASTER_2 noobs;        proc print data=UP_DATE_2 noobs;
run ;                                  run ;
```

Note: In the MASTER table, the ID "011" appears three times (the values under ACCOUNT are also identical). In UP_DATE, however, the ID "011" appears only twice (the values under ACCOUNT are different from each other and from MASTER).

Multiple entries reveal a remarkable functionality of the DATA step.

Program 9.4-7: Scenario III: Update Using DATA Step

```
/* III. Update using DATA Step */

data MASTER_BASE_2 (drop=COUNTER) ;
   update MASTER_2 UP_DATE_2;
   by ID COUNTER ;
run;
proc print data=MASTER_BASE_2 noobs ;
run;
```

Figure 9.4-4: Output: Scenario III: Updating a Table with COUNTER (DATA Step)

ID	ACCOUNT	
11	376	
11	377	
11	245	←
26	269	
28	374	
34	A	
57	.	←
(!)		
57	582	←

"26", "28", and "34" (unique): "26": Numerical values are *not* replaced by system-defined missing values. "28": Numerical values are replaced by numerical values. "34": Numerical values are replaced by user-defined missing values.

"011" (multiple): If the number of multiple IDs in MASTER and UP_DATE does not match, the surplus IDs in MASTER are *not* updated by values from UP_DATE (see value 245).

"057" (multiple): If the number of multiple IDs in MASTER and UP_DATE does not match, the excess IDs in MASTER are *not* updated by values from UP_DATE (see value 582). System-defined missing values are not replaced by user-defined missing values.

Scenario III shows that for the DATA step, considering an additional COUNTER variable that controls the number of levels of multiple IDs reduces the problem caused by multiple key entries. However, the COUNTER variable is not a solution to all problems.

Program 9.4-8: Scenario III: Update Using SQL (Variant I)

```
/* III. Update using PROC SQL (variant I) */

proc sql ;
update MASTER_2 as old
      set ACCOUNT=(select ACCOUNT from UP_DATE_2 as new
```

```
                where old.ID=new.ID and old.COUNTER=new.COUNTER)
                    where old.ID in (select ID from UP_DATE_2);
select ID, ACCOUNT, COUNTER from MASTER_2 ;
quit;


proc print data=MASTER_2 noobs ;
run ;
```

Figure 9.4-5: Output: Scenario III: Updating a Table with COUNTER (SQL I)

ID	ACCOUNT	COUNTER	
11	376	1	
11	377	2	
11	.	3	←
26	.	1	
28	374	1	
34	A	1	
57	_	1	←
57	.	2	←

"011" (multiple): If the number of multiple IDs in MASTER and UP_DATE does not match, the surplus IDs in MASTER are replaced by system-defined missing values (see the third entry in "011").

"057" (multiple): System-defined missing values are replaced by user-defined missing values. Surplus IDs are replaced by system-defined missing values (see the second entry in "057").

Scenario III shows for PROC SQL that the consideration of an additional COUNTER variable that controls the number of levels of multiple IDs also reduces the problem of multiple IDs. While the DATA step approach keeps surplus MASTER values, the PROC SQL approach replaces them by system-defined missing values. (See 245 in "011" and 582 in "057".) The following approach illustrates how surplus values from MASTER can be excluded during the update.

Variant II is based on data from SQL Variant I. To understand the functionality of SQL Variant II, the SAS code of SQL Variant I is also provided.

Program 9.4-9: Scenario III: Update Using SQL (Variant II)

```
/* III. Update using PROC SQL (variant II) */

* (a1) Filtering of data (MASTER) on the common key *;
proc sql ;
    create table MASTER_3 as
    select m.ID, m.ACCOUNT, m.COUNTER
    from MASTER_2 as m, UP_DATE_2 as u
    where m.COUNTER = u.COUNTER and m.ID = u.ID  ;
quit;

* (a2) Filtering of data (UPDATE) on the common key *;
proc sql ;
create table UP_DATE_3 as
    select u.ID, u.ACCOUNT, u.COUNTER
    from MASTER_2 as m, UP_DATE_2 as u
    where m.COUNTER=u.COUNTER and m.ID=u.ID  ;
quit;

* (b) Updating filtered partial data *;
proc sql ;
update MASTER_3 as old
    set ACCOUNT=(select ACCOUNT from UP_DATE_3 as new
        where old.ID=new.ID and old.COUNTER=new.COUNTER)
            where old.ID in (select ID from UP_DATE_3);
    select ID, ACCOUNT, COUNTER from MASTER_3 ;
create table MASTER_SQL_3 as
        select ID, ACCOUNT
    from MASTER_3 ;
create table EXL_UNION as
```

```
        (select * from MASTER_SQL_3
              except
        select ID, ACCOUNT from MASTER_2)
              union
        (select ID, ACCOUNT from MASTER_2
              except
        select * from MASTER_SQL_3) ;
quit;


* (c) Output of update result *;
proc print data=MASTER_3 noobs ;
run ;
```

Figure 9.4-6: Output: Scenario III: Updating a Table with COUNTER and Filter (SQL II)

ID	ACCOUNT	COUNTER
11	376	1
11	377	2
26	.	1
28	374	1
34	A	1
57	_	1

"011" / "057" (multiple): If the number of multiple IDs in MASTER and UP_DATE do not match, the surplus IDs in MASTER are excluded from the update result. The third entry in "011" and the second entry in "057" are missing in the update result.

Scenario III shows for PROC SQL that if the number of multiple IDs in MASTER and UP_DATE do not match, the surplus IDs in MASTER are excluded from the update result. Users are in the comfortable position of being able to decide for themselves which approach seems more technically appropriate to them: update while keeping the original values in surplus IDs in MASTER (DATA step approach), update with system-defined missing values instead of the original values in surplus IDs (SQL Approach I), or exclude the surplus IDs from MASTER (SQL Approach II).

Chapter 10: SAS Syntax–PROC SQL, SAS Functions, and SAS CALL Routines

A quick finder helps you to quickly find the right SQL statement for your desired purpose. Section 10.1 presents the syntax of PROC SQL in a schema. Then, SQL options for optimizing SQL and global SAS system options for PROC SQL are introduced, starting with SAS 9.4. Section 10.2 provides an overview of numerous SAS functions and CALL routines. A SAS function or CALL routine performs a calculation or manipulation on arguments and returns a value. Finally, Section 10.3 discusses special features of the pass-through facility for selected DBMS accesses. For more information, please refer to the SAS documentation (SAS Institute, 2019, 2016a,b).

Table 10-1: Quick Finder

To do this ...	use these SQL statements
Modify, add, or delete **columns.**	ALTER TABLE
Establish a connection with a **DBMS**, query of a DMBS, send DBMS-specific non-query SQL statements to a DBMS, and terminate the connection.	CONNECT TO, CONNECTION TO, EXECUTE, DISCONNECT FROM
Create an index, a PROC SQL table, a PROC SQL view.	CREATE INDEX, CREATE TABLE, CREATE VIEW
Select, add, delete, or execute **rows.**	DELETE, INSERT, SELECT
Display the definition of a table, a view, or of integrity constraints.	DESCRIBE TABLE, DESCRIBE VIEW, DESCRIBE TABLE CONSTRAINTS
Delete tables, views, or indexes.	DROP
Reset options that affect the procedure environment without having to restart the procedure itself.	RESET
Update values.	UPDATE
Check the correctness of a query including feedback.	VALIDATE

10.1 PROC SQL Syntax Overview

Subsection 10.1.1 gives an overview of the PROC SQL syntax of version SAS 9.4. New features for optimizing SQL are arranged according to the schematic overview. Subsection 10.1.2 presents a short description of PROC SQL options. It also introduces PROC SQL functions not currently documented, such as _METHOD and _TREE.

PROC SQL as a procedure is not supported on the CAS server. However, FedSQL expressions use many SQL statements and language elements such as clauses and options for CAS. In Chapter 7, you can find FedSQL syntax and examples for processing on CAS.

10.1.1 Schema of PROC SQL Syntax

This subsection gives a simplified overview of the PROC SQL syntax. PROC SQL is supported in SAS 9.4 and SAS Viya, here through the SAS Programming Environment. You can use it in code submitted by SAS Studio or in Compute Server jobs.

PROC SQL Schema

```
PROC SQL <Option(s) [see Subsection 10.1.2 for a description of the options] > ;
    ALTER TABLE Table name
            <ADD Column definition <, ... Column definition > >
            <ADD <CONSTRAINT> Constraint name+specs  <, ... Constraint name+specs > >
            <DROP Column  <, ... Column > >
            <DROP CONSTRAINT Constraint name  <, ... Constraint name > >
            <DROP FOREIGN KEY Constraint name >
            <DROP PRIMARY KEY>
            <MODIFY Column definition  <, ... Column definition > > ;

    CREATE <UNIQUE> INDEX Index name ON Table name (Column <, ... Column >) ;

    CREATE TABLE Table name
            (Column specification <, ... Column specification | Constraint specification >) ;

    CREATE TABLE Table name LIKE Table name2 ;

    CREATE TABLE Table name AS Query expression
            <ORDER BY Sorting variable <, ... Sorting variable > > ;

    CREATE VIEW PROC SQL View AS Query expression
            <ORDER BY Sorting variable <, ... Sorting variable > >
            <USING Libname clause <, ... Libname clause > > ;

    DELETE FROM Table name|PROC SQL View |SAS/ACCESS view
            <AS Alias>
            <WHERE SQL expression > ;

    DESCRIBE TABLE Table name <, ... Table name > ;

    DESCRIBE VIEW PROC SQL View  <, ... PROC SQL View > ;

    DESCRIBE TABLE CONSTRAINTS Table name <, ... Table name > ;

    DROP INDEX Index name  <, ... Index name> FROM Table name ;

    DROP TABLE Table name  <, ... Table name > ;

    DROP VIEW View name  <, ... View name > ;

    INSERT INTO Table name|PROC SQL view|SAS/ACCESS view < (Column <, ... Column>) >
            SET Column=SQL expression  <, ... Column=SQL expression  >
            <SET Column=SQL expression  <, ... Column=SQL expression  > > ;

    INSERT INTO Table name |PROC SQL view|SAS/ACCESS view < (Column <, ... Column>) >
            VALUES (Value <, ... Value>)
            <VALUES (Value <, ... Value>) > ;

    INSERT INTO Table name|SAS/ACCESS view|PROC SQL view  < (Column <, ... Column>) >
            Query expression ;

    RESET <Option(s)> ;
            SELECT <DISTINCT> Object Item   <, ... Object item >
                    <INTO Macro variable specification  <, ... Macro variable specification > >
                    FROM list
                            <WHERE SQL expression >
                            <GROUP BY Group item  <, ... Group item > >
                            <HAVING SQL expression>
                            <ORDER BY Sort item <, ... Sort item  > <ASC | DESC > ;

            UPDATE Table name |PROC SQL view|SAS/ACCESS view  <AS Alias>
                            SET Column=SQL expression  <, ... Column=SQL expression >
                            <SET Column=SQL expression  <, ... Column=SQL
                            expression > >
                            <WHERE SQL expression > ;

            VALIDATE Query expression ;
        <QUIT ;>
```

Schemas for Connecting to a DBMS

Use this schema to connect with a DBMS and retrieve the data from the DBMS (see **SELECT** and **CONNECTION TO**):

```
PROC SQL ;
            CONNECT TO  DBMS Name  < AS  Alias>
      < (Connection_parameter_1=Value   <... Connection_parameter_n=Value > ) >
      < (DB_connection_argument_1=Value <... DB_connection_argument_n=Value > ) > ;
            SELECT  Column list
            FROM CONNECTION TO  DBMS Name|Alias  (DBMS Query)
                                                optional PROC SQL clauses
;
            <DISCONNECT FROM  DBMS Name | Alias ; >
   < QUIT ; >
```

Use this schema to connect with a DBMS and send a DBMS-specific nonquery SQL statement (see **EXECUTE**):

```
PROC SQL ;
            CONNECT TO  DBMS name  < AS  Alias >
      < (Connection_parameter_1=Value   <... Connection_parameter_n=Value > ) >
      < (DB_connection_argument_1=Value <... DB_connection_argument_n=Value > ) > ;
            EXECUTE (DBMS SQL statement)  BY DBMS Name | Alias;
            <DISCONNECT FROM  DBMS Name | Alias ; >
   < QUIT ; >
```

10.1.2 Short Description of PROC SQL Options

Options for Displaying Internal Join Processes (Performance)

The two PROC SQL options _METHOD and _TREE are not currently documented in SAS. The author became aware of them through a SAS Global Forum presentation by Cheng (2003).

_METHOD
> In **SAS 9.4** and **PROC SQL**, this option describes the internal process hierarchy when selecting the join methods.
> **SAS Viya 3.1** adds a new _METHOD procedure option that prints a text description of the **FedSQL** query plan for executing the specified FedSQL statements; with SAS Viya 3.5, this description is without the stage query and number of threads used.

_TREE
> In **SAS 9.4** and **PROC SQL**, this option describes the internal processes of a join in detail in the form of a tree.

Options to Control Output

DICTDIAG | NODICTDIAG
> Create a Diagnostic column in the report.

DOUBLE | NODOUBLE
> Double-space the report.

FEEDBACK | NOFEEDBACK
> Write a statement to the SAS log that expands the query.

FLOW | NOFLOW
> Flow characters within a column.

NUMBER | NONUMBER
> Include a column of row numbers.

PRINT|NOPRINT
> Specify whether PROC SQL prints the query's result.

SORTMSG= sorting table | NOSORTMSG
> Specify whether PROC SQL should display sorting information.

SORTSEQ=| LINGUISTIC
> Specify a sorting table (collating sequence). LINGUISTIC specifies that the collating sequence is determined from the session locale. Do not use the SORTSEQ=LINGUISTIC option or SAS system option when a SORTKEY function is used in an ORDER BY clause. It produces unintended ordering.

UBUFSIZE=*n* | *n*K | *n*M | *n*G
> Specifies the internal transient buffer page size for the PROC SQL paged memory subsystem in byte and kilo-, mega-, and gigabyte. Replaces BUFFERSIZE used before SAS 9.4. BUFFERSIZE specified the permanent buffer size.

WARNRECURS | NOWARNRECURS
> Displays a warning in the SAS log for recursive references.

Options to Control Execution

CONSTDATETIME | NOCONSTDATETIME
> Replace references to the DATE, TIME, DATETIME, and TODAY functions in a query with their equivalent constant values before the query executes.

DQUOTE= ANSI | SAS
> Allows PROC SQL to use names other than SAS names.

ERRORSTOP | NOERRORSTOP
> Specify whether PROC SQL should stop executing after an error.

EXEC | NOEXEC
> Specify whether PROC SQL should execute statements.

EXITCODE
> Specify whether PROC SQL should clear an error code for any SQL statement. Error codes are assigned to the SQLEXITCODE macro variable.

IPASSTHRU | NOIPASSTHRU
> Specify whether implicit pass-through is enabled or disabled.

INOBS=
> Restricts the number of input rows.

LOOPS=
> Restricts PROC SQL to *n* iterations through its inner loop.

OUTOBS=

Restricts the number of output rows.

PROMPT | NOPROMPT

Prompt you to stop or continue if PROC SQL reaches a limit specified by INOBS=, OUTOBS=, or LOOPS=.

REDUCEPUT=ALL | NONE | DBMS | BASE

Specifies the engine type to use to optimize a PUT function in a query. The PUT function is replaced with a logically equivalent expression (same functionality as the SQLREDUCEPUT system option). See also the REDUCEPUTOBS= and REDUCEPUTVALUES= options.

REDUCEPUTOBS=n

If REDUCEPUT= DBMS, BASE, or ALL: Specify the minimum number of observations that must be in a table for PROC SQL to consider optimizing the PUT function in a query.

REDUCEPUTVALUES=n

If REDUCEPUT= DBMS, BASE, or ALL: Specify the maximum number of SAS format values that can exist in a PUT function expression for PROC SQL to consider optimizing the PUT function in a query.

REMERGE | NOREMERGE

Specify whether PROC SQL can process queries that use remerging of data. Remerging makes two passes through a table, using data in the second pass that was created in the first pass. When NOREMERGE is set, PROC SQL cannot process remerging of data (same functionality as the SQL-REMERGE system option).

STIMER | NOSTIMER

Specify whether PROC SQL writes timing information to the SAS log for each statement (for example, processing time).

STOPONTRUNC

Specifies to not insert or update a row that contains data larger than the column when a truncation error occurs.

THREADS | NOTHREADS

Specify that SAS uses threaded processing if available. The THREADS system option enables some legacy SAS processes that are thread-enabled to take advantage of multiple CPUs by threading the processing and I/O operations.

UNDO_POLICY=

Specifies how PROC SQL handles updates when there is an interruption.

Selected Options for Optimizing SQL

This subsection presents options from the sections to control execution and output. Interested readers will find more detailed information about how these SQL options work in Chapter 8 on performance and efficiency. See also Subsection 10.1.3 for SAS system options.

CONSTDATETIME | NOCONSTDATETIME

Specify whether the SQL procedure replaces references to DATE, TIME, DATETIME, and TODAY functions in a query with their equivalent constant values before the query executes. When the **NOCON-STDATETIME** option is set, PROC SQL evaluates these functions in a query **each time** it processes an observation.

INOBS= / OUTOBS=
> Restrict the number of input rows resp. output rows.

IPASSTHRU | NOIPASSTHRU
> Specify whether implicit pass-through is **enabled** or disabled.

REDUCEPUT=ALL | NONE | DBMS | BASE
> Specifies the engine type to use to **optimize a PUT function** in a query. The PUT function is replaced with a logically equivalent expression (same functionality as the SQLREDUCEPUT system option). See also the REDUCEPUTOBS= and REDUCEPUTVALUES= options.

REDUCEPUTOBS=n
> If REDUCEPUT= DBMS, BASE, or ALL: Specify the minimum number of observations that must be in a table for PROC SQL to consider **optimizing the PUT function** in a query.

REDUCEPUTVALUES=n
> If REDUCEPUT= DBMS, BASE, or ALL: Specify the maximum number of SAS format values that can exist in a PUT function expression for PROC SQL to consider **optimizing the PUT function** in a query.

REMERGE | NOREMERGE
> Specify whether PROC SQL can process queries that use remerging of data. When NOREMERGE is set, PROC SQL will not process the remerging of data. When referencing database tables, performance is enhanced because it enables more of the query to be passed down to the database.

SORTSEQ=| LINGUISTIC
> Specify a sorting table (collating sequence). LINGUISTIC specifies that the collating sequence is determined from the session locale. **Do not use** the SORTSEQ=LINGUISTIC option (or SAS system option) when a SORTKEY function is used in an ORDER BY clause. It produces unintended ordering.

THREADS | NOTHREADS
> Specify that SAS **uses threaded processing** if available. The THREADS system option enables some legacy SAS processes that are thread-enabled to take advantage of multiple CPUs by threading the processing and I/O operations.

UBUFSIZE=*n* | *n*K | *n*M | *n*G
> Specifies the internal transient **buffer page size** for the PROC SQL paged memory subsystem in byte and kilo-, mega-, and gigabyte. Replaces BUFFERSIZE used before SAS 9.4. BUFFERSIZE specified the permanent buffer size.

10.1.3 Selected SAS System Options for PROC SQL

This subsection presents SAS system options associated with SQL.

DBIDIRECTEXEC
> Lets the SQL pass-through facility optimize handling of SQL statements by passing them directly to the database for execution (for example, Amazon Redshift, Google BigQuery, Greenplum, SAP HANA, SAP IQ, Snowflake, Spark, Teradata, and Yellowbrick).

SQLCONSTANTDATETIME | NOSQLCONSTANTDATETIME
> Specifies whether the SQL procedure replaces references to the DATE, TIME, DATETIME, and TODAY functions in a query with their equivalent constant values before the query executes.

SQLGENERATION=

> Specifies whether and when SAS procedures generate SQL for in-database processing of source data. After DBMS=, you specify one or more SAS/ACCESS engines.

SQLIPONEATTEMPT | NOSQLIPONEATTEMPT

> Specifies whether PROC SQL allows an SQL query to continue processing when an implicit pass-through request fails.

SQLMAPPUTTO

> Specifies whether the PUT function is mapped to the SAS_PUT function for a database (for example, Aster, DB2 under UNIX and PC Hosts, Greenplum, Netezza, Teradata). Use this system option to specify an alternative database in which the SAS_PUT function is published.

SQLREDUCEPUT= ALL | NONE | DBMS | BASE

> Specifies the engine type to use to optimize a PUT function in a query. The PUT function is replaced with a logically equivalent expression.

SQLREDUCEPUTOBS=*n*

> Specifies the minimum number of observations that must be in a table for PROC SQL to optimize the PUT function in a query when the SQLREDUCEPUT= system option is set to DBMS, BASE, or ALL. *n* is an integer. 0 indicates that there is no minimum number of observations in a table for PROC SQL to optimize the PUT function. The current upper limit is approximately 9.2 quintillion.

SQLREDUCEPUTVALUES=*n*

> When the SQLREDUCEPUT= system option is set to DBMS, BASE, or ALL, specifies the maximum number of SAS format values that can exist in a PUT function expression for PROC SQL to optimize the PUT function in a query. *n* is an integer (range: 100–3000).

SQLREMERGE | NOSQLREMERGE

> Specifies whether PROC SQL can process queries that use remerged data.

SQLUNDOPOLICY= NONE | OPTIONAL | REQUIRED

> Specifies how PROC SQL handles updates when errors occur while you are updating data (see UNDO_POLICY= to control whether changes are permanent). Specifies whether PROC SQL can process queries that use remerged data.

SQL_PUSHTCINTOVIEW_FROM_OUTSIDE= YES | Y | NO | N

> Specifies that a truncated comparison operator in a WHERE clause returns all of the values in a view that begin with the specified value.

SYS_SQLSETLIMIT Macro Variable

> Specifies the maximum number of values that is used to optimize a hash join during DBMS processing.

10.2 SAS Functions and CALL Routines (Overview)

A SAS function performs a calculation or manipulation on arguments and returns a value. Most functions use arguments supplied by the user. A few SAS functions get their arguments from the operating environment. A SAS function can be used in DATA step, in a WHERE expression in macro programming, in PROC REPORT, and of course in PROC SQL. Some SAS functions (as well as other specific SAS language elements) might disable the implicit pass-through facility. (See Section 10.3.) Given the scope and variety of SAS functions and CALL routines, this greatly extends the functionality and flexibility of SQL.

SAS CALL routines are similar to SAS functions; they also perform calculations or manipulations on arguments and return values. One difference between SAS CALL routines and SAS functions is that they cannot be used in assignment statements. SAS CALL routines are also called with the CALL statement. So, if a CALL precedes a routine, it is a SAS CALL routine (the CALL represents the call of the routine). All others refer to SAS functions. For example, CAT is a SAS function for characters and strings, CALL CATS is a SAS CALL routine for characters and strings.

In some cases, there is both a SAS function and a SAS CALL routine for one and the same purpose; in view of the numerous functions and CALL routines, it is extremely difficult to qualify in two or three keywords. It could be argued, however, that functions tend to be somewhat easier to program, while CALL routines have a certain greater scope of performance. Function calls for random numbers, for example, allow greater control over the seeds than the functions. This brief characterization does not necessarily apply equally to all functions and function calls. In any case, users are required to weigh up the necessary programming and the actual scope of performance required.

This overview also flags which of the functions and CALL routines can also run on CAS servers.

10.2.1 A Small Selection of SAS Functions and SAS CALL Routines

Despite the scope of the overview, it reflects only a small selection of all SAS functions and CALL routines provided by SAS. SAS offers many more functions and CALL routines for certain products and for certain platforms (CAS, OpenVMS, UNIX, Windows, z/OS, and so on). For detailed information about these and other SAS functions and CALL routines, please refer to the SAS technical documentation. This is strongly recommended because many SAS software and modules provide additional, special SAS functions and CALL routines. They are only available when the correct product is licensed to use them.

- SAS provides functions and CALL routines in the **SAS Data Quality Server** software. These functions and CALL routines like DQCASE, DQGENDER, or DQIDENTIFY enable you to cleanse data and access DataFlux Data Management Servers and are available in the Expression Builder of SAS Data Integration Studio and SAS Enterprise Guide. The selection introduced below is valid in CAS, PROC SQL, and SAS 9.
- **SAS National Language Support (NLS)** ensures that users can write SAS applications that conform to local language conventions in regions such as Asia and Europe. The functions presented are mainly from the Double-Byte Character Sets (DBCS) category. DBCS are needed to present languages that have many characters such as East Asian languages that can have thousands of double bytes of information.
- SAS functions and CALL routines with a focus on analytics are distributed over several SAS modules. **SAS/QC** contains the functions AOQ2, CUSUMARL, EWMAARL, PROBMED, STDMED, and the CALL routine CALL BAYESACT. **SAS/IML** contains many more functions and CALL routines for linear algebra, numerical/statistical/graphical analysis, genetic algorithms (experimental), matrix calculation, wave transformations, as well as for time series analysis, and so on. Some SAS functions that can be mentioned are, for example, APPLY, ARMASIM, BLOCK, BRANKS, BTRAN, ECHELON, EIGVAL, HERMITE, POLYROOT and many more. Some *CALL routines* are CALL APPCORT, CALL ARMACOV, CALL ARMALIK, CALL CHANGE, CALL GDRAW, PGRAF, and many more CALL routines for nonlinear optimization, including CALL NLPCG, NLPDD, NLPFDD, NLPFEA, NLPHQN, NLPLM, NLPNMS, NLPNRA. The CALL routines SEQSCALE, SEQSHIFT, and SEQTESTS, for example, perform discrete sequential tests.
- SAS also provides many other functions for the **Macro Facility**, such as %BQUOTE, %EVAL, %INDEX, %LENGTH, %NRBQUOTE, %NRQUOTE, %QSCAN, %QUOTE, %SCAN, %SYSEVALF.
- In addition, **PROC FCMP** allows users to write, test, and save their own SAS functions, which can also be used in PROC SQL.

SAS can also move functions back and forth between the individual modules of the SAS software. For example, the former SAS/ETS functions INTCINDEX, INTCYCLE, INTFMT, INTINDEX as well as INTSEAS were moved to Base SAS since SAS 9.2 and are also ISO 8601 compliant. The SAS functions HOLIDAY and NWKDOM from the SAS High-Performance Forecasting are also available in Base SAS since SAS 9.2. Another reason is that in some cases the number of arguments needed might be platform-dependent, for example, with the EXP function.

SAS might grow faster than any documentation. In SAS 9.4, numerous functions and CALL routines are new, including COT, CSC, DOSUBL, or FCOPY, and many more CALL routines and functions were enhanced. Furthermore, an unknown number of undocumented SAS functions and SAS CALL routines can be assumed.

Calling SAS Functions and SAS CALL routines

The call of SAS functions and SAS CALL routines differs in one central aspect. To the left of a SAS function is an assignment statement, for example, "v1_3="; to the left of a SAS CALL routine is a call, "CALL".

Program 10.2-1: Calling a SAS Function

```
data _null_ ;
   v1='Hel';
   v2='lo wo';
   v3='rld!';
 v1_3=cat(v1,v2,v3);
   put v1_3 ;
run;
```

Program 10.2-2: Calling a SAS CALL routine

```
data _null_ ;
length v1_3 $ 12;
   v1='Hel';
   v2='lo wo';
   v3='rld!';
 call catt(v1_3,v1,v2,v3);
   put v1_3 ;
run;
```

Figure 10.2-1: Output in SAS Log

```
Hello world!
```

Figure 10.2-2: Output in SAS Log

```
Hello world!
```

The SAS documentation points out that values returned by SAS functions might be truncated, unlike SAS macro functions. For details about functions and CALL routines, please refer to the current SAS documentation.

10.2.2 Quick Finder: Categories of SAS Functions and CALL Routines

The following subsection briefly introduces selected SAS functions and CALL routines. The categories of the SAS functions and CALL routines are arranged in alphabetical order.

For additional details about which CALL routines and functions run on the CAS server and which run only on the client, and for other SAS functions and CALL routines, please refer to the SAS documentation.

Some SAS functions and CALL routines might appear in several categories such as TZONEID in "NLS" and "Date and Time", or HASHING in "Character" and "Hashing" (SAS Institute, 2016d).

Table 10.2-1: Overview: Categories of SAS Functions and CALL routines

Category	Description
Arithmetic	Returns the result of a division that handles special missing values for ODS output.
Array	Returns information about arrays.
Bitwise Logical Operations	Returns the bitwise logical result for an argument.
Character	Returns information based on character data.
Character String Matching	Returns information from Perl regular expressions.
Combinatorial	Generates combinations and permutations.
Data Quality	Enables data cleansing and access to DataFlux Data Management Servers. *Note:* Selection: Valid in CAS, PROC SQL, and SAS 9.
Date and Time	Returns date and time values, including time intervals.
Descriptive Statistics	Returns statistical values such as mean, median, and standard deviation.
Distance	Returns the geodetic distance.

(Continued)

Table 10.2-1: (*Continued*)

Category	Description
External Files	Returns information that is associated with external files.
External Routines	Returns a character or numeric value, or calls a routine without any return code.
Financial	Calculates financial values such as interest, periodic payments, depreciation, and prices for European options on stocks.
Git	Supports the libgit2 package for interacting with Git repositories.
Hashing	Transforms data (message) into a short fixed-length value (digest).
Hyperbolic	Performs hyperbolic calculations such as sine, cosine, and tangent.
Macro	Assigns a value to a macro variable, returns the value of a macro variable, determines whether a macro variable is global or local in scope, and identifies whether a macro variable exists.
Mathematical	Performs mathematical calculations such as factorials, absolute value, fuzzy comparisons, and logarithms.
National Language Support (NLS)	NLS ensures that users can process data successfully in their native languages and environments.
Numeric	Returns a numeric value based on whether an expression is true, false, or missing, or determines whether a software image exists in the installed version of SAS.
Probability	Returns probability calculations such as from a chi-square or Poisson distribution.
Quantile	Returns a quantile from specific distributions.
Random Number	Returns random variates from specific distributions.
SAS File I/O	Returns information about SAS files.
Search	Searches for character or numeric values.
Sort	Sorts the values of character or numeric arguments.
Special	Special functions and CALL routines that return and store memory addresses, suspend execution of a program, submit an operating-environment command for execution, specify formats and informats at run time, return the system return code, determine whether a product is licensed, as well as return other information about SAS processing.
State and ZIP Code	Returns ZIP codes, FIPS codes, state and city names, postal codes, and the geodetic distance between ZIP codes.
Trigonometric	Returns trigonometric values such as sine, cosine, and tangent.
Truncation	Truncates numeric values and returns numeric values, often using fuzzing or zero fuzzing.

10.2.3 Categories and Descriptions of SAS Functions and SAS CALL Routines

The following table introduces selected SAS functions and CALL routines in alphabetical order. In the column "CAS", the ● symbol shows which functions and CALL routines are supported for running on a CAS server (SAS Institute, 2016d).

Table 10.2-2: Categories and Descriptions of SAS Functions and SAS CALL routines

SAS Functions and CALL routines	Descriptions	CAS
Arithmetic		
DIVIDE	Returns the result of a division that handles special missing values for ODS output.	●
Array		
DIM	Returns the number of elements in an array.	●
HBOUND	Returns the upper bound of an array.	●
LBOUND	Returns the lower bound of an array.	●
Bitwise Logical Operations		
BAND	Returns the bitwise logical AND of two arguments.	●
BLSHIFT	Returns the bitwise logical left shift of two arguments.	●
BNOT	Returns the bitwise logical NOT of an argument.	●
BOR	Returns the bitwise logical OR of two arguments.	●
BRSHIFT	Returns the bitwise logical right shift of two arguments.	●
BXOR	Returns the bitwise logical EXCLUSIVE OR of two arguments.	●
Character		
ANYALNUM	Searches a character string for an alphanumeric character and returns the first position at which the character is found.	●
ANYALPHA	Searches a character string for an alphabetic character and returns the first position at which the character is found.	●
ANYCNTRL	Searches a character string for a control character and returns the first position at which that character is found.	●
ANYDIGIT	Searches a character string for a digit and returns the first position at which the digit is found.	●
ANYFIRST	Searches a character string for a character that is valid as the first character in a SAS variable name under VALIDVARNAME = V7 and returns the first position at which that character is found.	●
ANYGRAPH	Searches a character string for a graphical character and returns the first position at which that character is found.	●
ANYLOWER	Searches a character string for a lowercase letter and returns the first position at which the letter is found.	●
ANYNAME	Searches a character string for a character that is valid in a SAS variable name under VALIDVARNAME = V7 and returns the first position at which that character is found.	●
ANYPRINT	Searches a character string for a printable character and returns the first position at which that character is found.	●
ANYPUNCT	Searches a character string for a punctuation character and returns the first position at which that character is found.	●
ANYSPACE	Searches a character string for a whitespace character (blank, horizontal or vertical tab, carriage return, line feed, and form feed) and returns the first position at which that character is found.	●
ANYUPPER	Searches a character string for an uppercase letter and returns the first position at which the letter is found.	●

(*Continued*)

Table 10.2-2: (*Continued*)

SAS Functions and CALL routines	Descriptions	CAS
ANYXDIGIT	Searches a character string for a hexadecimal character that represents a digit and returns the first position at which that character is found.	●
BYTE	Returns one character in the ASCII or EBCDIC collating sequence.	●
CALL CATS	Removes leading and trailing blanks and returns a concatenated character string.	-
CALL CATT	Removes trailing blanks and returns a concatenated character string.	-
CALL CATX	Removes leading and trailing blanks, inserts delimiters, and returns a concatenated character string.	-
CALL COMPCOST	Sets the costs of operations for later use by the COMPGED function.	-
CALL MISSING	Assigns missing values to the specified character or numeric variables.	●
CALL SCAN	Returns the position and length of the nth word from a character string.	●
CAT	Does not remove leading or trailing blanks and returns a concatenated character string.	●
CATQ	Concatenates character and numeric values by using a delimiter to separate items and by adding quotation marks to strings that contain the delimiter.	●
CATS	Removes leading and trailing blanks and returns a concatenated character string.	●
CATT	Removes trailing blanks and returns a concatenated character string.	●
CATX	Removes leading and trailing blanks, inserts delimiters, and returns a concatenated character string.	●
CHAR	Returns a single character from a specified position in a character string.	-
CHOOSEC	Returns a character value that represents the results of choosing from a list of arguments.	●
CHOOSEN	Returns a numeric value that represents the results of choosing from a list of arguments.	●
COALESCEC	Returns the first nonmissing value from a list of character arguments.	●
COLLATE	Returns a character string in the ASCII or EBCDIC collating sequence.	●
COMPARE	Returns the position of the leftmost character by which two strings differ or returns 0 if there is no difference.	●
COMPBL	Removes multiple blanks from a character string.	●
COMPGED	Returns the generalized edit distance between two strings.	●
COMPLEV	Returns the Levenshtein edit distance between two strings.	●
COMPRESS	Returns a character string with specified characters removed from the original string.	●
COUNT	Counts the number of times that a specified substring appears within a character string.	●
COUNTC	Counts the number of characters that appear or do not appear in a list of characters.	●
COUNTW	Counts the number of words in a character string.	●
DEQUOTE	Removes matching quotation marks from a character string that begins with a quotation mark and deletes all characters to the right of the closing quotation mark.	●
FIND	Searches for a specific substring of characters within a character string.	●
FINDC	Searches a string for any character in a list of characters.	●

(*Continued*)

Table 10.2-2: (*Continued*)

SAS Functions and CALL routines	Descriptions	CAS
FINDW	Returns the character position of a word in a string or returns the number of the word in a string.	●
FIRST	Returns the first character in a character string.	-
HASHING	Transforms a message into a digest in hexadecimal representation.	●
HASHING_FILE	Transforms the entire contents of a file into a digest and returns the digest in hexadecimal representation.	●
HASHING_INT	Initializes a running hash.	●
IFC	Returns a character value based on whether an expression is true, false, or missing.	●
INDEX	Searches a character expression for a string of characters and returns the position of the string's first character for the first occurrence of the string.	●
INDEXC	Searches a character expression for any of the specified characters and returns the position of that character.	●
INDEXW	Searches a character expression for a string that is specified as a word and returns the position of the first character in the word. Since SAS 9.2, also with alternative separators. Thus, the end of a string is not automatically understood as the end of the data.	●
LEFT	Left-aligns a character string.	●
LENGTH	Returns the length of a non-blank character string, excluding trailing blanks, and returns 1 for a blank character string.	●
LENGTHC	Returns the length of a character string, including trailing blanks.	●
LENGTHM	Returns the amount of memory (in bytes) that is allocated for a character string.	●
LENGTHN	Returns the length of a character string, excluding trailing blanks. LENGTHN returns a value of 0 for blank character strings.	●
LOWCASE	Converts all uppercase single-width English alphabet letters in an argument to lowercase.	●
MISSING	Returns a numeric result that indicates whether the argument contains a missing value.	●
NLITERAL	Converts a character string that you specify to a SAS name literal (string in quotation marks, followed by an n; so-called n-literal).	-
NOTALNUM	Searches a character string for a nonalphanumeric character and returns the first position at which the character is found.	●
NOTALPHA	Searches a character string for a nonalphabetic character and returns the first position at which the character is found.	●
NOTCNTRL	Searches a character string for a character that is not a control character and returns the first position at which that character is found.	●
NOTDIGIT	Searches a character string for any character that is not a digit and returns the first position at which that character is found.	●
NOTFIRST	Searches a character string for an invalid first character in a SAS variable name under VALIDVARNAME=V7 and returns the first position at which that character is found.	●
NOTGRAPH	Searches a character string for a non-graphical character and returns the first position at which that character is found.	●

(*Continued*)

Table 10.2-2: (*Continued*)

SAS Functions and CALL routines	Descriptions	CAS
NOTLOWER	Searches a character string for a character that is not a lowercase letter and returns the first position at which that character is found.	●
NOTNAME	Searches a character string for an invalid character in a SAS variable name under VALIDVARNAME=V7 and returns the first position at which that character is found.	●
NOTPRINT	Searches a character string for a nonprintable character and returns the first position at which that character is found.	●
NOTPUNCT	Searches a character string for a character that is not a punctuation character and returns the first position at which that character is found.	●
NOTSPACE	Searches a character string for a character that is not a whitespace character (blank, horizontal and vertical tab, carriage return, line feed, and form feed) and returns the first position at which that character is found.	●
NOTUPPER	Searches a character string for a character that is not an uppercase letter and returns the first position at which that character is found.	●
NOTXDIGIT	Searches a character string for a character that is not a hexadecimal character and returns the first position at which that character is found.	●
NVALID	Checks the validity of a character string for use as a SAS variable name.	-
PROPCASE	Converts all words in an argument to proper case.	●
QUOTE	Adds double quotation marks to a character value.	●
RANK	Returns the position of a character in the ASCII collating sequence.	●
REPEAT	Returns a character value that consists of the first argument repeated n+1 times.	●
REVERSE	Reverses a character string.	●
RIGHT	Right-aligns a character expression.	●
SCAN	Returns the nth word from a character string.	●
SOUNDEX	Encodes a string to facilitate searching for the sound of words. The underlying algorithm was developed for the English language and might therefore be less useful for comparing strings from other languages.	-
SPEDIS	Determines the likelihood of two words matching, expressed as the asymmetric spelling distance between the two words.	-
STRIP	Returns a character string with all leading and trailing blanks removed.	●
SUBPAD	Returns a substring that has a length that you specify, using blank padding if necessary.	-
SUBSTR (left of =)	Replaces character value contents.	●
SUBSTR (right of =)	Extracts a substring from an argument.	●
SUBSTRN	Returns a substring, allowing a result with a length of zero.	●
TRANSLATE	Replaces specific characters in a character expression.	●
TRANSTRN	Replaces or removes all occurrences of a substring in a character string.	●
TRANWRD	Replaces all occurrences of a substring in a character string.	●
TRIM	Removes trailing blanks from a character string and returns one blank if the string is missing.	●
TRIMN	Removes trailing blanks from character expressions and returns a string with a length of zero if the expression is missing.	●

(*Continued*)

Table 10.2-2: (*Continued*)

SAS Functions and CALL routines	Descriptions	CAS
UPCASE	Converts all lowercase single-width English alphabet letters in an argument to uppercase.	●
VERIFY	Returns the position of the first character in a string that is not in specified data strings.	●
Character String Matching (please see note at the end of the table)		
CALL PRXCHANGE	Performs a pattern-matching replacement.	-
CALL PRXDEBUG	Enables Perl regular expressions in a DATA step to send debugging output to the SAS log.	-
CALL PRXFREE	Frees memory that was allocated for a Perl regular expression.	●
CALL PRXNEXT	Returns the position and length of a substring that matches a pattern and iterates over multiple matches within one string.	-
CALL PRXPOSN	Returns the start position and length for a capture buffer.	●
CALL PRXSUBSTR	Returns the position and length of a substring that matches a pattern.	-
PRXCHANGE	Performs a pattern-matching replacement.	-
PRXMATCH	Searches for a pattern match and returns the position at which the pattern is found.	-
PRXPAREN	Returns the last parenthesis (brackets) match for which there is a match in a pattern.	-
PRXPARSE	Compiles a Perl regular expression (PRX) that can be used for pattern matching of a character value.	-
PRXPOSN	Returns a character string that contains the value for a capture buffer.	-
Combinatorial		
ALLCOMB	Generates all combinations of the values of n variables taken k at a time in a minimal change order.	-
ALLPERM	Generates all permutations of the values of several variables in a minimal change order.	●
CALL ALLCOMB	Generates all combinations of the values of n variables taken k at a time in a minimal change order.	-
CALL ALLCOMBI	Generates all combinations of the indices of n objects taken k at a time in a minimal change order.	-
CALL ALLPERM	Generates all permutations of the values of several variables in a minimal change order.	-
CALL GRAYCODE	Generates all subsets of n items in a minimal change order.	-
CALL LEXCOMB	Generates all distinct combinations of the nonmissing values of n variables taken k at a time in lexicographic order.	-
CALL LEXCOMBI	Generates all combinations of the indices of n objects taken k at a time in lexicographic order.	-
CALL LEXPERK	Generates all distinct permutations of the nonmissing values of n variables taken k at a time in lexicographic order.	-
CALL LEXPERM	Generates all distinct permutations of the nonmissing values of several variables in lexicographic order.	-
CALL RANCOMB	Permutes the values of the arguments and returns a random combination of k out of n values.	-

(*Continued*)

Table 10.2-2: (*Continued*)

SAS Functions and CALL routines	Descriptions	CAS
CALL RANPERK	Permutes the values of the arguments and returns a random permutation of k out of n values.	-
CALL RANPERM	Randomly permutes the values of the arguments.	-
COMB	Computes the number of combinations of n elements taken r at a time.	•
GRAYCODE	Generates all subsets of n items in a minimal change order.	-
LCOMB	Computes the logarithm of the COMB function, which is the logarithm of the number of combinations of n objects taken r at a time.	•
LEXCOMB	Generates all distinct combinations of the nonmissing values of n variables taken k at a time in lexicographic order.	-
LEXCOMBI	Generates all combinations of the indices of n objects taken k at a time in lexicographic order.	-
LEXPERK	Generates all distinct permutations of the nonmissing values of n variables taken k at a time in lexicographic order.	-
LEXPERM	Generates all distinct permutations of the nonmissing values of several variables in lexicographic order.	-
LFACT	Computes the logarithm of the FACT (factorial) function.	•
LPERM	Computes the logarithm of the PERM function, which is the logarithm of the number of permutations of n objects, with the option of including r number of elements.	•
PERM	Computes the number of permutations of n items that are taken r at a time.	•
Data Quality (selection; valid in CAS, PROC SQL, and SAS 9)		
CALL DQPARSE	Returns a delimited string of parse token values and a status flag.	•
DQCASE	Returns a character value with standardized capitalization.	•
DQEXTINFOGET	Returns the names of the tokens that are supported by an extraction definition.	•
DQEXTRACT	Returns a delimited string of extraction token values from an input character value.	•
DQEXTTOKENGET	Returns an extraction token value from a delimited string of extraction token values.	•
DQEXTTOKENPUT	Inserts an extraction token value into a delimited string of extraction token values and returns the updated delimited string.	•
DQGENDER	Returns a gender determination from the name of an individual.	•
DQGENDERINFOGET	Returns the name of the parse definition that is associated with the specified gender definition.	•
DQGENDERPARSED	Returns the gender of an individual from a delimited string of parse token values.	•
DQIDENTIFY	Returns the highest-scoring identity for a character value.	•
DQIDENTIFYIDGET	Returns an identification analysis score for a given identity from a delimited string of identification analysis scores.	•
DQLOCALEGUESS	Returns the ISO code of the highest-scoring locale for a character value.	•
DQLOCALEINFOGET	Returns a list of the locales that are loaded into memory.	•
DQLOCALEINFOLIST	Returns a count of definitions and displays the names of definitions for a type of definition in a locale.	•
DQLOCALESCORE	Returns a locale confidence score for an input character value.	•

(*Continued*)

Table 10.2-2: (*Continued*)

SAS Functions and CALL routines	Descriptions	CAS
DQMATCH	Returns a matchcode from a character value.	●
DQMATCHINFOGET	Returns a list of the locales that are loaded into memory.	●
DQMATCHPARSED	Returns a matchcode from a delimited string of parse token values.	●
DQPARSE	Returns a delimited string of parse token values.	●
DQPARSEINFOGET	Returns the names of the tokens that are supported by a parse definition.	●
DQPARSETOKENGET	Returns a parse token value from a delimited string of parse token values.	●
DQPARSETOKENPUT	Inserts a parse token value into a delimited string of parse token values and returns the updated delimited string.	●
DQPATTERN	Returns a pattern analysis from a character value.	●
DQSTANDARDIZE	Standardizes the casing, spacing, and format of certain words and abbreviations and returns an updated character value.	●
DQTOKEN	Returns the value of a token from an input character value.	●
DQVER	Returns the version of the SAS Data Quality engine.	●
DQVERQKB	Returns the version of the currently loaded QKB.	●
Date and Time		
CALL ISO8601_ CONVERT	Converts an ISO 8601 interval to datetime and duration values or converts datetime and duration values to an ISO 8601 interval.	-
DATDIF	Returns the number of days between two dates after computing the difference between the dates according to specified day count conventions. Since SAS 9.2 with reference to a document of the Securities Industry Association.	●
DATE	Returns the current date as a SAS date value.	●
DATEJUL	Converts a Julian date to a SAS date value.	●
DATEPART	Extracts the date from a SAS datetime value.	●
DATETIME	Returns the current date and time of day as a SAS datetime value.	●
DAY	Returns the day of the month from a SAS date value.	●
DHMS	Returns a SAS datetime value from date, hour, minute, and second values.	●
HMS	Returns a SAS time value from hour, minute, and second values.	●
HOLIDAY	Returns a SAS date value of a specified holiday for a specified year.	●
HOLIDAYCK	Returns the number of occurrences of the holiday value between date1 and date2.	-
HOLIDAYCOUNT	Returns the number of holidays defined for a SAS date value.	-
HOLIDAYNAME	Returns the name of the holiday that corresponds to the SAS date or a blank string if a holiday is not defined for the SAS date.	-
HOLIDAYNX	Returns the nth occurrence of the holiday relative to the date argument.	-
HOLIDAYNY	Returns the nth occurrence of the holiday for the year.	-
HOUR	Returns the hour from a SAS time or datetime value.	●
INTCK	Returns the number of interval boundaries of a given kind that lie between two dates, times, or datetime values. Since SAS 9.2 also ISO 8601 compliant.	●
INTFIT	Returns a time interval that is aligned between two dates.	●
INTFMT	Returns a recommended SAS format when a date, time, or datetime interval is specified.	●

(*Continued*)

Table 10.2-2: (*Continued***)**

SAS Functions and CALL routines	Descriptions	CAS
INTGET	Returns a time interval based on three date or datetime values.	•
INTNX	Increments a date, time, or datetime value by a given time interval, and returns a date, time, or datetime value; since SAS 9.2 also ISO 8601 compliant.	•
INTSHIFT	Returns the shift interval that corresponds to the base interval.	•
INTTEST	Returns 1 if a time interval is valid and returns 0 if a time interval is invalid.	•
JULDATE	Returns the Julian date from a SAS date value.	•
JULDATE7	Returns a seven-digit Julian date from a SAS date value.	•
MDY	Returns a SAS date value from month, day, and year values.	•
MINUTE	Returns the minute from a SAS time or datetime value.	•
MONTH	Returns the month from a SAS date value.	•
QTR	Returns the quarter of the year from a SAS date value.	•
SECOND	Returns the seconds and milliseconds from a SAS time or datetime value.	•
TIME	Returns the current time of day as a numeric SAS time value.	•
TIMEPART	Extracts a time value from a SAS datetime value.	•
TODAY	Returns the current date as a numeric SAS date value.	•
TZONEID	Returns the current time zone ID.	-
TZONENAME	Returns the current standard or Daylight Saving Time, time zone name.	-
TZONEU2S	Converts a UTC date time value to a SAS date time value.	-
WEEK	Returns the week-number value.	•
WEEKDAY	From a SAS date value, returns an integer that corresponds to the day of the week	•
YEAR	Returns the year from a SAS date value.	•
YRDIF	Returns the difference in years between two dates according to specified day count conventions; returns a person's age.	•
YYQ	Returns a SAS date value from year and quarter year values.	•
Descriptive Statistics		
CMISS	Counts the number of missing arguments.	•
CSS	Returns the corrected sum of squares.	•
CV	Returns the coefficient of variation.	•
EUCLID	Returns the Euclidean norm of the nonmissing arguments.	-
GEOMEAN	Returns the geometric mean.	•
GEOMEANZ	Returns the geometric mean, using zero fuzzing.	•
HARMEAN	Returns the harmonic mean.	•
HARMEANZ	Returns the harmonic mean, using zero fuzzing.	•
IQR	Returns the interquartile range (IQR).	•
KURTOSIS	Returns the kurtosis.	•
LARGEST	Returns the kth largest nonmissing value.	•
LPNORM	Returns the Lp norm of the second argument and subsequent nonmissing arguments.	-
MAD	Returns the median absolute deviation from the median.	•

(Continued)

Table 10.2-2: (*Continued*)

SAS Functions and CALL routines	Descriptions	CAS
MAX	Returns the largest value.	●
MEAN	Returns the arithmetic mean (average).	●
MEDIAN	Returns the median value.	●
MIN	Returns the smallest value.	●
MISSING	Returns a numeric result that indicates whether the argument contains a missing value.	●
N	Returns the number of nonmissing numeric values.	●
NMISS	Returns the number of missing numeric values.	●
ORDINAL	Returns the kth smallest of the missing and nonmissing values.	●
PCTL	Returns the percentile that corresponds to the percentage.	●
RANGE	Returns the range of the nonmissing values.	●
RMS	Returns the root mean square of the nonmissing arguments.	●
SKEWNESS	Returns the skewness of the nonmissing arguments.	●
SMALLEST	Returns the kth smallest nonmissing value.	●
STD	Returns the standard deviation of the nonmissing arguments.	●
STDERR	Returns the standard error of the mean of the nonmissing arguments.	●
SUM	Returns the sum of the nonmissing arguments.	●
SUMABS	Returns the sum of the absolute values of the nonmissing arguments.	●
USS	Returns the uncorrected sum of squares of the nonmissing arguments.	●
VAR	Returns the variance of the nonmissing arguments.	●
Distance		
GEODIST	Returns the geodetic distance between two latitude and longitude coordinates.	●
ZIPCITYDISTANCE	Returns the geodetic distance between two ZIP code locations.	-
External Files		
DCLOSE	Closes a directory that was opened by the DOPEN function.	-
DCREATE	Returns the complete pathname of a new, external directory.	-
DINFO	Returns information about a directory.	-
DNUM	Returns the number of members in a directory.	-
DOPEN	Opens a directory and returns a directory identifier value.	-
DOPTNAME	Returns directory attribute information. The number, names, and type of directory information depend on the respective operating environment.	-
DOPTNUM	Returns the number of information items that are available for a directory.	-
DREAD	Returns the name of a directory member.	-
DROPNOTE	Deletes a note marker from a SAS data set or an external file.	-
FAPPEND	Appends the current record to the end of an external file.	-
FCLOSE	Closes an external file, directory, or directory member.	-
FCOL	Returns the current column position in the File Data Buffer (FDB).	-
FDELETE	Deletes an external file or an empty directory.	-
FEXIST	Verifies the existence of an external file that is associated with a fileref.	-

(*Continued*)

Table 10.2-2: (*Continued*)

SAS Functions and CALL routines	Descriptions	CAS
FGET	Copies data from the File Data Buffer (FDB) into a variable.	-
FILEEXIST	Verifies the existence of an external file by its physical name.	-
FILENAME	Assigns or deassigns a fileref to an external file, directory, or output device.	-
FILEREF	Verifies whether a fileref has been assigned for the current SAS session.	-
FINFO	Returns the value of a file information item.	-
FNOTE	Identifies the last record that was read and returns a value that the FPOINT function can use.	-
FOPEN	Opens an external file and returns a file identifier value.	-
FOPTNAME	Returns the name of an item of information about an external file.	-
FOPTNUM	Returns the number of information items, such as file name or record length, which are available for an external file.	-
FPOINT	Positions the read pointer on the next record to be read.	-
FPOS	Sets the position of the column pointer in the File Data Buffer (FDB).	-
FPUT	Moves data to the File Data Buffer (FDB) of an external file, starting at the FDB's current column position.	-
FREAD	Reads a record from an external file into the File Data Buffer (FDB).	-
FREWIND	Positions the file pointer to the start of the file.	-
FRLEN	Returns the size of the last record that was read, or if the file is opened for output, returns the current record size.	-
FSEP	Sets the token delimiters for the FGET function; since SAS 9.2 also as hexadecimal value.	-
FWRITE	Writes a record to an external file.	-
MOPEN	Opens a file by directory ID and member name, and returns either the file identifier or a 0.	-
PATHNAME	Returns the physical name of an external file or a SAS library, or returns a blank.	-
RENAME	Renames a member of a SAS library, an entry in a SAS catalog, an external file, or a directory.	-
SYSMSG	Returns error or warning message text from processing the last data set or external file function.	-
SYSRC	Returns a system error number.	-
External Routines		
CALL MODULE	Calls an external routine without any return code.	-
MODULE	Calls a specific routine or module that resides in an external dynamic link library (DLL).	-
MODULEC	Calls an external routine and returns a character value.	-
CALL MODULEI	Calls an external routine that does not return a value. Supported only by the IML procedure.	-
MODULEIC	Calls an external routine that returns a character value. Supported only by the IML procedure.	-
MODULEIN	Calls an external routine that returns a numerical value. Supported only by the IML procedure.	-
MODULEN	Calls an external routine and returns a numeric value.	-

(*Continued*)

Table 10.2-2: (*Continued*)

SAS Functions and CALL routines	Descriptions	CAS
Finance		
BLACKCLPRC	Calculates call prices for European options on futures, based on the Black model.	●
BLACKPTPRC	Calculates put prices for European options on futures, based on the Black model.	●
BLKSHCLPRT	Calculates call prices for European options on stocks, based on the Black-Scholes model.	-
BLKSHPTPRT	Calculates put prices for European options on stocks, based on the Black-Scholes model.	-
COMPOUND	Returns compound interest parameters.	●
CONVX	Returns the convexity for an enumerated cash flow.	●
CONVXP	Returns the convexity for a periodic cash flow stream such as a bond.	●
DACCDB	Returns the accumulated declining balance depreciation.	-
DACCDBSL	Returns the accumulated declining balance with conversion to a straight-line depreciation.	-
DACCSL	Returns the accumulated straight-line depreciation.	-
DACCSYD	Returns the accumulated sum-of-years-digits depreciation.	-
DACCTAB	Returns the accumulated depreciation from specified tables.	-
DEPDB	Returns the declining balance depreciation.	-
DEPDBSL	Returns the declining balance with conversion to a straight-line depreciation.	-
DEPSL	Returns the straight-line depreciation.	-
DEPSYD	Returns the sum-of-years-digits depreciation.	-
DEPTAB	Returns the depreciation from specified tables.	-
DUR	Returns the modified duration for an enumerated cash flow.	●
DURP	Returns the modified duration for a periodic cash flow stream, such as a bond.	●
FINANCE	Computes numerous financial calculations such as depreciation, maturation, accrued interest, net present value, periodic savings, and internal rates of return.	●
GARKHCLPRC	Calculates call prices for European options on stocks, based on the Garman-Kohlhagen model.	●
GARKHPTPRC	Calculates put prices for European options on stocks, based on the Garman-Kohlhagen model.	●
INTRR	Returns the internal rate of return as a fraction.	●
IRR	Returns the internal rate of return as a percentage.	●
MARGRCLPRC	Calculates call prices for European options on stocks, based on the Margrabe model.	●
MARGRPTPRC	Calculates put prices for European options on stocks, based on the Margrabe model.	●
MORT	Returns amortization parameters.	●
NETPV	Returns the net present value as a percent.	●
NPV	Returns the net present value with the rate expressed as a percentage.	●
PVP	Returns the present value for a periodic cash flow stream (such as a bond), with repayment of principal at maturity.	●
SAVING	Returns the future value of a periodic saving.	●

(*Continued*)

Table 10.2-2: (*Continued*)

SAS Functions and CALL routines	Descriptions	CAS
SAVINGS	Returns the balance of a periodic savings by using variable interest rates.	•
YIELDP	Returns the yield-to-maturity for a periodic cash flow stream, such as a bond.	•
Git		
GIT_BRANCH_MERGE	Merges a Git branch into the currently checked-out branch.	-
GIT_BRANCH_NEW	Creates a Git branch.	-
GIT_CLONE	Clones a Git repository into a directory on the SAS server.	-
GIT_COMMIT	Commits staged files to the local repository.	-
GIT_COMMIT_GET	Returns the specified attribute of the nth commit object that is associated with the local repository.	-
GIT_DELETE_REPO	Deletes a local Git repository and all content within the repository.	-
GIT_DIFF_FREE	Clears the diff record object associated with a local repository.	-
GIT_DIFF	Returns the number of diffs between two commits in the local repository and creates a diff record object for the local repository.	-
GIT_FETCH	Fetches updates from the remote repository.	-
GIT_INDEX_REMOVE	Unstages 1 to n number of files to commit to the local repository.	-
GIT_PULL	Pulls changes from the remote repository into the local repository.	-
GIT_PUSH	Pushes the committed files in the local repository to the remote repository.	-
GIT_REBASE	Rebases your current branch to a specified commit ID.	-
GIT_RESET	Resets the local repository to a specified commit.	-
GIT_SET_URL	Sets the remote repository URL for a local repository.	-
GIT_STASH	Stashes file changes that have not been committed.	-
GIT_STATUS	Returns the status objects for files in the local repository and creates a status record.	-
GIT_VERSION	Specifies whether libgit2 is available and if available, specifies the version that is being used.	-
GITFN_CLONE	Clones a Git repository into a directory on the SAS server.	-
GITFN_COMMIT_LOG	Returns the number of commit objects that are associated with the local repository.	-
GITFN_COMMIT	Commits staged files to the local repository.	-
GITFN_DEL_REPO	Deletes a local Git repository and its contents.	-
GITFN_DIFF_GET	Returns the specified attribute of the nth diff object in the local repository.	-
GITFN_DIFF	Returns the number of diffs between two commits in the local repository and creates a diff record object for the local repository.	-
GITFN_IDX_ADD	Stages 1 to n number of files to commit to the local repository.	-
GITFN_NEW_BRANCH	Creates a Git branch.	-
GITFN_PULL	Pulls changes from the remote repository into the local repository.	-
GITFN_PUSH	Pushes the committed files in the local repository to the remote repository.	-
GITFN_RESET	Resets the local repository to a specified commit.	-
GITFN_STATUS	Returns the status objects for files in the local repository and creates a status record.	-

(*Continued*)

Table 10.2-2: (*Continued*)

SAS Functions and CALL routines	Descriptions	CAS
GITFN_VERSION	Specifies whether libgit2 is available and if available, specifies the version that is being used.	-
Hashing		
HASHING	Transforms a message into a digest in hexadecimal representation.	●
HASHING_FILE	Transforms the entire contents of a file into a digest and returns the digest in hexadecimal representation.	●
HASHING_INT	Initializes a running hash.	●
MD5	Returns the MD5 digest for a specified message string.	●
SHA256	Returns the SHA256 digest for a specified message string.	●
SHA256HEX	Returns the SHA256 digest for a specified message, and the digest is provided in hexadecimal representation.	●
Hyperbolic		
ARCOSH	Returns the inverse hyperbolic cosine.	●
ARSINH	Returns the inverse hyperbolic sine.	●
ARTANH	Returns the inverse hyperbolic tangent.	●
COSH	Returns the hyperbolic cosine.	●
SINH	Returns the hyperbolic sine.	●
TANH	Returns the hyperbolic tangent.	●
Macro		
CALL EXECUTE	Resolves the argument and issues the resolved value for execution at the next step boundary.	-
CALL SYMPUT	Assigns a value produced in a DATA step to a macro variable. Not supported by the CAS engine.	-
CALL SYMPUTX	Assigns a value to a macro variable and removes both leading and trailing blanks.	-
DOSUBL	Enables the immediate execution of SAS code after a text string is passed.	-
RESOLVE	Returns the resolved value of the argument after the argument has been processed by the macro facility.	-
SYMEXIST	Returns an indication of the existence of a macro variable.	-
SYMGET	Returns the value of a macro variable during DATA step execution.	-
SYMGLOBL	Returns an indication of whether a macro variable is in global scope to the DATA step during DATA step execution.	-
SYMLOCAL	Returns an indication of whether a macro variable is in local scope to the DATA step during DATA step execution.	-
Mathematical		
ABS	Returns the absolute value.	●
AIRY	Returns the value of a differential equation.	●
BETA	Returns the value of the beta function.	●
CALL LOGISTIC	Applies the logistic function to each argument.	●
CALL SOFTMAX	Returns the softmax value.	-
CALL STDIZE	Standardizes the values of one or more variables.	-
CALL TANH	Returns the hyperbolic tangent.	-

Table 10.2-2: (*Continued*)

SAS Functions and CALL routines	Descriptions	CAS
CNONCT	Returns the noncentrality parameter from a chi-square distribution.	●
COALESCE	Returns the first nonmissing value from a list of numeric arguments.	●
COMPFUZZ	Performs a fuzzy comparison of two numeric values.	●
CONSTANT	Computes machine and mathematical constants.	●
DAIRY	Returns the derivative of the AIRY function.	●
DEVIANCE	Returns the deviance based on a probability distribution.	●
DIGAMMA	Returns the value of the digamma function.	●
ERF	Returns the value of the (normal) error function.	●
ERFC	Returns the value of the complementary (normal) error function.	●
EXP	Returns the value of the exponential function.	●
FACT	Computes a factorial.	●
FNONCT	Returns the value of the noncentrality parameter of an F distribution.	●
GAMMA	Returns the value of the gamma function.	●
GCD	Returns the greatest common divisor for one or more integers.	●
IBESSEL	Returns the value of the modified Bessel function.	●
JBESSEL	Returns the value of the Bessel function.	●
LCM	Returns the least common multiple.	●
LGAMMA	Returns the natural logarithm of the Gamma function.	●
LOG	Returns the natural (base e) logarithm.	●
LOG10	Returns the logarithm to the base 10.	●
LOG1PX	Returns the log of 1 plus the argument.	●
LOG2	Returns the logarithm to the base 2.	●
LOGBETA	Returns the logarithm of the beta function.	●
MOD	Returns the remainder from the division of the first argument by the second argument, fuzzed to avoid most unexpected floating-point results.	●
MODZ	Returns the remainder from the division of the first argument by the second argument, using zero fuzzing.	●
MSPLINT	Returns the ordinate of a monotonicity-preserving interpolating spline.	-
SIGN	Returns the sign of a value.	●
SQRT	Returns the square root of a value.	●
TNONCT	Returns the value of the noncentrality parameter from the Student's t distribution.	●
TRIGAMMA	Returns the value of the trigamma function.	●
National Language Support (NLS)		
BASECHAR	Converts characters to base characters.	-
ENCODCOMPAT	Verifies the transcoding compatibility between two encodings.	-
EUROCURR	Converts one European currency to another.	-
GETLOCENV	Returns the current locale/language environment.	-
KCOMPARE	Returns the result of a comparison of character expressions.	-
KCOMPRESS	Removes specified characters from a character expression.	-

(*Continued*)

Table 10.2-2: (*Continued*)

SAS Functions and CALL routines	Descriptions	CAS
KCOUNT	Returns the number of double-byte characters in an expression.	-
KCVT	Converts data from one type of encoding data to another type of encoding data.	-
KINDEX	Searches a character expression for a string of characters.	-
KINDEXC	Searches a character expression for specified characters and returns character-based values.	-
KLEFT	Left-aligns a character expression by removing unnecessary leading DBCS blanks and SO/SI.	-
KLENGTH	Returns the length of an argument.	-
KLOWCASE	Converts the uppercase alphabetic letters to lowercase letters.	-
KREVERSE	Reverses a character expression.	-
KRIGHT	Right-aligns a character expression by trimming trailing DBCS blanks and SO/SI.	-
KSCAN	Selects a specified word from a character expression.	-
KSTRCAT	Concatenates two or more character expressions.	-
KSUBSTR	Extracts a substring from an argument.	-
KSUBSTRB	Extracts a substring from an argument according to the byte position of the substring in the argument.	-
KTRANSLATE	Replaces specific characters in a character expression.	-
KTRIM	Removes trailing DBCS blanks and SO/SI from character expressions.	-
KTRUNCATE	Truncates a string to a specified length in byte unit without breaking multibyte characters.	-
KUPCASE	Converts the lowercase alphabetic letters to uppercase letters.	-
KUPDATE	Inserts, deletes, and replaces character value contents.	-
KUPDATEB	Inserts, deletes, and replaces the contents of the character value according to the byte position of the character value in the argument.	-
KVERIFY	Returns the position of the first character (character-based value) that is unique to an expression.	-
NLDATE	Converts the SAS date value to the date value of the specified locale by using the date format descriptors.	-
NLDATM	Converts the SAS datetime value to the time value of the specified locale by using the datetime-format descriptors.	-
NLTIME	Converts the SAS time or the datetime value to the time value of the specified locale by using the NLTIME descriptors.	-
TRANTAB	Transcodes data by using the specified translation table.	-
TZONEID	Returns the current time zone ID.	-
VARTRANSCODE	Returns the transcode attribute of a SAS data set variable.	-
VTRANSCODE	Returns a value that indicates whether transcoding is enabled for the specified character variable.	-
VTRANSCODEX	Returns a value that indicates whether transcoding is enabled for the specified argument.	-
Numeric		
IFN	Returns a numeric value based on whether an expression is true, false, or missing.	●

(*Continued*)

Table 10.2-2: (*Continued*)

SAS Functions and CALL routines	Descriptions	CAS
MODEXIST	Determines whether a software image exists in the version of SAS that you have installed.	-
Probability		
CDF	Returns a value from a cumulative probability distribution, for example, Bernoulli, Beta, Binomial, Cauchy, Conway-Maxwell-Poisson, Exponential, F, Generalized Poisson, Normal Mixture, Pareto, Tweedle, Weibull, and so on.	•
LOGCDF	Returns the logarithm of a left cumulative distribution function.	•
LOGPDF	Returns the logarithm of a probability density (mass) function.	•
LOGSDF	Returns the logarithm of a survival function.	•
PDF	Returns a value from a probability density (mass) distribution.	•
POISSON	Returns the probability from a Poisson distribution.	•
PROBBETA	Returns the probability from a beta distribution.	•
PROBBNML	Returns the probability from a binomial distribution.	•
PROBBNRM	Returns a probability from a bivariate normal distribution.	•
PROBCHI	Returns the probability from a chi-square distribution.	•
PROBF	Returns the probability from an F distribution.	•
PROBGAM	Returns the probability from a gamma distribution.	•
PROBHYPR	Returns the probability from a hypergeometric distribution.	•
PROBMC	Returns a probability or a quantile from various distributions for multiple comparisons of means.	•
PROBMED	Computes cumulative probabilities for the sample median.	•
PROBNEGB	Returns the probability from a negative binomial distribution.	•
PROBNORM	Returns the probability from the standard normal distribution.	•
PROBT	Returns the probability from a t distribution.	•
SDF	Returns a survival function.	•
Quantiles		
BETAINV	Returns a quantile from the beta distribution.	•
CINV	Returns a quantile from the chi-square distribution.	-
FINV	Returns a quantile from the F distribution.	-
GAMINV	Returns a quantile from the gamma distribution.	•
PROBIT	Returns a quantile from the standard normal distribution.	•
QUANTILE	Returns the quantile from a distribution when you specify the left probability (CDF).	•
SQUANTILE	Returns the quantile from a distribution when you specify the right probability (SDF).	•
TINV	Returns a quantile from the t distribution.	•
Random		
CALL RANBIN	Returns a random variate from a binomial distribution.	-
CALL RANCAU	Returns a random variate from a Cauchy distribution.	-
CALL RANEXP	Returns a random variate from an exponential distribution.	-

(*Continued*)

Table 10.2-2: (*Continued*)

SAS Functions and CALL routines	Descriptions	CAS
CALL RANGAM	Returns a random variate from a gamma distribution.	-
CALL RANNOR	Returns a random variate from a normal distribution.	-
CALL RANPOI	Returns a random variate from a Poisson distribution.	-
CALL RANTBL	Returns a random variate from a tabled probability distribution.	-
CALL RANTRI	Returns a random variate from a triangular distribution.	-
CALL RANUNI	Returns a random variate from a uniform distribution.	-
CALL STREAM	Specifies a random-number stream to use for subsequent calls to the RAND function.	-
CALL STREAMINIT	Specifies a random-number generator and seed value for generating random numbers.	-
CALL STREAMREWIND	Rewinds a stream to its initial state for subsequent random-number generation.	-
NORMAL	Returns a random variate from a normal, or Gaussian, distribution.	-
RANBIN	Returns a random variate from a binomial distribution.	-
RANCAU	Returns a random variate from a Cauchy distribution.	-
RAND	Generates random numbers from a distribution that you specify.	●
RANEXP	Returns a random variate from an exponential distribution.	-
RANGAM	Returns a random variate from a gamma distribution.	-
RANNOR	Returns a random variate from a normal distribution.	-
RANPOI	Returns a random variate from a Poisson distribution.	-
RANTBL	Returns a random variate from a tabled probability distribution.	-
RANTRI	Returns a random variate from a triangular distribution.	-
RANUNI	Returns a random variate from a uniform distribution.	-
UNIFORM	Returns a random variate from a uniform distribution.	-
SAS File I/O		
ATTRC	Returns the value of a character attribute for a SAS data set.	-
ATTRN	Returns the value of a numeric attribute for a SAS data set.	-
CEXIST	Verifies the existence of a SAS catalog or SAS catalog entry.	-
CLOSE	Closes a SAS data set.	-
CUROBS	Returns the observation number of the current observation.	-
DROPNOTE	Deletes a note marker from a SAS data set or an external file.	-
DSNAME	Returns the SAS data set name that is associated with a data set identifier.	-
ENVLEN	Returns the length of an environment variable.	-
EXIST	Verifies the existence of a SAS library member.	-
FETCH	Reads the next non-deleted observation from a SAS data set into the Data Set Data Vector (DDV).	-
FETCHOBS	Reads a specified observation from a SAS data set into the Data Set Data Vector (DDV).	-
GETVARC	Returns the value of a SAS data set character variable.	-
GETVARN	Returns the value of a SAS data set numeric variable.	-
IORCMSG	Returns a formatted error message for _IORC_.	-

(*Continued*)

Table 10.2-2: (*Continued*)

SAS Functions and CALL routines	Descriptions	CAS
LIBNAME	Assigns or clears a libref for a SAS library.	-
LIBREF	Verifies that a libref has been assigned.	-
NOTE	Returns an observation ID for the current observation of a SAS data set.	-
OPEN	Opens a SAS data set.	-
PATHNAME	Returns the physical name of an external file or a SAS library, or returns a blank.	-
POINT	Locates an observation that is identified by the NOTE function.	-
RENAME	Renames a member of a SAS library, an entry in a SAS catalog, an external file, or a directory.	-
REWIND	Positions the data set pointer at the beginning of a SAS data set.	-
SYSEXIST	Returns a value that indicates whether an operating-environment variable exists in your environment.	-
SYSMSG	Returns error or warning message text from processing the last data set or external file function.	-
SYSRC	Returns a system error number.	-
VARFMT	Returns the format that is assigned to a SAS data set variable.	-
VARINFMT	Returns the informat that is assigned to a SAS data set variable.	-
VARLABEL	Returns the label that is assigned to a SAS data set variable.	-
VARLEN	Returns the length of a SAS data set variable.	-
VARNAME	Returns the name of a SAS data set variable.	-
VARNUM	Returns the number of a variable's position in a SAS data set.	-
VARTYPE	Returns the data type of a SAS data set variable.	-
Search		
WHICHC	Searches for a character value that is equal to the first argument and returns the index of the first matching value.	●
WHICHN	Searches for a numeric value that is equal to the first argument and returns the index of the first matching value.	●
Sort		
CALL SORTC	Sorts the values of character arguments.	●
CALL SORTN	Sorts the values of numeric arguments.	●
Special		
ADDR	Returns the memory address of a variable on a 32-bit platform.	-
ADDRLONG	Returns the memory address of a variable on 32-bit and 64-bit platforms.	-
CALL POKE	Writes a value directly into memory on a 32-bit platform (floating point since SAS 9.2).	-
CALL POKELONG	Writes a value directly into memory on 32-bit and 64-bit platforms (floating point since SAS 9.2).	-
CALL SLEEP	For a specified period of time, suspends the execution of a program that invokes this CALL routine.	●
CALL SYSTEM	Submits an operating-environment command for execution.	-
CALL TSO	Executes a TSO command, emulated USS command, or MVS program.	-
DIF	Returns differences between an argument and its nth lag.	-

(*Continued*)

Table 10.2-2: (*Continued*)

SAS Functions and CALL routines	Descriptions	CAS
FMTINFO	Retrieves information about a format or informat.	-
GETOPTION	Returns the value of a SAS system or graphics option.	-
INPUT	Returns the value that is produced when SAS converts an expression by using the specified informat.	●
INPUTC	Enables you to specify a character informat at run time.	●
INPUTN	Enables you to specify a numeric informat at run time.	●
LAG	Returns values from a queue.	-
PEEK	Stores the contents of a memory address in a numeric variable on a 32-bit platform.	-
PEEKC	Stores the contents of a memory address in a character variable on a 32-bit platform.	-
PEEKCLONG	Stores the contents of a memory address in a character variable on 32-bit and 64-bit platforms.	-
PEEKLONG	Stores the contents of a memory address in a numeric variable on 32-bit and 64-bit platforms.	-
PTRLONGADD	Returns the pointer address as a character variable on 32-bit and 64-bit platforms.	-
PUT	Returns a value using a specified format.	●
PUTC	Enables you to specify a character format at run time.	●
PUTN	Enables you to specify a numeric format at run time.	●
SLEEP	Suspends the execution of a program that invokes this function for a period of time.	●
SYSEXIST	Returns a value that indicates whether an operating-environment variable exists in your environment.	-
SYSGET	Returns the value of the specified operating-environment variable.	-
SYSPARM	Returns the system parameter string.	-
SYSPROCESSID	Returns the process ID of the current process.	-
SYSPROCESSNAME	Returns the process name that is associated with a given process ID or returns the name of the current process.	-
SYSPROD	Determines whether a product is licensed.	-
SYSTEM	Issues an operating-environment command during a SAS session and returns the system return code.	-
UUIDGEN	Returns the short or binary form of a Universally Unique Identifier (UUID).	●
State and ZIP Code		
FIPNAME	Converts two-digit FIPS codes to uppercase state names.	-
FIPNAMEL	Converts two-digit FIPS codes to mixed case state names.	-
FIPSTATE	Converts two-digit FIPS codes to two-character state postal codes.	-
STFIPS	Converts state postal codes to FIPS state codes.	-
STNAME	Converts state postal codes to uppercase state names.	-
STNAMEL	Converts state postal codes to mixed case state names.	-

(*Continued*)

Table 10.2-2: (*Continued*)

SAS Functions and CALL routines	Descriptions	CAS
ZIPCITY	Returns a city name and the two-character postal code that corresponds to a ZIP code.	-
ZIPCITYDISTANCE	Returns the geodetic distance between two ZIP code locations.	-
ZIPFIPS	Converts ZIP codes to two-digit FIPS codes.	-
ZIPNAME	Converts ZIP codes to uppercase state names.	-
ZIPNAMEL	Converts ZIP codes to mixed-case state names.	-
ZIPSTATE	Converts ZIP codes to two-character state postal codes incl. Army Post Office (APO) und Fleet Post Office (FPO).	-
Trigonometric		
ARCOS	Returns the arccosine.	●
ARSIN	Returns the arcsine.	●
ATAN	Returns the arc tangent.	●
ATAN2	Returns the arc tangent of the ratio of two numeric variables.	●
COS	Returns the cosine.	●
CSC	Returns the cosecant.	●
SEC	Returns the secant.	●
SIN	Returns the sine.	●
TAN	Returns the tangent.	●
Truncation		
CEIL	Returns the smallest integer that is greater than or equal to the argument, fuzzed to avoid unexpected floating-point results.	●
CEILZ	Returns the smallest integer that is greater than or equal to the argument, using zero fuzzing.	●
FLOOR	Returns the largest integer that is less than or equal to the argument, fuzzed to avoid unexpected floating-point results.	●
FLOORZ	Returns the largest integer that is less than or equal to the argument, using zero fuzzing.	●
FUZZ	Returns the nearest integer if the argument is within $1E{-}12$ of that integer.	●
INT	Returns the integer value, fuzzed to avoid unexpected floating-point results. If the argument's value is within $1E{-}12$ of an integer, the function results in that integer. If the value of argument is positive, the INT function has the same result as the FLOOR function. If the value of argument is negative, the INT function has the same result as the CEIL function.	●
INTZ	Returns the integer portion of the argument, using zero fuzzing.	●
ROUND	Rounds the first argument to the nearest multiple of the second argument or to the nearest integer when the second argument is omitted.	●
ROUNDE	Rounds the first argument to the nearest multiple of the second argument and returns an even multiple when the first argument is halfway between the two nearest multiples.	●
ROUNDZ	Rounds the first argument to the nearest multiple of the second argument, using zero fuzzing.	●
TRUNC	Truncates a numeric value to a specified number of bytes, for example, as specified by LENGTH.	●

(*Continued*)

Table 10.2-2: (*Continued*)

SAS Functions and CALL routines	Descriptions	CAS
Variable Control		
CALL LABEL	Assigns a variable label to a specified character variable.	●
CALL SET	Links SAS data set variables to DATA step or macro variables that have the same name and data type.	-
CALL VNAME	Assigns a variable name as the value of a specified variable.	●
Variable Information		
CALL VNEXT	Returns the name, type, and length of a variable that is used in a DATA step.	●
VARRAY	Returns a value that indicates whether the specified name is an array.	●
VARRAYX	Returns a value that indicates whether the value of the specified argument is an array.	●
VFORMAT	Returns the format that is associated with the specified variable.	●
VFORMATD	Returns the decimal value of the format that is associated with the specified variable.	●
VFORMATDX	Returns the decimal value of the format that is associated with the value of the specified argument.	●
VFORMATN	Returns the format name that is associated with the specified variable.	●
VFORMATNX	Returns the format name that is associated with the value of the specified argument.	●
VFORMATW	Returns the format width that is associated with the specified variable.	●
VFORMATWX	Returns the format width that is associated with the value of the specified argument.	●
VFORMATX	Returns the format that is associated with the value of the specified argument.	●
VINARRAY	Returns a value that indicates whether the specified variable is a member of an array.	●
VINARRAYX	Returns a value that indicates whether the value of the specified argument is a member of an array.	●
VINFORMAT	Returns the informat that is associated with the specified variable.	●
VINFORMATD	Returns the decimal value of the informat that is associated with the specified variable.	●
VINFORMATDX	Returns the decimal value of the informat that is associated with the value of the specified variable.	●
VINFORMATN	Returns the informat name that is associated with the specified variable.	●
VINFORMATNX	Returns the informat name that is associated with the value of the specified argument.	●
VINFORMATW	Returns the informat width that is associated with the specified variable.	●
VINFORMATWX	Returns the informat width that is associated with the value of the specified argument.	●
VINFORMATX	Returns the informat that is associated with the value of the specified argument.	●
VLABEL	Returns the label that is associated with the specified variable.	●
VLABELX	Returns the label that is associated with the value of the specified argument.	●
VLENGTH	Returns the compile-time (allocated) size of the specified variable.	●

(*Continued*)

Table 10.2-2: (*Continued*)

SAS Functions and CALL routines	Descriptions	CAS
VLENGTHX	Returns the compile-time (allocated) size for the variable with a name that is the same as the value of the argument.	●
VNAME	Returns the name of the specified variable.	●
VNAMEX	Validates the value of the specified argument as a variable name.	●
VTYPE	Returns the type (character or numeric) of the specified variable.	●
VTYPEX	Returns the type (character or numeric) for the value of the specified argument.	●
VVALUE	Returns the formatted value that is associated with the variable that you specify.	●
VVALUEX	Returns the formatted value that is associated with the argument that you specify.	●
Web Service		
SOAPWEB	Calls a web service by using basic web authentication; credentials are provided in the arguments.	-
SOAPWEBMETA	Calls a web service by using basic web authentication; credentials for the authentication domain are retrieved from metadata.	-
SOAPWIPSERVICE	Calls a SAS web service by using WS-Security authentication; credentials are provided in the arguments.	-
SOAPWIPSRS	Calls a SAS web service by using WS-Security authentication; credentials are provided in the arguments.	-
SOAPWS	Calls a web service by using WS-Security authentication; credentials are provided in the arguments.	-
SOAPWSMETA	Calls a web service by using WS-Security authentication; credentials for the provided authentication domain are retrieved from metadata.	-
Web Tools		
HTMLDECODE	Decodes a string that contains HTML numeric character references or HTML character entity references and returns the decoded string.	-
HTMLENCODE	Encodes characters using HTML character entity references and returns the encoded string.	-
URLDECODE	Returns a string that was decoded using the URL escape syntax.	●
URLENCODE	Returns a string that was encoded using the URL escape syntax.	●

Note: Since SAS 9.2, some RX functions and CALL routines have been replaced with PRX versions that were previously available in SAS. According to SAS, the following RX versions are therefore no longer listed in the documentation: RXMATCH and RXPARSE, and the CALL routines RXCHANGE, RXFREE, and RXSUBSTR. By the way, the PRX prefix indicates that the function or CALL routine is a Perl Regular Expression. Perl regulations might also have to be observed when programming.

Note: The SCANQ function and the CALL SCANQ routine have been replaced since SAS 9.2 by the more powerful SCAN function and CALL SCAN routine, respectively, and are therefore no longer listed in the documentation.

10.3 Pass-Through Facility (Features)

The pass-through facility is one of the most important ways SAS can connect to DBMS. The pass-through facility supports sending DBMS-specific SQL syntax directly to the respective DBMS. The type and scope of the functionality varies from DBMS to DBMS and might also depend on the host. The following section introduces the pass-through facility. Special advantages of the pass-through facility are:

- uses server performance for processing
- enables you to specify any SQL syntax that the DBMS understands, even SQL syntax that is not valid for PROC SQL
- translates your PROC SQL statements into DMS-specific statements
- optimizes queries for joins, the handling of indexes, or aggregation functions like AVG, GROUP BY clauses, or columns created by expressions

Processing on the server side is usually much more efficient than passing data back and forth between the SAS client and server for processing. SAS offers two different approaches to execute as many queries and perform as many calculations as possible on the server than on the SAS client side:

- **Implicit pass-through** is the process of translating PROC SQL query code into equivalent data source-specific SQL code so that it can be passed directly to the data source for processing. That means that you do not need to be familiar with DBMS-specific SQL syntax. Implicit pass-through optimizes the processing in several aspects, for example, reducing the volume of the transferred data transfer and leveraging of data source-specific capabilities such as massively parallel processing and advanced join techniques.
- **Explicit pass-through** sends SQL code as written. There is no translation or optimization done. Use this functionality when you want to control exactly which commands are sent from PROC SQL to the server's SQL processor, or when you want to optimize the SQL yourself. All in all, the explicit pass-through offers more control over the query code and its processing on the server, including locking and spooling.

Implicit Pass-Through

Simply specifying the IP=YES option in the LIBNAME statement enables the implicit pass-through from PROC SQL. When IP=YES is set (see bold in the SAS code below), PROC SQL internally analyzes the SELECT requests for eligibility for pass-through and passes requests that are eligible through to the server for processing. When you also want to create server tables with PROC SQL, you also need set the DBIDIRECTEXEC= system option; it optimizes CREATE TABLE operations so that they can be passed to the server SQL processor for processing. No other options need to be set. Implicit pass-through generally requires less SAS code. Implicit pass-through is easy to code and works like a charm when the SELECT requests are eligible, that is, ANSI-compatible.

Program 10.3-1: Example for Implicit Pass-Through

```
libname spdsip sasspds 'conversion_area' server=husky.5400 user="siteusr1"
        password="xxxxxxxx" ip=yes;
options dbidirectexec=yes;
proc sql;
 create table spdsip.lotterywin (ticketno num, winname char(30));
 insert into spdsip.lotterywin values (1, 'Wishu Weremee');
quit;
```

This PROC SQL program uses an implicit pass-through from PROC SQL to create a new table named Lotterywin on the server. The SASSPDS LIBNAME statement includes **IP**=YES to enable implicit pass-through to the server. The **DBIDIRECTEXEC**=YES system option enables the pass-through of table creation requests. Below you find an explicit pass-through version of this program.

If the implicit pass-through is used, the advantage is that the DBMS-specific SQL need not be known. As appealing as it seems, the automatic translating and optimizing of the PROC SQL query code into equivalent data source-specific SQL code has its trade-offs. A disadvantage of the implicit pass-through is that the translation process can be at the expense of performance. A second disadvantage of the implicit pass-through statement is that, depending on the DBMS, not all or the same SAS functions can be translated into DBMS-specific SQL.

- **Interferences:** Non-ANSI SQL, a few SAS functions, and other specific SAS language elements might disable implicit pass-through from PROC SQL, which might turn queries into a resource-intensive processing. For example, specifying SAS functions on **PROC SQL**'s SELECT clause can prevent joins from being passed. Other functions such as MOD or statements might yield different results depending on where they run. Selected examples for **FedSQL** are aggregate statistics such as SKEWNESS, STUDENTS_T, or NMISS, or mathematical functions such as SIN, COS, and TAN. For other FedSQL limitations that might prevent implicit pass-through, please see the SAS documentation.

- **Retrying the processing on the client:** An important requirement is that the server SQL engine successfully parses the submitted SQL statements. If the server cannot successfully parse the statements, especially the case of non-ANSI, SQL quietly prevents implicit pass-through from happening (why "quietly"? see below), resulting in the consumption of excess resources. PROC SQL retries the query on the *client*—a resource-intensive approach which you tried to avoid in the first place.
- **Simplifying is not always optimizing:** As already mentioned, the server SQL engine must successfully parse the submitted SQL statement. If the server cannot *process* the query, PROC SQL *simplifies* that query until it succeeds. An undesired simplification might lead to a final result that differs from what would have been expected from the original code.
- **Lack of knowledge and control:** The advantages of the implicit pass-through might turn into its disadvantages. Users have to re-gain control over "translating," "optimizing," and the location of processing the queries. These are not always obvious, as the server SQL engine does not provide feedback in the *SAS log about* which implicit pass-through queries could not be handled.
- **Tables:** The tables referenced in the SQL statement must be server tables (see DBIDIRECTEXEC).

Explicit Pass-Through

The explicit pass-through uses the CONNECT statement or the CONNECT TO component. Remember, implicit pass-through is translating and optimizing automatically. Because it is not translating, explicit pass-through sends SQL code very efficiently *exactly as written* to the server's SQL processor. Use explicit pass-through if you want to be in control of which commands are sent to the server's SQL processor, and the way you want to optimize the SQL yourself using the optimization features of the server.

Program 10.3-2: Example for Explicit Pass-Through

```
proc sql;
connect to sasspds (dbq='conversion_area'
                    server=husky.5400
                    user='anonymous');
execute (create table lotterywin(ticketno num, winname char(30))) by sasspds;
execute (insert into lotterywin values (1, 'Wishu Wereme')) by sasspds;
disconnect from sasspds;
quit;
```

This PROC SQL program uses an explicit pass-through from PROC SQL to create a new table named Lotterywin on the server. The authorizations to access the server are passed using the CONNECT TO component. Compare its parameters with those of the LIBNAME statement in the implicit pass-through. Above you find an implicit pass-through version of this program.

Which one to choose? See implicit pass-through (IP) and explicit pass-through (EP) as complementary approaches. From a *skill* focus, EP expects you to be familiar with translating and optimizing from the start; IP only if something didn't work as expected. So, use IP if you process standard SQL code, want SAS to do all the translating, and no "hiccups" are expected. Use EP if you process nonstandard SQL code, and you are able to do all the programming in SAS and optimizing on the server yourself. From a *performance* focus, use EP also if you pass standard SQL code. It's efficient because no translations are required and no "hiccups" are expected.

10.3.1 Implicit Pass-Through: Options

The implicit pass-through consists of only a few options, for example:

- Use the **IP**=YES in the LIBNAME statement to enable implicit pass-through to the SAS SPD Server session.
- Use the **DBIDIRECTEXEC**=YES system option to enable the pass-through of table creation requests.
- By default, errors are reported in the *server log*. To turn on the error reporting in the SAS log (as notes, not errors), set the macro variable as follows: %let SPDSIPDB=YES. Feedback about query failures are returned as notes, not errors.

Users will find various notes in connection with optimizing the implicit pass-through performance in Section 8.6 including some LIBNAME options like SQL_FUNCTIONS= or DIRECT_SQL.

- By default, only DBMS functions are called that produce the same results as the comparable function in SAS. **SQL_FUNCTIONS**=ALL allows functions to be passed to the DBMS that produce slightly different results than those that a comparable SAS function generates.
- If you want to access a DBMS column that has NULL values by a WHERE clause, some DBMSs remove the NULL values before processing, whereas SAS does not. In the end, your results might differ based on whether the WHERE processing occurs in the DBMS or in SAS. **DIRECT_SQL**=NOWHERE prevents the SQL code from being passed to the DBMS.

10.3.2 Explicit Pass-Through: Statements and Component

The explicit pass-through consists of a few, but powerful statements and a component that enables you to perform the following tasks, for example:

- You can use the **CONNECT** statement to connect with a DBMS and terminate this connection with the **DISCONNECT** statement.
- You can use the **EXECUTE** statement to pass dynamic DBMS-specific SQL statements to the DBMS that are not restricted to queries.
- You can use the **CONNECTION TO** component (specified in the FROM clause of a PROC SQL SELECT statement) to retrieve data directly from a DBMS. Using CONNECTION TO, you can specify any SQL syntax that the DBMS understands, even SQL syntax that is not valid for PROC SQL.

The pass-through facility does not support stored procedures with output parameters. For details about the individual statements and the CONNECTION TO component of the pass-through facility, see the current technical documentation from SAS.

CONNECT Statement: Syntax

```
PROC SQL ;
     CONNECT TO  DBMS name  < AS  alias >
        < (CONNECT_statement_1=Value <... CONNECT_statement_n=Value > ) >
        < (DB_connection_argument_1=Value <... DB_connection_argument_n=Value > ) > ;
```

Program 10.3-3: Example for CONNECT Statement

```
proc sql ;
connect to SAPASE as SAPASE_CONN
   (server=MY_SRVR database=BCBS239 user=CFGS
password=MY_PWD connection=GLOBAL) ;
```

CONNECT establishes a connection to the SAPASE database using an alias (see AS) and passes several credentials to authorize DB access including user and password.

Description of the CONNECT Syntax

The connection established by CONNECT remains in effect until you issue a DISCONNECT statement or terminate the SQL procedure.

DBMS name:

Name of the DBMS; you can specify DB2, NETEZZA, ODBC, ORACLE, SAPASE, TERADATA, and so on.

Alias (optional):

Defines an alias for the connection. If an alias is defined, AS must be specified before it. If no alias is specified, the DBMS name is used as the name of the pass-through connection.

CONNECT TO: DBMS name | alias

Specifies values for arguments that indicate whether you can make multiple connections, shared or unique connections, and so on, to the database. With these arguments, the SQL pass-through facility can use some of the connection management features of the LIBNAME statement or of SAS system options. These arguments are optional, but if they are specified, they must be enclosed in parentheses.

CONNECTION= SHARED | GLOBAL

Shows whether multiple CONNECT statements can use the same connection for a DBMS.

CONNECTION_GROUP=Name of the connection group; default: no name.

Defines a connection that can be shared by multiple CONNECT statements in the pass-through facility.

DBCONINIT= User-defined initialization command; default: No command

Defines a user-defined initialization command that is executed immediately *after* the connection to the DBMS.

DBCONTERM= User-defined initialization command; default: No command

Defines a user-defined termination command that is executed *before* the connection to the DBMS is interrupted.

DBGEN_NAME= DBMS | SAS

Specifies whether the names of DBMS columns that contain characters that SAS does not allow (for example, $) should automatically be converted to valid SAS variable names.

DBMAX_TEXT= *integer*.

Sets the maximum length value for reading, appending, or updating very long DBMS columns of type String. DBMAX_TEXT= does not apply when you are creating a table.

DBPROMPT=YES | NO; default: NO.

Specifies whether SAS opens a window in which the user enters the required connection information before connecting to the DBMS.

DEFER=NO | YES; default: NO.

Determines when the connection to the DBMS occurs. If YES, the connection occurs when the first pass-through statement is executed; if NO, the connection occurs when the CONNECT statement is executed.

VALIDMEMNAME=COMPATIBLE | EXTEND; default: COMPATIBLE.

Specifies the rules for naming SAS data sets, SAS data views, and item stores. If COMPATIBLE, the names can have up to 32 characters, begin with a letter of the Latin alphabet or an underscore, and cannot contain blanks or special characters except the underscore; iF EXTEND, the names can be up to 32 *bytes* in length, can include national and special characters (with few exceptions), and cannot begin with a blank or a period.

VALIDVARNAME=V7 (default)

Indicates the compatibility mode for variable names for the SQL pass-through facility. V7 indicates that only variable names that are compatible with SAS version 7 are valid.

DISCONNECT Statement: Syntax

PROC SQL ;
      DISCONNECT FROM   DBMS name < **AS** alias >
          < (CONNECT_statement_1=Value <... CONNECT_statement_n=Value >) >
          < (DB_connection_argument_1=Value<... DB_connection_argument_n=Value >) > ;
< **QUIT** ; >

Program 10.3-4: Example for DISCONNECT Statement

```
proc sql ;
        disconnect from SAPASE_CONN ;
quit ;
```

DISCONNECT terminates the connection established by CONNECT (demonstrated in the CONNECT code).

Description of the Syntax of the DISCONNECT Statement

DISCONNECT terminates the connection established by CONNECT.

DBMS name:

Name of the DBMS; you can specify DB2, NETEZZA, ODBC, ORACLE, SAPASE, TERADATA, and so on.

Alias (optional):

Alias that was defined in the CONNECT statement.

DISCONNECT FROM: DBMS name | Alias

Terminates a connection with a DBMS. If you do not include the DISCONNECT statement, SAS performs an implicit DISCONNECT when PROC SQL terminates.

EXECUTE Statement: Syntax

      EXECUTE *(DBMS-specific SQL statements)* **BY** DBMS name | alias ;

Program 10.3-5: Example for EXECUTE Statement

```
proc sql ;
connect to SAPASE as SAPASE_CONN
   (server=MY_SRVR database=BCBS239 user=CFGS
password=MY_PWD connection=GLOBAL) ;
%put &sqlxrc ;
%put &sqlxmsg ;
      execute (create view MRKTD_CMPGNE as
                   select CID, PID, F_NAME, L_NAME, MRKT_PREFS, ADDRESS, STREET,
                         TPHONE, EMAIL
             from CUSTOMER, POLICES
                where CUSTOMER.C_ID= POLICES.P_ID) by SAPASE_CONN;
      execute (grant select on MRKTD_CMPGNE to CFGS) by SAPASE_CONN;
      disconnect from SAPASE_CONN ;
%put &sqlxmsg ;
%put &sqlxrc ;
quit ;
```

Two EXECUTE statements are passing dynamic DBMS-specific SQL statements to SAPASE. The first EXECUTE creates the view MRKTD_CMPGNE using selected columns from the tables CUSTOMER and POLICES. The second EXECUTE grants SELECT privilege on the created view to the specified user. The SAS macro variables SQLXRC and SQLXMSG contain provider-specific return codes and messages that identify and describe the generated error. The contents of these two macro variables can be written to the SAS log using %PUT.

Description of the Syntax of the EXECUTE Statement

Passes dynamic DBMS-specific SQL statements to a DBMS that are not restricted to queries.

DBMS name:

Name of the DBMS; you can specify DB2, NETEZZA, ODBC, ORACLE, SAPASE, TERADATA, and so on.

Alias (optional):

Alias that was defined in the CONNECT statement. **BY** must precede it.

(dbms-specific SQL statements)

This argument is required and must be enclosed in parentheses. The SQL statement is passed to the DBMS exactly as you enter it; it must not contain semicolons. Depending on the DBMS, the SQL statement is case-sensitive. (See Teradata.) The following statements can be passed to the DBMS: **CREATE, DELETE, DROP, GRANT, INSERT, REVOKE**, and **UPDATE**. The functionality of these statements is DBMS-specific and cannot be compared to SQL commands, some of which have the same name.

DBMS name:

Specifies the DBMS to which you direct the DBMS-specific SQL statement. **BY** must precede it.

Alias (optional):

Alias that was defined in the CONNECT statement. **BY** must precede it.

CONNECTION TO Component: Syntax

CONNECTION TO DBMS name | alias *(DBMS query)* ;

Program 10.3-6: Example for CONNECTION TO Component

```
proc sql ;
connect to SAPASE as SAPASE_CONN
   (server=MY_SRVR database=BCBS239 user=CFGS
password=MY_PWD connection=GLOBAL) ;
      %put &sqlxmsg;
select * from connection to SAPASE
   (select * from CUSTOMER where C_ID='01051978');
```

CONNECTION TO establishes a connection to the SAPASE database using an alias (see AS) and passes several credentials to authorize DB access (for example, user and password). You can also take advantage of the macro variables SQLXRC and SQLXMSG. (See EXECUTE above.) For example, SQLXMSG returns provider-specific error messages that identify and describe a generated error in more detail. The SELECT statement selects all columns from CUSTOMER where C_ID equals that specified value.

Description of the Syntax of the CONNECTION TO Component

Retrieves and uses DBMS data in a PROC SQL query or view.

DBMS name:

Name of the DBMS; you can specify DB2, NETEZZA, ODBC, ORACLE, SAPASE, TERADATA, and so on.

Alias (optional):

Alias that was defined in the CONNECT statement.

(DBMS-query):

Defines the query or view to be passed to the DBMS. The query can contain **any DBMS-specific SQL** syntax or syntax that is valid for the DBMS. The query cannot contain semicolons and must be enclosed in parentheses. The query is passed to the DBMS exactly as it is written, so if the DBMS is case-sensitive, the query must be written accordingly. You can store the pass-through facility component and the enclosed statements in an SQL view. If the SQL view is used in a SAS program, SAS can connect to the DBMS using the defining arguments stored in the view.

10.3.3 Explicit Pass-Through: Examples for DBMS-specific Features

This subsection introduces DBMS-specific features of the explicit pass-through facility for selected DBMSs: MySQL, Netezza, ODBC, Oracle and Teradata. Many DBMSs are not presented like DB2 UNIX/PC, DB2 z/OS, Informix, and OLE DB. For these and other DBMSs, please refer to the technical documentation of SAS 9.4. This chapter assumes you do not process on the CAS server. PROC SQL is not supported on the CAS server. Chapter 7 contains FedSQL examples for CAS. The examples in this subsection are taken from the SAS documentation.

1. MySQL

PROC SQL connects to MySQL and passes two EXECUTE statements for processing. The first EXECUTE creates the WHOTOOKORDERS table from the selected columns from ORDERS and EMPLOYEES. The second EXECUTE passes a GRANT SELECT to the DBMS. The user TESTUSER is granted a SELECT authorization, that is, permission to apply the specified SELECT statement to the WHOTOOKORDERS table. DISCONNECT terminates the connection.

Program 10.3-7: Example for MySQL Specific Features

```
proc sql;
   connect to mysql (user=testuser password=testpass
                     server=mysqlserv
                     database=mysqldb port=9876) ;
   execute (create table whotookorders as
      select ordernum, takenby,
             firstname, lastname, phone
         from orders, employees
         where orders.takenby=employees.empid)
      by mysql ;
   execute (grant select on whotookorders
            to testuser) by mysql ;
   disconnect from mysql ;
quit ;
```

Description of Connection Specifics

The arguments of the CONNECT statement are identical to the options of its LIBNAME connection. The arguments of the CONNECT statement for connecting to MySQL are:

- **USER=** Optional MySQL login ID. If USER= is specified, PASSWORD= must also be used and vice versa.
- **PASSWORD=** Passes the password that belongs to the MySQL Login ID.
- **SERVER=** Name or IP address of the MySQL server to connect to.
- **DATABASE=** Name of the MySQL database.
- **PORT=** Server port used for the TCP/IP connection.

Due to a current limitation in the MySQL client library, you cannot run MySQL stored procedures when SAS is running on AIX.

2. Netezza

PROC SQL connects to Netezza. The first SELECT statement selects all columns from CUSTOMERS that meet a WHERE clause. The second SELECT statement uses SQLTables as SQLAPI to query for a list of tables that match the specified arguments. The result returns the table MY_TEST (including all its columns) from the catalog TEST and the schema "".

Program 10.3-8: Example for Netezza Specific Features

```
proc sql;
   connect to netezza
   (server=mysrv1 database=test user=myuser password=mypwd);
select * from connection to netezza
   (select * from customers where customer like '1%');
select * from connection to netezza
        (NETEZZA::SQLTables "test","","my_test") ;
quit;
```

Description of Connection Specifics

The CONNECT statement is required. The arguments of the CONNECT statement are identical to the options of its LIBNAME connection. The arguments of the CONNECT statement for connecting to Netezza are:

- **USER=** Optional Netezza login ID. If USER= is specified, PASSWORD= must also be used and vice versa.
- **PASSWORD=** Passes the password that belongs to the Netezza login ID.
- **SERVER=** Name or IP address of the Netezza server to connect to.
- **DATABASE=** Name of the Netezza database.
- **Netezza::SQLAPI:** In Netezza, you can use application programming interfaces (APIs) in queries. The general format of the special queries is: *Netezza::SQLAPI "parameter 1","parameter n"* ("parameter n" is a quoted string that is delimited by commas). *As* **SQLAPI,** *you could specify parameters for different special queries such as* SQLTables for a list of tables, SQLColumns for a list of columns, or SQLPrimaryKeys for a list of primary key columns. **Netezza::** is required. *SQLAPI* is the specific API being called. For example, the convention to query tables is: **Netezza::SQLTables** <"Catalog", "Schema", "Table-name", "Type"> ; if no arguments are specified, all accessible table names and information are returned.

3. ODBC

PROC SQL connects to the ODBC interface.

- **Example 1** uses SQLColumns as SQLAPI to query for a list of columns that match the specified arguments. The result returns a list of the columns in the table CUSTOMERS.
- **Example 2** enables you to select any data source that is configured on your machine.
- **Example 3** connects to an ODBC Server, configured under the data source name "ODBC Server", and sends an SQL query to the database for processing. SELECT queries all rows for the specified columns from SASDEMO.EMPLOYEES that meet the requirement that the HIREDATE value is higher and more recent than 31.12.1988.

Program 10.3-9: Examples for ODBC Specific Features

```
/* Example 1 */

proc sql;
   connect to odbc as mydb
      (datasrc="SQL Server" user=testuser password=testpass) ;
   select * from connection to mydb
      (ODBC::SQLColumns (, , "CUSTOMERS")) ;
quit ;
```

```
/* Example 2 */
proc sql ;
   connect to odbc (required) ;
        %put &sqlxmsg ;
 quit ;
/* Example 3 */
proc sql ;
connect to odbc as odbcconn
    (datasrc="ODBC Server" user=myusr1 password=mypwd1 readbuff=250 trace=yes) ;
select *
   from connection to odbcconn
        (select empid, lastname, firstname, hiredate, salary
         from sasdemo.employees
         where hiredate>='31.12.1988'
         ) ;
disconnect from odbcconn ;
quit ;
```

Description of Connection Specifics

The **CONNECT** statement is required. The arguments of the CONNECT statement are identical to the options of its LIBNAME connection. Not all ODBC drivers fully support the CONNECT statement options. PROC SQL also supports multiple connections to ODBC. The arguments of the CONNECT statement for connecting to ODBC are:

- **USER=** Optional ODBC login ID. If USER= is specified, PASSWORD= must also be used and vice versa.
- **PASSWORD=** Passes the password that belongs to the ODBC login ID.
- **DATASRC=** Name of server and database (ODBC), configured under a data source name, for example, "SQL Server". The data source, user and password specifications are enclosed in parentheses.
- **READBUFF=** (aliases: BUFF=, BUFFSIZE=). Specifies the number of rows to retrieve from a table or view with each fetch. Using this option can improve the performance of any query. Default: 250 rows per fetch. 250 rows is the minimum value that READBUFF= accepts.
- **TRACE= YES** |NO. Specifies whether to turn on tracing information for use in debugging.
- **ODBC::SQLAPI:** In ODBC, you can use application programming interfaces (APIs) in queries. The general format of the special queries is: *ODBC::SQLAPI "parameter 1","parameter n"* ("parameter n" is a quoted string that is delimited by commas). *As* **SQLAPI,** *you could specify parameters for different special queries,* SQLTables for a list of tables, SQLColumns for a list of columns, SQLColumnPrivileges for a list of column privileges or SQLPrimaryKeys for a list of primary key columns. **ODBC::** is required. *SQLAPI* is the specific API being called. For example, the convention to query columns is: **ODBC::SQLColumns** <"Catalog", "Schema", "Table-name", "Column-name">; if no arguments are specified, all accessible column names and information are returned.

4. Oracle

PROC SQL connects to Oracle.

- **Example 1** passes two EXECUTE statements for processing. The **first** EXECUTE passes a specific Oracle SQL statement to the DBMS. The WHOTOOKORDERS view is created from the selected columns from the ORDERS and EMPLOYEES tables. The **second** EXECUTE also passes a specific Oracle SQL statement to the DBMS. Using **GRANT SELECT**, the user TESTUSER is granted a SELECT authorization, that is, permission to apply the specified SELECT statement to the WHOTOOKORDERS table. DISCONNECT terminates the connection.
- **Example 2** passes additional information ("hints") to the DBMS for processing: **PRESERVE_COMMENTS** passes the hints, for example, how to choose the best processing method, to Oracle's Query Optimizer. The hints are specified in the parenthesis expression of the subquery (CONNECTION TO component). **INDX** identifies the index for the Oracle Query Optimizer. **ALL_ROWS** requests all lines. Hints are separated with blanks.

Program 10.3-10: Examples for Oracle Specific Features

```
/* Example 1 */

proc sql ;
   connect to oracle as dbcon
       (user=testuser password=testpass
        readbuff=100 path='myorapath') ;
   execute (create view whotookorders as
      select ordernum, takenby, firstname, lastname, phone
         from orders, employees
         where orders.takenby=employees.empid)
      by oracle ;
   execute (grant select on whotookorders
           to testuser) by oracle ;
   disconnect from dbcon ;
quit ;


/* Example 2 */
proc sql ;
   connect to oracle as dbcon
       (user=testuser password=testpass preserve_comments);
   select *
      from connection to dbcon
        (select     +indx(empid) all_rows
            count(*) from employees);
quit ;
```

Description of Connection Specifics

The **CONNECT** statement is optional. When you omit a CONNECT statement, an implicit connection is performed when the first EXECUTE statement or CONNECTION TO component is passed to Oracle. The arguments of the CONNECT statement for connecting to Oracle are:

- **USER=** Optional Oracle login ID. If USER= is specified, PASSWORD= must also be used and vice versa.
- **PASSWORD=** Passes the password that belongs to the Oracle login ID. The Oracle password can be encoded using PROC PWENCODE, if necessary.
- **READBUFF=** (aliases: BUFF=, BUFFSIZE=) Specifies the number of rows to retrieve from an Oracle table or view with each fetch. Using this option can improve the performance of any query to Oracle. Default: 250 rows per fetch. The maximum this depends on available memory. The theoretical maximum is 2.147.483.647 rows per fetch.
- **PATH=** identifies the Oracle driver, nodes, and the database. Aliases are required if SQL*Net Version 2.0 or later is used. In some operating environments the information required for the PATH= statement can be passed before invoking SAS.
- **PRESERVE_COMMENTS:** Passes processing information ("hints") to Oracle's Query Optimizer suggesting the best processing method.

5. Teradata

PROC SQL connects to Teradata.

- **Example 1** connects to Teradata and passes two EXECUTE statements for processing. The first EXECUTE performs an update in the SALARY table. For each NAME 'Irma L.', the entry in CURRRENT _SALARY is set to 45000. An update is regarded as a transaction and is therefore completed with a COMMIT statement in the second EXECUTE; otherwise, the update is rolled back. A COMMIT statement makes the changes a permanent part of the database.
- **Example 2** passes seven EXECUTE statements for processing. The first EXECUTE deletes the table SALARY. The second EXECUTE passes a first COMMIT statement for processing; the connection is therefore in ANSI

mode. The third EXECUTE creates (DDL statement) another table SALARY specifying the columns CURRENT_ SALARY and NAME. The fourth EXECUTE passes another COMMIT statement to makes changes that are so far permanent. The fifth and sixth EXECUTE insert each two rows into the columns CURRENT_SALARY and NAME (transactions). The seventh EXECUTE passes a final COMMIT statement to turn the changes by these transactions into a permanent part of the database.

- **Example 3** operates in Teradata mode. In this mode, data processing is not case-sensitive, and no explicit COMMIT statements need to be specified. This example is made of two PROC SQL queries. The first PROC SQL example passes two EXECUTE statements for processing and creates and populates table CASETEST in Teradata mode. Table CASETEST is created with the string variable "X" and filled with the value "Case Insensitivity Desired". Since the connection runs in Teradata mode, no explicit COMMIT statements need to be specified. The second PROC SQL queries the table CASETEST (see (1) and the string variable "X"). Due to the Teradata mode, the query succeeds because the WHERE clause is satisfied because of case-insensitivity. The processing ignores differences in upper- and lowercase. (See INSERT INTO versus WHERE.)

Program 10.3-11: Examples for Teradata Specific Features

```
/* Example 1 */

proc sql ;
    connect to teradata as tera (user=testuser password=testpass) ;
        execute (update salary set current_salary=45000
                    where (name='Irma L.')) by tera ;
        execute (commit) by tera ;
    disconnect from tera ;
quit ;


/* Example 2 */
proc sql ;
    connect to teradata as tera (user=testuser password=testpass) ;
    execute (drop table salary) by tera ;
    execute (commit) by tera ;
    execute (create table salary (current_salary float, name char(10)))
            by tera ;
    execute (commit) by tera ;
    execute (insert into salary values (35335.00, 'Dan J.')) by tera ;
    execute (insert into salary values (40300.00, 'Irma L.')) by tera ;
    execute (commit) by tera ;
    disconnect from tera ;
quit ;


/* Example 3 */
proc sql ;
/* (1) Create and populate table in Teradata Mode (case insensitive) */
    connect to teradata (user=testuser pass=testpass mode=teradata) ;
    execute(create table casetest(x char(28))) by teradata ;
    execute(insert into casetest values('Case Insensitivity Desired')) by teradata ;
quit ;
proc sql ;
/* (2) Query table in Teradata Mode (for case insensitive match)   */
    connect to teradata (user=testuser pass=testpass mode=teradata) ;
    select * from connection to teradata
            (select * from casetest where x='case insensitivity desired');
quit ;
```

Description of Connection Specifics

The **CONNECT** statement is required. The arguments of the CONNECT statement are identical to the options of its LIBNAME connection. By default, SAS/ACCESS opens Teradata in ANSI mode. In ANSI mode, every DDL statement like DROP TABLE or CREATE TABLE and every transaction like INSERT must end with an explicit **COMMIT** statement, otherwise Teradata rolls back updates and inserts. If data is only read, that is, no changes are made to the database, no COMMIT statements need to be specified. If you add MODE=TERADATA to your CONNECT statement, you do not specify explicit COMMIT statements. (See below.) Arguments of the CONNECT statement for connecting to Teradata are:

- **USER=** Optional Teradata login ID. If USER= is specified, PASSWORD= must also be used and vice versa.
- **PASSWORD=** Passes the password that belongs to the Teradata login ID.
- **MODE= TERADATA** Data processing is not case-sensitive. No explicit COMMIT statements need be specified. In default MODE= **ANSI**, the data processing is case-sensitive.

If you must change the default Teradata database, add the **DATABASE=** option to your **CONNECT** statement. Do not issue the Teradata DATABASE statement within the EXECUTE statement in PROC SQL.

References

American National Standards Institute (ANSI) Document ANSI X3.135-1992. International Organization for Standardization (ISO): Database Languages - SQL. Document ISO/IEC 9075:1992.

Barnett, Vic & Lewis, Toby (1994). Outliers in statistical data. New York: John Wiley & Sons.

Burlew, Michele M. 2001. Debugging SAS Programs: a Handbook of Tools and Techniques. Cary, NC: SAS Institute Inc.

Burlew, Michele M. (1998). SAS Macro Programming Made Easy. Cary, NC: SAS Institute Inc.

Cheng, Wei (2004). Helpful undocumented features in SAS. *Proceedings of the Twenty-Ninth Annual SAS Users Group International Conference,* Cary, NC: SAS Institute., Paper 40-29.

Dickstein, Craig; Pass, Ray; Davis, Michael L. (2007). DATA Step vs. PROC SQL: What's a neophyte to do? *Proceedings of the SAS Global Forum 2007 Conference,* Cary, NC: SAS Institute, Paper 237-2007.

Dorfman, Paul M.; Vyverman, Koen; Dorfman, Victor (2010). Black Belt Hashigana. *Proceedings of the SAS Global Forum 2010 Conference,* Cary, NC: SAS Institute, Paper 023-2010.

Dorfman, Paul M.; Vyverman, Koen (2009). The SAS Hash Object in Action. *Proceedings of the SAS Global Forum 2009 Conference,* Cary, NC: SAS Institute, Paper 153-2009.

Dorfman, Paul M.; Shajenko, Lessia S.; Vyverman, Koen (2008). Hash Crash and Beyond. *Proceedings of the SAS Global Forum 2008 Conference,* Cary, NC: SAS Institute, Paper 037-2008.

Droogendyk, Harry & Dosani, Faisal (2008). Joining Data: Data Step Merge or SQL? *Proceedings of the SAS Global Forum 2008 Conference,* Cary, NC: SAS Institute, Paper 178-2008.

Ebbesmeyer, Curtis C. & Ingraham Jr., W. James (1994). Pacific Toy Spill Fuels Ocean Current Pathways Research, Eos, v.75, no.37, p. 425-430: https://agupubs.onlinelibrary.wiley.com/doi/abs/10.1029/94EO01056

Feder, Steven (2003). Comparative efficiency of SQL and Base code when reading from database tables and existing data Sets. *Proceedings of the Twenty-Eighth Annual SAS Users Group International Conference,* Cary, NC: SAS Institute, Paper 76-28.

Foley, Malachy J. (2005). MERGING vs. JOINING: Comparing the DATA Step with SQL. *Proceedings of the Thirtieth Annual SAS Users Group International Conference,* Cary, NC: SAS Institute Presentation at SUGI 30, Paper 249-30.

Hadden, Louise S. & Zdeb, Mike S. (2010). ZIP Code 411: Decoding SASHELP.ZIPCODE and Other SAS® Maps Online Mysteries. *Proceedings of the SAS Global Forum 2010 Conference,* Cary, NC: SAS Institute, Paper 219-2010.

Hadden, Louise S., Zdeb, Mike S. & Allison, Robert (2007). Wow! You Did That Map with SAS/GRAPH? *Proceedings of the Twentieth Annual NorthEast SAS Users Group Conference.*

Hartung, Joachim, Elpelt, Barbel, and Klosener, Karl-Heinz. (1999). Statistik. München Wien: R.Oldenbourg Verlag.

Hughes, Troy Martin (2019). User-Defined Multithreading with the SAS DS2 Procedure: Performance Testing DS2 Against Functionally Equivalent DATA Steps. *Proceedings of the Annual Pharmaceutical SAS Users Group Conference,* Paper AD-228.

Ivis, Frank (2006). Calculating geographic distance: Concepts and methods. *Proceedings of the Nineteenth Annual NorthEast SAS Users Group Conference* Presentation at NESUG 26, Paper DA15-2006.

Jordan, Mark (2018). Mastering the DS2 Procedure: Advanced Data Wrangling Techniques. Cary NC: SAS Institute Inc. (2nd ed.).

Lafler, Kirk P. (2004). PROC SQL: Beyond the Basics Using SAS. Cary, NC: SAS Institute Inc.

Langston, Rick (2005). Efficiency Considerations Using the SAS® System. *Proceedings of the Thirtieth Annual SAS Users Group International Conference,* Cary, NC: SAS Institute. Presentation at SUGI 30, Paper 002-30.

Little, Roderick J.A. & Rubin, Donald B. (2002). Statistical Analysis with Missing Data. Hoboken, NJ: John Wiley & Sons. Second Edition.

McJones, Paul (ed.); Bamford, Roger; Blasgen, Mike; Chamberlin, Don; Cheng, Josephine; Daudenarde, Jean-Jacques; Finkelstein, Shel; Gray, Jim; Jolls, Bob; Lindsay, Bruce; Lorie, Raymond; Mehl, Jim; Miller, Roger; Mohan, C.; Nauman, John; Pong, Mike; Price, Tom; Putzolu, Franco; Schkolnick, Mario; Selinger, Bob; Selinger, Pat; Slutz, Don; Traiger, Irv; Wade, Brad; Yost, Bob (1997). The 1995 SQL Reunion: People, Projects, and Politics. SRC Technical Note 1997 – 018. Palo Alto, CA 94301, Systems Research Center (August 20, 1997).

Mohammed, Zabiulla; Gangarajula, Ganesh Kumar; Kalakota, Pradeep (2015). Working with PROC FEDSQL in SAS 9.4. *Proceedings of the SAS Global Forum 2015 Conference,* Cary, NC: SAS Institute, Paper 3390-2015.

Pendergrass, Jerry (2017). The Architecture of the SAS Cloud Analytic Services in SAS Viya. *Proceedings of the SAS Global Forum 2017 Conference,* Cary, NC: SAS Institute, Paper 309-2017.

Rasch, Dieter, Herrendörfer, Günter, Bock, Jürgen, Victor, Norbert und Guiard, Volker (Hsg.) (1996). Verfahrensbibliothek: Versuchsplanung und -auswertung. Band I. München Wien: R. Oldenbourg Verlag.

Ray, Robert & Secosky, Jason (2008). Better Hashing in SAS® 9.2. *Proceedings of the SAS Global Forum 2008 Conference,* Cary, NC: SAS Institute, Paper 306-2008.

RiskNet (2020). Data error inflated Wells Fargo's op risk capital by $5 billion. Risk.net (Woodall, Louie, 03.06.2020).

Sadof, Michael G. (2000). Keeping Your Data in Step – Utilizing Efficiencies. *Proceedings of the Twenty-Fifth Annual SAS Users Group International Conference,* Cary, NC: SAS Institute, Paper 60-25.

Sadof, Michael G. (1999). Keeping Your Data in Step – Utilizing Efficiencies. *Proceedings of the Twenty-Fourth Annual SAS Users Group International Conference,* Cary, NC: SAS Institute, Paper 37.

SAS Institute (2019). SAS 9.4 Language Reference: Concepts. Sixth editon. Cary, NC: SAS Institute (2019.11).

SAS Institute (2016a). Base SAS 9.4 Procedures Guide: Statistical Procedures. 5th edition. Cary, NC: SAS Institute (2020.08).

SAS Institute (2016b). SAS 9.4 SQL Procedure User's Guide. Fourth edition. Cary, NC: SAS Institute (2021.02).

SAS Institute (2016c). SAS 9.4 FedSQL Language Reference, 5th Ed. Cary, NC: SAS Institute (2020.11).

SAS Institute (2016d). SAS 9.4 Functions and CALL Routines: Reference, 5th Ed. Cary, NC: SAS Institute (2020.07).

SAS Institute (2016e). SAS 9.4 Macro Language: Reference, 5th Ed. Cary, NC: SAS Institute Inc. (2020.08).

SAS Institute (1994). SAS Macro Facility: Tips & Techniques. Cary, NC: SAS Institute Inc.

SAS Institute (1990). SAS Programming Tips: A Guide to Efficient SAS Processing. Cary, NC: SAS Institute Inc.

Schendera, Christian FG (2020). Data Quality with SPSS. Amazon Paperback. ISBN-13: 979-8573296067.

Schendera, Christian FG (2012). SQL mit SAS: Band 2: PROC SQL für Fortgeschrittene. München: Oldenbourg Wissenschaftsverlag.

Schendera, Christian FG (2011). SQL mit SAS: Band 1: PROC SQL für Einsteiger. München: Oldenbourg.

Schendera, Christian FG (2010). Clusteranalyse mit SPSS. München: Oldenbourg Wissenschaftsverlag.

Schendera, Christian FG (2008). Regressionsanalyse mit SPSS. München: Oldenbourg.

Schendera, Christian FG (2007). Datenqualität mit SPSS. München: Oldenbourg Wissenschaftsverlag.

Schendera, Christian FG (2005). Datenmanagement mit SPSS. Heidelberg: Springer.

Schendera, Christian FG (2004). Datenmanagement und Datenanalyse mit dem SAS-System. München: Oldenbourg Wissenschaftsverlag.

Slocum, Terry A., McMaster, Robert B., Kessler, Fritz C. & Howard, Hugh H. (2009). Thematic Cartography and Geographic Visualization. Upper Saddle River, NJ: Pearson Prentice Hall. Third Edition.

Smith, Kevin J.; Khan, Muhammad Z.; Zhang, Yadong (2003). PROC SQL vs. Merge. The Miller Lite Question of 2002 and Beyond. *Proceedings of the Twenty-Eighth Annual SAS Users Group International Conference,* Cary, NC: SAS Institute, Paper 96-28.

Süddeutsche (2014). Behörden-Panne: Fiskus verteilte Steueridentifikationsnummern falsch; Süddeutsche (Bohsem, Guido, 13.02.2014).

Theuwissen, Henri (2011). Don't Waste Too Many Resources to Get Your Data in a Specific Sequence. *Proceedings of the SAS Global Forum 2011 Conference,* Cary, NC: SAS Institute, Paper 242-2011.

Theuwissen, Henri (2010). Complex Data Combinations on Large Volumes: How to get the Best Performance. *Proceedings of the SAS Global Forum 2010 Conference,* Cary, NC: SAS Institute, Paper 027-2010.

Theuwissen, Henri (2009). You Need to Consolidate Massive Amounts of Data? Select the Fastest Method or Go for Minimal Memory Requirements to Build Your Result. *Proceedings of the SAS Global Forum 2009 Conference,* Cary, NC: SAS Institute, Paper 040-2009.

Warner-Freeman, Jennifer K. (2007). I cut my processing time by 90% using hash tables – You can do it too! *Proceedings of the Twentieth Annual NorthEast SAS Users Group Conference*, Paper BB16.

Williams, Christianna S. (2008). PROC SQL for DATA Step Die-hards. *Proceedings of the SAS Global Forum 2008 Conference,* Cary, NC: SAS Institute, Paper 185-2008.

Wothke, Werner (2000). Longitudinal and multigroup modeling with missing data. In T.D. Little, Kai U. Schnabel, & Jürgen Baumert (Eds.) Modeling longitudinal and multilevel data: Practical issues, applied approaches, and specific examples. Mahwah, NJ: Lawrence Erlbaum Associates.

Wilcox, Andrew (2000). Efficiency Techniques for Accessing Large Data Files. *Proceedings of the Twenty-Fifth Annual SAS Users Group International Conference,* Cary, NC: SAS Institute, Paper 115-25.

Yi, Danbo & Zhang, Lei (1998). Handling missing values in the SQL Procedure. *Proceedings of the Eleventh Annual NorthEast SAS Users Group Conference*, Pittsburgh.

Syntax Index

Subject Index